Commercial Crop Technology

Commercical Crop Technology

C.K. Jain

Oxford Book Company
Jaipur ▪ New Delhi

Published By

Oxford Book Company

Regd. Off. "Bony Residency", Gate No. 2
Opp. Tilak Public School, Vishweswariya Nagar,
Gopalpura Road, Jaipur - 302018, India
Ph: 0141 - 2761280, Telefax: 2761381
email - oxfordbookjpr@gmail.com
website - oxfordbookcompany.com

Delhi Office: 102, 1st Floor, Satyam House,
4327/3, Ansari Road, Darya Ganj,
New Delhi - 110002, India
Ph: 011-45652440

Distributed by

KEYA INTERNATIONAL PRESS LLP
E-mail: keyainternationalpressllp@gmail.com

Published in India, 2025

ISBN : 9789380179315

Typeset by Shivangi Computers, Jaipur

Printed and bound in India by Replika Press Pvt. Ltd.

Preface

Commercial crops are crops grown solely for sale rather than for the farmer's own use, for example, oilseed, tea, coffee, cotton, or sugarcane. Many developing world countries grow commercial crops to meet their debt repayments rather than grow food for their own people. India is noted for several plantation cash crops, of which the most important are tea, rubber, oilseed, coconut, and sugarcane. Jute, a commercial fibre, though it has decreased in significance, remains a major export crop of India.

The increase in post -independence agricultural production has been brought about by bringing additional area under cultivation, extension of irrigation facilities, use of better seeds, better techniques, water management, and plant protection. Dependence on India agricultural imports in the early 1960s convinced planners that India's growing population, as well as concerns about national independence, security, and political stability, required self-sufficiency in food production. This perception led to a programme of agricultural improvement called the green revolution, to a public distribution system, and to price supports for farmers. The growth in food-grain production is a result of concentrated efforts to increase all the green revolution inputs needed for higher yields: better seed, more fertiliser, improved irrigation, and education of farmers.

The present book provides latest technological advances in the production of some of the major commercial crops of India. It is organised into 11 chapters with an introductory chapter and one chapter each for different commercial crops such as oilseed, sugarcane, fiber crops, tea, coffee, black pepper, cashew, coconut, rubber and tobacco. The introductory chapter gives an insight into the present scenario of plantation crop production, geographical distribution, nutrient management, crop protection strategies etc. Individual chapters on crops covers updated information on crop improvement, biotechnology, production and management and post harvest handling with emphasis on integrated nutrient and pest management. This book will be useful to all who associated with agriculture—farmers, estate managers, extension workers, technologists, policy makers, researchers and students.

C.K. Jain

Contents

1

Introduction

India ranks second worldwide in farm output. Agriculture and allied sectors like forestry, logging and fishing accounted for 16.6% of the GDP in 2007, employed 60% of the total workforce and despite a steady decline of its share in the GDP, is still the largest economic sector and plays a significant role in the overall socio-economic development of India. India is the largest producer in the world of milk, cashew nuts, coconuts, tea, ginger, turmeric and black pepper. It also has the world's largest cattle population It is the second largest producer of wheat, rice, sugar, groundnut and inland fish. It is the third largest producer of tobacco. India accounts for 10% of the world fruit production with first rank in the production of banana and sapota.

India has many farm insurance companies that insure wheat, fruit, rice and rubber farmers in the event of natural disasters or catastrophic crop failure, under the supervision of the Ministry of Agriculture. One notable company that provides all of these insurance policies is agriculture insurance company of India and it alone insures almost 20 million farmers. India's population is growing faster than its ability to produce rice and wheat. The most important structural reform for self-sufficiency is the ITC Limited plan to connect 20,000 villages to the Internet by 2013. This will provide farmers with up to date crop prices for the first time, which should minimise losses incurred from neighbouring producers selling early and in turn facilitate investment in rural areas.

India ranks second worldwide in farm output. Agriculture and allied sectors like forestry, logging and fishing accounted for 18% of the GDP in 2007, employed 70% of the total workforce and despite a steady decline of its share in the GDP, is still the largest economic sector and plays a significant role in the overall socio-economic development of India. Yields per unit area of all crops have grown since 1950, due to the special emphasis placed on agriculture in the five-year plans and steady improvements in irrigation, technology, application of modern agricultural practices and provision of agricultural credit and subsidies since Green revolution in India.

However, international comparisons reveal that the average yield in India is generally 30% to 50% of the highest average yield in the world. Indian Punjab is called the "Granary of India" or "India's bread-basket." It produces 14% of India's cotton, 20% of India's wheat, and 9% of India's rice.

COMMERCIAL CROP PRODUCTION

India has a variety of crops grown in irrigated and rainfed areas. Rice, wheat, jowar, maize and bajra are the major foodgrains. Oilseed, sugarcane, cotton and jute are the important cash crops. Likewise, tea and coffee are important plantation crops. India's agricultural production systems also include a wide range of horticultural crops.

Rice is widely grown in Southern, Eastern and Northeastern states. Wheat is mainly grown in Punjab and Haryana. Jowar and Bajra are important foodgrains in dryland areas. Assam and West Bengal are famous for tea, whereas Karnataka, Tamil Nadu and Kerala are coffee and coconut producing states.

In the highly diversified Indian context, no simple gender division of labour exists with regard to crop production. In certain areas in India women play a key role as seed selectors and in seedling production. Their knowledge on seeds and seed storage contribute to the viability of the agricultural diversity and production. As weeders, women contribute to crop management. Women prepare and apply green and farmyard manure. As integrated pest management practices are introduced, it could be expected that women's work will increase due to more labour-intensive activities.

On the eve of Independence, Indian agriculture had experienced many years of relatively low growth and by many accounts even deceleration in both productivity and output growth. From annual agricultural output growth of greater than 1 percent in the late 1800s, the first half of the 1900s saw annual growth plummet to about 0.3 percent per annum. Growth in grain yields has been estimated at between -0.4 to -0.6 percent annually from 1916 to 1946.

It is generally agreed that a combination of (1) structure of land rights and (2) resource endowments generated a set of incentives that worked against increases in productivity. The argument is that those who had ownership rights and access to capital did not have the right set of incentives to invest in productivity enhancing measures. Why this occurred has been the topic of much debate and will not be revisited here. The more important question is why this continued in the period following Independence. The answer is that class structure and resource endowments take time to change. Subsequently Indian agriculture has witnessed significant positive changes. Growth of net domestic products from agriculture averaged 1.87 percent from 1950 to 1967, increased to about 2.25 percent in 1966 to 1981 and further increased to about 3.3 percent in 1980 to 1999. However it now appears to be stagnating. The growth in production can be broadly ascribed to (1) growth in factor inputs and (2) growth in productivity. The first phase was more a result of an increase in factor inputs and the latter was driven by increases in productivity. It is debatable whether the early twenty-first century will witness stagnation or a continued increase.

At the same time the Indian economy overall is also growing faster. Moreover the dependence of the overall economy on agricultural growth is now much lower. According to estimates from Indian Council for Research on International Economic Relations (ICRIER) about 45 percent of the historical variation in growth can be accounted for by variations in agricultural output. But growth evidence from the early 2000s suggests that high levels of dependence belong to the past.

However, high growth is essential, with increases in population, incomes and nutritional requirements being various aspects. The ability to advance progress and the lifestyles of a large mass of the population is an important component. The experience of China is relevant here. A recent International Food Policy Research Institute (IFPRI) report summarises the issue well: "The comparative study shows that agriculture-led reform and development not only provide the necessary conditions for the manufacturing and service sectors to grow and reform, but also lead to larger reductions in poverty given the same rate of growth. The Chinese reform starting with agriculture ensured that the majority of the population benefited from the initial reforms, as the population was overwhelmingly dependent on the primary sector. The breathtaking reduction of poverty, from 33 percent of the population in 1978 to 3 percent in 2001, is primarily attributed to the acceleration of agricultural growth during the same time period."

India's experience however has been rather tame in comparison. The reasons are apparent. For one, the governance structure is quite different. There have been few effective policy measures in agriculture despite policy documents stating otherwise. This does not mean that policy-makers do not appreciate the importance of agriculture. But a combination of factors ranging from the fiscal situation, India's infamous infrastructure bottlenecks, manifold laws and coordination failures have contributed to the situation. The politico economic argument of entrenched lobbies is one of the more frequently cited reasons. But not the only one. India had fewer price distortions, and fewer controls on deliveries at the very outset, and therefore agricultural reforms were not as critical as they may have been in China.

AGRICULTURAL ECONOMY

India has a land area of about 328 million ha and a coastline of about 7 000 km. Of the total land area about 165 million ha are arable and 54 million ha belong to the net irrigated area. Ranging from the Himalayas in the north to the Thar desert in the west and the tropical climate in the south, there is a wide range of climatic conditions within the country.

Growth

There are various ways to disaggregate the temporal trend of India's growth experience, for instance, the decadal system. Alternatively, one can use the five-year plan framework, as India has a system of five-year plans. Real annual average GNP growth was 3.7 percent during the First Plan (1951-1956), 4.2 percent during the Second Plan (1956-1961), 2.8 percent during the Third Plan (1961-1966) and 3.4 percent during the Fourth Plan (1969-1974) (GOI 2004-2005). The period up till the mid-1970s is often described by the term "Hindu rate of growth", a term coined for an annual average real growth of 3.5 percent, a rate that India never seemed to cross till the mid-1970s. From the second half of the 1970s, growth rates inched up, especially in the 1980s, to an annual average real growth approaching 5.5 percent. For instance, the Fifth Plan (1974-1979) produced average growth of 5.0 percent, the Sixth Plan (1980-1985) 5.5 percent and the Seventh Plan (1985-1990) 5.8 percent.

As a trend, the new Hindu rate of growth became 5.5 percent. Some reforms were introduced in the late 1970s and 1980s, particularly in the second half of the 1980s. Faced with a balance of payments crisis in 1990 to 1991, more substantive economic reforms were introduced in the 1990s, beginning with 1991. Following these reforms, the Eighth Plan (1992-1997) produced average growth of 6.8 percent and the Ninth Plan (1997-2002) 5.6 percent. The Tenth Plan (2002-2007) has a target growth rate of 8 percent and the NCMP (National Common Minimum Programme) of the United Progressive Alliance (UPA) Government mentions growth of between 7 and 8 percent. The 8 percent target during the Tenth Plan is now impossible to achieve. The first three years have produced 4.2 percent in 2002 to 2003, 8.5 percent in 2003 to 2004 and 6.8 percent in 2004 to 2005, with around 7.5 percent expected in 2005 to 2006. Growth in 2006 to 2007 will have to cross 10 percent for the 8 percent goal of the Tenth Plan to be reached - an unlikely situation.

Poverty

How does growth affect poverty and unemployment in a relatively poor country like India beyond direct antipoverty programmes? Income poverty is certainly not the only indicator of poverty, but it is an important one. Table 1 shows the incidence of poverty in India. The poverty ratio or head count ratio is based on the percentage of population below the national poverty line. This national poverty line is not the same as the international PPP (purchasing power parity) standard of US$1.00/day, but is close to it. Data for computing poverty ratios are obtained through surveys and the National Sample Survey (NSS) organisation conducts sizeable and reliable surveys approximately once every five years. The NSS collects data on consumer expenditure, not income. Similar to surveys elsewhere in the world, the aggregate of consumption expenditure obtained in household surveys falls short of the aggregate of consumption expenditure obtained through national accounts. Ideally, one ought to make adjustments for this gap, but ultimately all such adjustments tend to be subjective. Nationwide, there are significant inter-regional variations.

Income poverty has dropped significantly, especially since the early 1980s (Table 1). This is understandable, and has largely been a function of growth. Income and expenditure distributions have not deteriorated. Depending on the year, the Gini coefficient of rural consumption expenditure distributions has varied between 0.25 and 0.30, while the Gini coefficient for urban consumption expenditure distributions has varied between 0.32 and 0.34. Therefore, because distributions have not declined, growth has induced trickle-down poverty reduction effects. In addition, because income distributions are typically log normal, sharp drops are possible when the thick part of the distribution passes above the poverty line. Arguably, that is precisely what has been happening in the 1990s and this trend should continue for the next decade also.

Table 1. Incidence of poverty

Year	*Rural poverty ratio (%)*	*Urban poverty ratio (%)*	*Combined poverty ratio (%)*	*Absolute number of rural poor (million)*	*Absolute number of urban poor (million)*
1977-78	53.1	45.2	51.3	264.3	64.6
1983	45.7	40.8	44.5	252.0	70.9
1987-88	39.1	38.2	38.9	231.9	75.2
1993-94	37.3	32.4	36.0	244.0	76.3
1999-2000	27.1	23.6	26.1	193.2	67.1

The decline in the combined poverty ratio from 36.0 percent in 1993 to 1994 to 26.1 percent in 1999 to 2000 was one of the sharpest drops ever witnessed. But despite the reduction, the absolute numbers of the poor were quite large, particularly in rural India, where, in 1999 to 2000, almost 200 million individuals were Below Poverty Line (BPL). In addition, the rural poverty ratio was higher than the urban poverty ratio. There are significant regional variations within India, such as between states or between districts. Also there are states where the rural poverty ratio is lower than the urban ratio. This was evidenced in Andhra Pradesh, Gujarat, Haryana, Karnataka, Madhya Pradesh, Rajasthan and Tamil Nadu in 1999 to 2000. However there were other states where the rural poverty ratio in 1999 to 2000 exceeded 30 percent, for example Assam, Bihar, Madhya Pradesh, Orissa, Uttar Pradesh and West Bengal as well as some northeastern states.

All of this evidence indicates that high growth is a significant force behind high poverty reduction. Neither government poverty alleviation programmes nor other direct policy instruments appear to have made a significant dent in poverty levels on a large enough scale. At the same time, there are too many state-level and within state variations in both poverty levels and their trends, indicating that if poverty has to be impacted through other means, then location-specific models need to be built around the socio-economic and other peculiarities of specific villages and urban communities.

Unemployment

The last NSS survey was held in 1999 to 2000 and is somewhat dated now. It revealed a labour force of 363.33 million and a workforce of 336.75 million, with an unemployment rate of 7.2 percent. With a significant share of the workforce employed in the rural sector, unemployment figures may not reveal a great deal. The problem is more of underemployment. Again, there are significant variations in unemployment rates across states. Unemployment rates exceeded 10 percent in states such as Kerala, Tamil Nadu and West Bengal. What is worrying is that the annual average growth in employment has slowed down. While the average annual growth rate in employment was 2.89 percent between 1983-1987-1988 and 2.50 percent between 1987-1988 and 1993-1994, it was 1.07 percent between 1993-1994 and 1999-2000.

This deceleration is largely because in the 1990s agriculture failed to create jobs. While this is recognised and increasing rural employment is important in the present UPA government's agenda, notwithstanding the emphasis on agricultural reforms, it is fair to say that the emphasis is not on creating employment through crop output. Instead, the emphasis is on creating employment through off-farm activities and within farm activities, through non-crop diversification.

Broad Sectoral Trends

Sectoral compositions of the GDP are not constant. They change overtime. In 2003 to 2004, agriculture and allied activities, accounted for 21.71 percent of the GDP, compared to 59.20 percent in 1950 to 1951, 54.75 percent in 1960 to 1961, 48.12 percent in 1970 to 1971, 41.82 percent in 1980 to 1981 and 34.93 percent in 1990 to 1991. This relative decline, particularly marked in the 1990s, has been due to the increased importance of the services sector, rather than increasing importance of the secondary sector. This classification of the primary sector also includes mining and quarrying. In 2003 to 2004, mining and quarrying accounted for 2.3 percent of the GDP. That is, with mining and quarrying excluded, agriculture, forestry and logging and fishing accounted for 19.8 percent of the GDP. The forestry and logging share was 0.01 percent of the GDP and the fishery share was also 0.01 percent. Thus, agriculture contributed 19.6 percent of the GDP.

Trends in the Agriculture Sector

There were three distinct periods in the evolution of Indian agriculture and agricultural policy after Independence. The first period was before the Green Revolution up to the mid-1960s. The second period was the Green Revolution from the mid-1960s till the late 1980s. The third phase continues from then to the present day.

Phase 1

The first period saw the government focusing policy on the following:

- irrigation,
- land reforms,
- community development, and
- restructuring of rural credit institutions.

Land reforms involved abolition of intermediaries, tenancy reforms, acquisition of surplus land through ceilings on holdings and redistribution and consolidation of holdings; success varied across states. Land reforms quite successfully altered agrarian structures in states like Kerala, West Bengal and Karnataka, but failed to do so in states such as Bihar, Orissa and Rajasthan. Changes in agrarian structures occurred in Punjab, Haryana, Western Uttar Pradesh, Tamil Nadu and Andhra Pradesh.

A Community Development Programme was started in 1952, for integrated village development by coordinating the development of agriculture, animal husbandry, infrastructure and extension at the block level. Also in 1952 the National Extension Programme was spliced into the Community Development Programme, incorporating provision of technical inputs. The Intensive Agricultural District Programme (IADP), started in selected potential districts in 1960 and 1961, aimed to provide a package of high-yielding inputs to farmers. The Intensive Agricultural Area Programme (IAAP), started in selected potential blocks in 1964 and 1965 aimed to provide technological inputs for identified crops.

Whether these changes were successful or not depends upon how one benchmarks success. Compared to the low or negative growth in the period immediately preceding Independence, they were successful. Both productivity and output increased during most of the period. However, progress in productivity improvements was not commensurate with the requirements even during good monsoon years. There were institutional constraints in providing inputs like high-yielding varieties, fertilisers and irrigation, coupled with a switch during the Second Plan, with reduced investments in agriculture as fallout. In other words, the government was unable to inject the required financial resources, or develop adequate institutional back-up to enable widespread and sustained implementation. Moreover, the government was also unable to effect improvements in technology and the usage of improved seeds, inputs and methods did not occur until much later.

On the positive side the first phase did succeed in other aspects that facilitated later changes. Land reforms, assigning ownership rights to tenants, consolidation of holdings, among others, were important measures that started during this period, and have continued to a varying extent to the present day. The establishment of many agricultural technology institutions and the birth of a widespread extension services system also transpired during this time. Most importantly, administrative systems were developed during this period with the capacity to take on the challenge of the Green Revolution when the technologies finally became available.

The mid-1960s witnessed a succession of exogenous negative shocks that had a major impact on agricultural output: a war with China in 1962, a war with Pakistan in 1965 and successive monsoon failures in 1965 to 1966 and 1966 to 1967. India had to resort to large-scale imports of food.

Phase 2

The Green Revolution package was based on increased availability/access to:

- high-yielding varieties (HYV) of seeds,
- fertilisers,
- irrigation,
- biochemicals,
- extension services,
- availability of credit,
- establishment of marketing and price-support mechanisms for farmers.

This was spliced with a price-support policy. The government's role in ensuring the availability of complementary inputs for HYV seeds was quite significant. Ahluwalia summarises it well, "[it] required a comprehensive strategy for agricultural change requiring active Government intervention in many dimensions...".

A sustained effort was needed to expand irrigation with a shift from major to medium and minor irrigation projects. It was necessary to move the banking system into rural areas to provide credit for the purchase of biochemical inputs needed for HYVs. Nationalised banks were given the task of upgrading their rural operations and they succeeded to a large extent. Primary agriculture markets were regulated and some of the usurious practices of traders stopped. Extension services were set up and backed by numerous agricultural research establishments. Development of appropriate varieties was critical given the heterogeneity in agroclimatic conditions. The ability of the government to effect adequate coordinating mechanisms was also quite important. Provision of credit was facilitated through the nationalisation of banks. Fertiliser availability and accessibility was expanded through subsidies and was juxtaposed by public and private sector expansion in manufacturing. Moreover price support was instituted at remunerative prices. This coordination role cannot be understated. While subsidised inputs were becoming available and productivity increases were rapid and concentrated, the price-support mechanism ensured that incomes for farmers increased. The three key characteristics of the mandate were: simultaneous handling of constraints in supply and marketing, a coordinated approach and recognition of heterogeneity in agroclimatic conditions. This approach can be assigned most of the credit for the success of the Green Revolution.

From 1949-1950 to 1964-1965, non-foodgrain was the major component of crop production and for both foodgrain and non-foodgrain, production growth primarily transpired through area expansion. This was reversed from 1967-1968 to 1980-1981. During this period, expansion of production for both foodgrain and non-foodgrain occurred through productivity increases, area expansion having slowed down. This trend continued from 1979-1980 to 1989-1990, with productivity increases becoming sharper and the area under foodgrain actually declining. However, in the last few years, productivity growth has slackened, and so has growth in production, particularly of foodgrain.

However the Green Revolution had several positive aspects: (1) As foodgrain production increased faster than the population growth rate, per capita availability of food increased. Per capita net availability of cereals and pulses was 394.9 grams in 1951, 468.7 grams in 1961, 468.8 grams in 1971, 454.8 grams in 1981, 510.1 grams in 1991 and 416.2 grams in 2001. (2) Per capita income generation in agriculture increased. (3) Agriculture was better protected from the ravages of drought. (4) There was greater commercialisation and diversification, with cropping patterns changing in favour of commercial crops and moving away from coarse cereals, even for small and marginal farmers. (5) Capital accumulation rose, including investments from the private sector. (6) While there may have been some reservations about the initial distributional impact of the Green Revolution package, the benefits became more broad-based in subsequent years.

Many holdings are marginal or small. Barring pulses and coarse cereals, all other crops seem to have benefited from the new technologies. In addition, there was a paradigm shift from foodgrain towards commercial crops and even fruits and vegetables, this diversification being at the expense of pulses and coarse cereals. Although a large percentage of small and marginal farmers is still dependent on foodgrain, there was some diversification among small and marginal farmers. For instance, the share of cereals in area cultivated by small and marginal farmers declined from 71.44 percent in 1970 to 1971 to 70.57 percent in 1980 to 1981 and 66.22 percent in 1990 to 1991; the share of fruit and vegetables increased in area from 2.43 percent in 1970 to 1971 to 3.25 percent in 1980 to 1981 and 3.71 percent in 1990 to 1991. Thus it can be argued that there has been a degree of commercialisation and diversification among small and marginal farmers but this could have been greater. However there are some qualifiers. The first is that the coordinated approach did not lead to the spread of the benefits of the new technologies across India. Only a few areas benefited. It is difficult for a central government to introduce measures and mechanisms that are fine-tuned to the requirements of each state and district, Success was based on the term "provision" - inter alia provision of credit, HYVs, subsidised fertilisers and extension services. These had a negative long-term impact on the development of market-backed services and commodities. Lastly, the new technologies were biased towards high usage of water for irrigation. As many parts of the country were not irrigated, power-operated motorised tubewells sprung up across the country. Crops such as rice in Punjab and sugar cane in Maharashtra have become quite important even though surface water is limited in some of these areas. The result has been a rapid fall in water levels.

Phase 3

In the 1980s, when the balance between demand and supply of cereals was in sight, the overall goal of agricultural strategy/policy shifted from "maximisation of production of foodgrain" to "evolution of a production pattern in line with the demand pattern". The implication of this shift is that the emphasis of the policy shifted from foodgrain to other agricultural commodities like oilseed, fruit and vegetables. The shift helped to increase the output of non-cereal food items.

Diversification has been occurring in India since Phase 1, however, it became more prominent in the 1980s. There is little agreement on how much of this was due to the government efforts and how much transpired because of changing economic conditions. It would be fair to argue for oilseed that the shift from coarse cereals to oilseed in rain-fed areas was aided by (i) a protective trade environment; (ii) favourable price policy; and (iii) the connecting of the Technology Mission on Oilseeds. At the same time diversification was not merely attributable to greater area under high-value commodities; the improvements in yield also would have been generated by government-aided supply and technology-related factors.

Table 2 shows the improvements in both production and yield for the crop sector. There were significant improvements for livestock in the 1980s - the value of the livestock sector in total agricultural output value increased from about 18 percent in the triennium ending (TE) 1980 to 1981 to 23 percent in TE 1990 to 1991 (although it remained stagnant at that level by TE 1997 to 1998. Almost 70 percent of the livestock sector's production is accounted for by the milk and products subsector; the role of "Operation Flood" cannot be understated.

The value of the fisheries subsector swelled by about 50 percent during the 1990s, although as a share of the total it declined from 1.3 to 1.0 percent of the total agriculture sector value during the period. There were significant and coordinated government efforts to increase fish production. The central government's outlays towards the fishery sector as a share of the total agriculture sector outlays were more than doubled from 2 to 3 percent in the 1970s to about 5.5 percent in the 1980s and 1990s. Production and development programmes were instituted in both marine and inland areas. Farmers' development agencies were established both for fresh and brackish water areas. These programmes included technology upgrading components, encouragement and involvement of the private sector in activities such as seed, feed and other inputs and also creation of suitable infrastructure for storage, transport, marketing and credit. By 1998/1999 more than 50 seed hatcheries at the national level had been established, fishery industrial estates had been created and 30 minor fishing harbours and 130 fish-landing centres had evolved.

Table 2. Compounded annual rates of growth of production and yield

	1980-81 to 1989-90 (production)	*1980-81 to 1989-90 (yield)*	*1990-91 to 1999-2000 (production)*	*1990-91 to 1999-2000 (yield)*
Non-foodgrain	3.77	2.31	2.78	1.04
Foodgrain	2.85	2.74	1.94	1.52
Pulses	1.52	1.61	0.61	0.96
Oilseed	5.20	2.43	2.13	1.25
Cotton	2.80	4.10	1.73	-0.61
Sugar cane	2.70	1.24	2.78	0.95
Tobacco	-1.05	1.79	1.05	-0.23

Diversification in favour of high-value commodities is considered by some to have only partly benefited through government efforts. Both econometric and GIS-based studies also suggest that the presence of infrastructure aided diversification. Rao et al. found that in their sample, the areas close to large urban centres could diversify more to high-value commodities. They concluded that urbanisation is a strong demand driver for high-value commodities. In an earlier econometric study, Joshi et al. came to a similar conclusion - that diversification into high-value commodities was demand driven, unlike the supply driven Green Revolution.

Between 1990-1991 and 1998-1999, the annual average increase in yields was only 1.79 percent and 1.31 percent for wheat and rice, respectively. Despite the successes of the Green Revolution, there continue to be concerns with agricultural performance during Phase 3, the present phase of globalisation and diversification, initiated in the 1990s, and some macro issues emerge.

The first macro point is the one already alluded to indirectly. Compared to the 1980s, there was deceleration of agricultural growth in the 1990s. This is evident from Table 2 and cuts across all crops and both production and yields. The quality of the system for collecting agricultural statistics and the extent of deceleration can be debated, but the conclusion that growth in the crop sector decelerated in the 1990s is fairly obvious. In any case, deficiencies in the quality of statistics are constant factors and cannot explain the deceleration. Indeed, growth in non-food and non-crop output was also faster in the 1980s than in the 1990s and growth in animal husbandry, poultry, dairy, horticulture and fisheries also slowed down. This deceleration happened despite the reversal of historical discrimination against agriculture. In the 1990s, the terms of trade moved in favour of agriculture and this was independent of the terms of trade measures used. There were several reasons for this reversal of trend. There were higher prices for rice and wheat through support/procurement policies and prices of other crops also increased. There is a positive correlation between procurement prices, open market prices and higher prices in the PDS (public distribution system). Simultaneously because protection granted to manufacturing declined, the relative prices of manufactured products also declined. A priori, one should therefore have expected a positive supply response and increased capital accumulation.

However farm profitability declined in the 1990s. Not all farmers had access to higher support prices for rice, wheat or sugar cane, there was a deceleration in yields and real input prices also increased, thanks to prices for fertilisers, power and diesel that were closer to market-determined prices. Higher cereal prices also contributed to increases in wage costs. Consequent to per capita income growth, NSS data show changes in consumption patterns, with a decline in consumption of cereals in both rural and urban India, despite cereals still being important in the food basket. There has been a shift from consumption of coarse cereals to rice and wheat and also a shift towards consumption of fruit and vegetables and even fish and eggs. Table 3 shows some of these changes.

Table 3. Per capita consumption

Commodity	*1955-56*	*1975-76*	*1990-91*	*1997-98*
Foodgrain (kg/year)	155.6	158.5	180.6	179.3
Edible oils and vanaspati (kg/year)	3.2	4.2	6.5	8.2
Sugar (kg/year)	5.0	6.2	12.5	14.1
Textiles (cotton equivalent, m/year	14.4	17.6	24.8	28.0
Tea (kg/year)	0.36	0.45	0.61	0.64
Milk (l/year)	4.7	4.6	6.3	7.2
Eggs (no./year)	5.3	15.5	26.0	31.0

Consumption patterns are changing, even for the bottom 30 percent of the population and the shares of non-cereal food (fruit and vegetables) and non-food items have been increasing in consumption baskets. Increased cereal prices and lower rural incomes may have depressed demand for both cereal and non-cereal food, but there is no denying that preferences are also changing.

FUTURE EXPECTATIONS

Because of deceleration in the 1990s, expanding population and rapid economic growth, food security concerns have been raised. Food security at the national level need not be equated with 100 percent self-sufficiency in foodgrain production. Adequate foreign exchange reserves should instead be the issue. The mindset is largely a reflection of the food import scare India confronted before the Green Revolution. Future demand projections are contingent on assumptions made about population, per capita income growth, the time frame and hunger removal. The time frame tends to be 2020. Population projections for 2020 vary with 1.315 billion being closer to the mark now.Consider the basic trends first. With economic growth, population increases and changing age distributions, food requirements will be higher. How they match up with expected production given current trends, and what implications this has for international markets are discussed later.

Population

India's population in 2001 was 1.02 billion according to the Census of India, making it the second most populous country in the world after China. It has been estimated that in the next 15 years, the population will increase by about 23 percent to 1.24 billion persons. But estimates vary, some being as high as 1.4 billion. High growth rates can be ascribed to death rates being lower than birth rates. Between 2001 and 2010, almost 150 million more people need to be provided with adequate nutrition and between 2010 and 2015, another 83 million people need to be covered. Most GOI estimates do not go beyond 2015, but trends till 2020 are not likely to be any different.

Incomes and Expenditures

According to India's national accounts, total GDP in 2003 to 2004 was about US$600 billion. Of this, personal disposable income was in the range of US$512 billion and 24.6 percent of this income was directed into savings by the household sector. During the postliberalisation decade, from 1993-1994 to 2003-2004 the average annualised growth rate of India's GDP has been around 6.2 percent and this has been accelerating in the early 2000s. Table 4 records the growth rates of some of the key indicators of national accounts of India. Poverty has declined significantly, and is expected to continue to fall. The poverty line is defined by the GOI as the cost of a package of commodities that can provide about 2 400 calories and 2 100 calories to an average Indian citizen living in rural and urban areas, respectively. The HCR (head count ratio) in India had fallen to about 26 percent in 1999 to 2000 from 39 percent in 1987 to 1988. Our own estimates are that if the GDP growth is sustained, it will fall to about 14 percent by 2010 and then to 8 percent by 2015. These are not very different from other estimates discussed earlier. Two further points need to be mentioned in this connection.

Table 4. Growth of macroeconomic variables

National accounts of India, 2003-2004	*Annualised growth rate between 1993-94 and 2003-04*
GDP	6.18
National income	6.41
Net national disposable income	6.55
Private income	6.73
Personal income	6.59
Personal disposable income	6.57
Domestic saving for household sector	9.77
Wholesale price index	5.82

First, income distributions are typically log normal and as the thick part of the distribution passes through the poverty line, it is possible for sharp reductions in HCRs to occur. Second, when the Indian poverty line evolved in India in the early 1960s, the presumption was that health and education would be merit goods, if not public goods. In either event, they would be provided by the government and need not be ingredients in private consumption expenditure. Hence, these are not included in defining or computing the poverty line. The 1990s, however, witnessed a switch from public expenditure to private expenditure in both health and education. It is thus possible that the Indian poverty line might be redefined at some time in the future. Notwithstanding this possibility, the proposition about decline in HCRs, assuming an unchanged poverty line, remains valid.

Both rural and urban areas have a very similar poverty scenario. However, India has a long way to go before the whole population has the means to consume the minimum required calories per day. During 1999 to 2000, more than one-fourth of India's population was below the poverty line and this amounted to about 260 million people at that time. If the rapid rate decline in poverty after 1991 were to continue, not only the percentage, but also the actual number of people below the poverty line is expected to be lower. The net impact on the demand for greater nutrition is obvious.

Changing Food Basket

As incomes, and as a consequence expenditures, increase, not only are expenditures on food and agricultural commodities as a whole likely to increase, but the consumption basket characteristics are also likely to change. In both rural and urban areas, dairy products, meat, fruit and vegetables are likely to have a greater share of the additional demand being generated. This is the standard picture across different countries, and is also reflected in the differing elasticities observed by various studies both for India as well as other countries. The income elasticities for five different income groups (quintiles 1-5) in urban and rural areas of India are shown in Table 5.

Table 5. Income elasticities across quintiles

	Q1	Q2	Q3	Q4	Q5
Rural					
Wheat	0.50	0.49	0.48	0.48	0.47
Rice	0.72	0.47	0.31	0.12	-0.21
Maize	-0.63	-0.51	-0.42	-0.33	-0.17
Milk and products	2.35	1.96	1.70	1.40	0.88
Eggs	2.44	2.17	1.98	1.77	1.39
Chicken	1.35	1.32	1.30	1.28	1.25
Other meat	1.35	1.32	1.30	1.28	1.25
Urban					
Wheat	0.32	0.25	0.19	0.13	0.04
Rice	0.39	0.22	0.09	-0.05	-0.29
Maize	-0.60	-0.63	-0.65	-0.67	-0.71
Milk and products	1.64	1.38	1.19	0.97	0.61
Eggs	1.72	1.56	1.44	1.29	1.06
Chicken	0.48	0.45	0.43	0.41	0.37
Other meat	0.48	0.45	0.43	0.41	0.37

The food requirement is dependent on the consumption behaviour, which in turn is a direct outcome of income. With development, incomes are expected to rise and would thus impact consumption behaviour. The income elasticities are a measure of the future demand of various food items.

Quintile one (Q1) denotes the poorest 20 percent of the population and Q5 denotes the richest 20 percent. Income elasticities across all commodities, as well as quintiles, are higher for the rural areas as compared to urban areas. The elasticities also fall as income rises, both within urban and rural quintiles. Wheat shows an elasticity of 0.50 and 0.32 for rural and urban areas respectively, in Q1. However, the elasticity decreases to 0.47 and 0.04 for the same areas in the case of Q5. Similarly, the income elasticity of rice in rural areas drops from 0.72 to a negative 0.21 as we move up from Q1 to Q5. A similar pattern is observed in the urban areas as well.

The elasticities of milk and milk products, along with eggs, show very high values of greater than 1 for the lower quintiles, implying that the proportionate change in consumption demand is greater than the proportionate change in income. However, these elasticities also follow the general decreasing trend as we ascend to higher income groups. Elasticities for chicken and other meat are lower than those of milk and eggs. For rural areas, income elasticities for meat are greater than 1 across all income groups (1.25 for Q5), while they are less than 1 for all income groups falling in urban areas. Meat eating is common across India, however a significant share of the population is vegetarian. Even among those who consume meat, at certain times of the year a largely vegetarian diet is followed. Consequently, the income elasticities of meat products may not be as high as in other countries.

In both rural and urban areas, the own price elasticity declines in absolute terms as we move from the very poor to the non-poor section of the population. For cereals in rural areas, elasticity declines from 0.7 to 0.1, while it changes from 0.5 to 0.1 for the urban areas. Milk, edible oils and meat show similar trends in both urban and rural areas, although demand for milk shows a higher sensitivity to price changes. Price elasticity of demand for sugar in rural areas increases marginally from 0.7 to 0.8, before declining to 0.6 for the non-poor. In urban areas, however, the elasticity shows a gradual decline as we move to higher income groups.

Consumption of Agricultural Products

Others have estimated the growth in food consumption and most have similar insights. Milk and milk products will see the largest increase up to 2020. Fruit and vegetables, sugar and meat and fish consumption will also increase significantly. The reasons are obvious in light of the previous discussion. Income increases matched by the high income responsiveness of these commodities will be the driving force. Among cereals, the highest percentage growth will be in wheat, followed by rice. Coarse grains are not likely to grow as much.

The demand for rice is projected to grow from 78.3 million tonnes/year in 2000 to 118.9 million tonnes/year in 2020, showing a compounded annual growth rate of 2.1 percent. The main factor driving this demand will be the increase in population. Wheat demand rises from 54.2 million tonnes in 2000 to 72.1 and 92.4 million tonnes in 2010 and 2020 respectively, with a growth rate of 2.7 percent. The responsible factors will be the rising population and positive income elasticity for wheat. The demand for other cereals will grow by 0.9 percent, primarily on account of change in preferences as incomes grow. On the whole, consumption demand for all cereals will show a growth of 76 million tonnes from 2000 to 2020 at the rate of 2.1 percent, which is marginally higher than the projected growth rate of the population (Tables 6 and 7).

Demand for food items such as milk products, meat and fruit is expected to grow at much higher levels of 3 to 5 percent in the next few years. This is a natural result of growth in incomes following economic development. The estimated demand for foodgrain in 2020 is 240.6 million tonnes/year, growing from 155.6 tonnes/year in 2000 at the rate of 2.2 percent per annum. The income effect will manifest itself to raise the demand for milk and milk products to 166 tonnes, edible oils to 11 tonnes, meat and fish to 11 tonnes and sugar and gur (jaggery) to 25 tonnes/year (Tables 6 and 7).

Table 6. Consumption estimates (million tonnes/year)

Commodity	*2000*	*2005*	*2010*	*2015*	*2020*
Rice	78.3	88.1	98.0	108.5	118.9
Wheat	54.2	63.1	72.1	82.2	92.4
Other cereals	13.1	13.6	14.1	14.8	15.6
All cereals	145.1	163.1	181.1	201.1	221.1
Pulses	10.6	12.6	14.6	17.1	19.5
Foodgrain	155.6	175.7	195.7	218.2	240.6
Milk & milk products	64.2	85.3	106.4	136.1	165.8
Edible oils	5.3	6.5	7.7	9.3	10.9
Meat & fish	4.7	6.0	7.3	9.0	10.8
Sugar & gur†	11.5	14.3	17.2	21.2	25.1

Table 7. Growth in consumption

Commodity	*CARG (2000-2020), %*	*Increase between 2000-2020 (million tonnes)*
Rice	2.10	40.7
Wheat	2.70	38.2
Other cereals	0.90	2.5
All cereals	2.10	76.0
Pulses	3.10	9.0
Foodgrain	2.20	85.0
Milk & milk products	4.90	101.7
Edible oils	3.70	5.7
Meat & fish	4.30	6.1
Sugar & gur	4.00	13.6

Production of Agricultural Commodities

India has one of the largest land masses of any country of the world, and a high proportion of its land mass is arable. Although most of the land depends upon the monsoons for irrigation, irrigated areas have been increasing and comprise about 40 percent of the gross cropped area. However, yields in India are among the lowest in the world, and therefore there is significant scope for increases in the coming future.

A yield comparison across countries shows how India's crop production fares vis-à-vis international benchmarks. India's figures are contrasted with crop yields in the United States, China and an average for the world. While these can be compared for a number of different crops, the results are remarkably similar, with a few exceptions. For cereals, coarse grains and pulses, both the United States and China are well above the world average. India has extremely low yields, even by developing country standards. For oil crops and primary fibre crops as well, India has yield levels below the United States, China and aggregate world levels. Across different product segments, a similar picture emerges.

While the static picture of India's crop production is not encouraging, the indicators for the future provide grounds for optimism. The potential improvements made possible by improvements in technology, organisational expertise and policy redirection have lifted the yield levels of many of India's crops. From 1961 to 2004, commodities with the largest increases in yield were wheat, maize and jute, and rice to a lesser extent. Pulses, oil crops and primary fibre crops have had muted increases in yield.

On an even more cautionary note, in the last ten years, improvements in yield have plateaued for Green Revolution heavyweights like wheat and cereals. From 2001 to 2004, several crops even experienced an absolute decline in their yields. Given these indications, prospects for the future of Indian agriculture production are mixed. While yield levels seem to have plateaued, they are still significantly below competitive levels for most crops. While countries at the peak of the agricultural technology curve must pour funds into R&D to invent new ways to increase production, countries on the other side of the curve stand to benefit from implementing those methods that have already been pioneered. Catching up is always easier and for this reason alone, India should be able to realise fast and significant gains in the near future.

Price Competitiveness

Currently, there are some estimates of price competitiveness, the results often being functions of whether an importable or an exportable hypothesis assumption is used and whether one uses the nominal protection coefficient (NPC), the effective protection coefficient (EPC), the effective subsidy coefficient (ESC) or the domestic resource cost (DRC). A widely referred to study by Bhalla revealed that:

- Most crops except oilseed, some coarse cereals and sugar are internationally competitive;
- More crops would be competitive if developed countries withdrew domestic support to agriculture;
- However, import competitiveness becomes reduced overtime because:
    - Successive price hikes have led to very high prices for many agricultural commodities. For example major crops such as rice and wheat became non-competitive during recent years because of a fall in their international price combined with a major increase in their domestic price owing to increase in MSP.

- Productivity growth has decelerated, moreover production efficiency is being affected by lack of implementation of technological innovations.
- A major thrust is needed in infrastructure investment in general and investment in science and technology in particular for India to maintain and enhance its competitiveness.

Production, Consumption and Surpluses

The GTAP model has been used to predict India's trade with the rest of the world including India's own production and consumption. The quantitative assumptions behind the model are given in Table 8.

Table 8. Assumptions: the growth rate of exogenous variables of India in four stages (per year %)

	2001-2005	2006-2010	2011-2015	2015-2020
GDP	6.73	7.00	7.50	8.00
Population	1.80	1.70	1.60	1.50
Cultivated land area				
- net	0.00	0.00	0.00	0.00
- gross	0.50	0.75	1.00	1.00
Labour supply	2.70	2.71	2.08	1.73
- unskilled labour	2.46	2.48	1.85	1.58
- skilled labour	7.24	6.40	5.32	3.63
Capital stock	6.46	6.67	6.92	7.20

India's GDP is expected to grow at more than 7 percent for much of the period under consideration. Moreover, the growth rate is expected to increase for the coming decade. As population growth has been falling owing to falling birth rates for the past two decades, the growth in labour supply would decline over the next decade and a half. Irrigated area is expected to grow, although not too significantly, as further increases will require significant public investments that show no signs of accelerating. Private efforts are likely to be the main driving force behind the 1 percent annual increase in irrigated area. Skilled and highly educated labour is expected to rise much faster than unskilled labour; this reflects the latest improvements in the educational achievements of a large segment of the population. The capital stock would have to grow rapidly as well, and is expected to increase as labour supply growth tapers off to sustain the high growth rates of 8 percent.

Although long-term projections are rare, when they are made, most experts expect real GDP growth of 7 to 8 percent for India in the period leading up to 2020. Notwithstanding reservations about the speed of economic reforms, consequent to debates about liberalisation in a democratic policy, these estimates of 6, 7 or 8 percent are probably underestimates, even if one ignores the exchange rate aspect. There are different ways to argue this out and all of these trends reinforce one another. Looking forward to 2015, an average savings rate of 30 percent is certainly plausible given the current rate of 28.1 percent rising. Foreign capital inflows are also increasing. There is no reason why the average investment rate should not therefore be 32 percent. The present incremental capital/output ratio (ICOR) is around 4. Whatever the figure, there is no reason why this should increase significantly in the near future. Indeed, with reforms, competition and resultant efficiency improvements, the ICOR should decline. But even with an ICOR of 4, there is growth of 8 percent.

It is obvious that these quantitative forecasts have not considered the possibilities of future environmental degradation, unforeseen changes in economic and trade policy and of course technological changes.

The growth in rice is going to be in the order of 1 percent per annum till 2010, for both production and consumption. For wheat however, the rise in incomes is going to create a marginally higher growth in consumption than expected production increases. For both oilseed and processed foods, consumption increases will outstrip production increases. Given current conditions and trends, similarly, for forestry products, as well as other foods, consumption increases are going to be far higher than likely production increases.

Overall, the patterns are quite unambiguous. Although production levels are likely to increase significantly, they are not going to be able to match the increases in consumption, at current and expected overall economic growth.

This does not imply a fall in export earnings, as overall expected price increases due to the opening of international markets, as well as reduction of subsidies will tend to have a positive impact on prices. This is regardless of the temporary impasse at WTO on agricultural liberalisation, which hopefully, will be temporary.

India is likely to maintain the status of an exporting country for rice and wheat. However for oilseed, India is likely to be a significant importing country in another decade and a half. It will also remain as a marginal exporter of sugar and coarse grains. Despite being an exporter of jute and cotton, India will be a net importer of plant-based fibre. Moreover, it will also start to become a net importer of other horticultural crops. Although overall milk production is expected to increase rapidly, consumption increases will prevent India from becoming a large dairy product exporter. Cattle and red meat exports are likely to increase, although other animal products, such as leather, are likely to see a net increase in imports.

Expectations for India

Overall, India will remain a marginal exporter of wheat, coarse grains, sugar, cattle and red meat, fish and other foods. However, it will become a significant importer of oilseed, forestry products, other animal products and plant fibre. For products such as plant fibre and animal products, its position as a net exporter of manufactured items will be facilitated by larger imports of raw materials. For the other segments, rising domestic consumption not matched by domestic supply increases, will be the driving principle.

Table 9. Net export share of world export in value terms (%)

	2001	2005	2010	2015	2020
Rice	10.0	9.7	9.8	10.2	10.8
Wheat	2.6	1.9	1.4	1.1	1.0
Coarse grains	0.2	0.2	0.2	0.2	0.2
Oilseed	-3.3	-5.0	-7.2	-9.8	-12.8
Sugar	2.7	2.2	1.8	1.3	0.9
Plant-based fibre	-6.6	-9.4	-12.6	-15.5	-18.1
Other crops	1.1	0.3	-0.6	-1.3	-2,1
Cattle and meat	1.0	0.9	1.0	1.2	1.7
Other agricultural products	-0.9	-1.8	-3.2	-5.3	-8.4
Milk	0.2	0.1	0.0	-0.1	-0.3
Fish	0.2	0.2	0.1	0.1	0.2
Other food	1.1	1.1	1.1	1.2	1.3
Forestry	-4.9	-6.1	-8.2	-11.1	-14.6

Forestry products are likely to be a significant import item, the bulk being related to wood. Industrial and furniture requirements are the important components that will drive forestry product shortfalls. Although India has large forest cover, commercial forestry is insignificant and is unlikely to expand in a big way given its current environmental protection laws. The statistics for forestry products have large gaps. According to some FAO estimates, production in India for industrial roundwood has been falling, while fuelwood has stagnated at about 2.9 million m^3. Pulp and matchwood have also been showing a negative trend. On the whole, it is expected that India will be substituting imports for domestic production more and more in the near future.

The overall position is quite unambiguous. For Indian policy-makers, the most worrying aspects are stagnating yields. This in itself is not surprising, given the low investments in agriculture, as well as the constraints on agricultural trade. If India is to become a significant exporter of other agricultural products apart from rice, emphasis will have to be placed on improving yields. Investments in rural areas are therefore essential.

How would these estimates change if growth assumptions were different? India would continue to be a large importer of forestry products whether economic growth is 6 or 8 percent. Although the quantum may differ. Similarly, its position as a significant textile and garment producer may require it to import plant-based fibre, in spite of plus or minus 1 percent variation in growth. India's low production of oilseed and high requirements will also drive its oilseed imports. But this needs to be qualified. Given its land area, it is conceivable that productivity enhancements could lead to a lower shortfall in oilseed and pulses. However, current trends indicate this to be a remote possibility. Now rice remains. Compared to the past, rice is not a preferred cereal; income elasticity measures also indicate that at the middle and upper income levels, the parameters are lower. In other words, estimates of rice surpluses are not as sensitive to economic growth assumptions.

AGRICULTURAL REFORMS

One way to categorise the reform agenda could be in terms of objectives: (a) protection of natural resources, (b) rise in productivity, (c) encouragement for value-added agriculture and (d) disseminating the benefits of agricultural growth for overall rural development. However as many of the necessary actions span these objectives, the requirements can be grouped under: (i) land, (ii) water, (iii) credit, (iv) markets, (v) diversification and (vi) technology.

Numerous reports and studies within various organs of the government have identified the critical issues that need to be addressed. However, it will take some years for the political establishment to tackle many of them even if there is a consensus on the issue.

Land

The arguments in the Approach Paper to the Tenth Five Year Plan (APTFYP) are almost generic. Both in agriculture and in forestry, uncultivated land must be brought into productive use. This requires a proper land-use policy, including tenurial reforms. Leasing and sharecropping of uncultivated agricultural land needs to be regularised.

- There is unequal access to land and this adversely affects farmers at the bottom of the farm size hierarchy. Despite land redistribution attempts, the land base of marginal and near-landless farmers has not improved much over time; land redistribution gains having been restricted mostly to middle levels of the peasantry. Currently, however, land redistribution does not appear to be on the agenda of any political entity.
- Fragmentation of holdings has led to reduced economies of scale, low allocative efficiency and low investments and mechanisation. In many cases, this leads to subsistence farming. One should review the policy on land availability and ideally, repeal land ceiling acts. Consolidation of landholdings is important to exploit economies of scale, as has happened in Punjab. Though land rights become complicated because of rights and subrights, it is possible to perform the consolidation on the basis of primary rights and legalise leasing of land. Some states

are in the process of improving their land records systems and this is expected to gather momentum in coming years. While it might not address fragmentation directly, it would at the very least enable functioning of land markets and enable mutually beneficial land transactions.

— There is underutilisation of available agricultural land. If land is limited, there must be land-augmenting or land-saving technologies. Agricultural holdings of socially disadvantaged groups like "scheduled castes" and "scheduled tribes" also exhibit greater underutilisation of land, particularly in states such as Bihar, Uttar Pradesh, Gujarat, Assam and Kerala and this may be due to the poor quality of land. The solution lies in the development of appropriate technologies and provision of support services to enable the small and disadvantaged farmer to utilise this resource.

— There is increasing land degradation because of natural and human-induced factors. Much land has become unusable owing to land degradation, salinisation or waterlogging. There is no transparent policy to make this land available to agro-industry. Panchayats (village local bodies) can be used to identify unusable, saline or waterlogged land, which can be reclaimed through joint cooperatives of landless farmers and agro-industry for cultivation, afforestation and plantation crops. With greater powers and funds being devolved to the Panchayats, it is expected that the local communities will be better motivated to take up these tasks collectively. Moreover there are some measures for growing biofuel-yielding plants that are gathering momentum.

— Notwithstanding official bans on leasing in and leasing out of land, tenancy continues de facto, although the form has changed from sharecropping to fixed rent tenancy. Given economies of scale in production, this suggests a conversion of the de facto practice to de jure legal changes, so that there are incentives to invest in land and land can also act as collateral. However, these changes are likely to be highly sensitive politically and unlikely to be changed in the near future.

Somewhat in the same vein, provisions are needed for contract farming, with explicit clauses on contract enforcement and dispute resolution.

Water

Of the total agricultural land, only 35 to 40 percent is irrigated and agriculture is therefore still dependent on monsoons. Moreover, during the 1990s , the area under canal irrigation declined, despite an increase in expenditure on major and medium irrigation projects. Simultaneously, there was also a decline in the use of tank irrigation; an increasing source of irrigation has been groundwater irrigation, mostly owned and managed by farmers.

There are several issues connected with irrigation and water usage: (a) water conservation and appropriate pricing of water; (b) diversion of water from surplus basins to deficit basins; (c) decline of traditional water management structures and institutions; (d) disparity in the spread of irrigation across regions; (e) absence of sufficient attention to the software of irrigation systems, rules and procedures, allocation and distribution of water and relationships between users and the bureaucracy; (f) inequity in groundwater usage and the absence of legal frameworks and enforcement mechanisms to check overexploitation; (g) insufficient usage of drip and sprinkler irrigation; and (h) declining user management and participation in irrigation systems. Nor should one forget that parts of India are still characterised by rain-fed agriculture, such as parts of Andhra Pradesh, Karnataka, Maharashtra, Madhya Pradesh and Rajasthan, producing coarse cereals and oilseed. These drought-prone areas often witness farmer trauma and institutional and policy support for dryland agriculture is weak or non-existent.

The solutions are well-known: community-based water conservation measures, revamping of traditional water-harvesting mechanisms, greater investments in maintenance of irrigation systems and appropriate pricing of electricity for agricultural purposes. Water will emerge as a serious consideration over the next few years, and it is likely that many, if not most, of the above solutions will have to be put in place.

Credit

The objective of the Agricultural Credit Policy in India since Independence has been gradual replacement of moneylenders by cooperatives and lowering of interest rates. Until banks were nationalised, cooperative institutions were the only source of institutional credit in rural areas. Since nationalisation, scheduled commercial banks and regional rural banks (RRBs) have also been part of the formal credit system. As percentage of agricultural GDP, institutional credit to agriculture increased from 2.56 percent in 1970-1971 to 7.11 percent in 1980-1981 to 12.14 percent in 2000-2001, although studies do suggest asymmetries in distribution of credit across farm size and across regions.

But small farmers continue to resort to informal lenders, as the current system of institutional credit to farmers suffers from non-farmer friendly practices, delays in credit delivery and collateral problems. This forces farmers to resort to non-institutional sources of credit at high rates of interest. There should be a shift in attention from the cost of credit to availability of credit. There are a number of strands in the reform package:

- First, improvements in the supply of credit by simplifying loan application and documentation procedures, easing collateral requirements and recapitalising regional rural banks (RRBs) and cooperative banks.

- Second, there are issues in reforming cooperatives and freeing them from government intervention.
- Third, there are questions of stimulating demand for credit by pushing the formation of self-help groups (SHGs) and NGO involvement, especially in the context of microcredit. There is a government policy to link SHGs with banks and up to December 2004, 1.28 million SHGs have been linked with banks.
- Fourth, the agricultural credit issue cannot be delinked from insurance. There is poor protection for farmers because the current Comprehensive Crop Insurance Scheme is too complicated; it does not cover all crops and has a complex system for loss assessment.
- A simplified crop insurance scheme should have farmer friendly administration, cover all crops, cover farmers who do not take loans, charge actuarial rates and provide subsidies for small and marginal farmers. In all fairness, one should point out that this is easier said than done.

Markets

In much of the agricultural reform agenda, there is questioning of what has historically believed to have been the government's role in pushing agricultural development. At a generic level, these government policies include agrarian reforms; the broad objective of what was called self-sufficiency in foodgrain production; a price-support policy; government interventions in production and distribution of agricultural products, inputs and services; technology generation and dissemination; and promotion of cooperatives. For instance, the self-sufficiency in foodgrain production objective led to a system of price-support purchases through the FCI (Food Corporation of India) and subsidised distribution through the PDS. This price-support system for rice and wheat created incentives in favour of rice and wheat, and favoured states that had such surpluses - Punjab, Haryana, the western parts of Uttar Pradesh and Andhra Pradesh. Quite often, the MSPs were increased far more than what the CACP (Commission on Agricultural Costs and Prices) had recommended. The FCI procured both for buffer stocks and the PDS and, in both cases, inefficiencies in FCI procurement have been documented. It is being suggested that while the FCI might continue to procure for buffer stock purposes, for the PDS, procurement should be thrown open to state governments, the private and the cooperative sector.

Nor has the PDS worked well. The APTFYP comments on the targeted PDS, in an attempt to reform the PDS: "There is 36 percent diversion of wheat, 31 percent diversion of rice and 23 percent diversion of sugar from the system at the national level. ....Due to poor off-take by the states and even poorer actual lifting by the BPL families, the scheme has not made any impact on the nutrition levels in [some] states." The

Expenditure Reform Commission's (ERC) Report on Food Subsidy is a devastating critique of the PDS, highlighting FCI inefficiencies as well as the pro-urban bias of the PDS. The reform agenda therefore encompasses a revamp of the PDS system as well and some states have now begun to experiment with a system of food stamps.

In addition to procurement and PDS, the APMC Acts and the Essential Commodities Act (ECA) are perceived to be unnecessary impediments towards developing a national and common market for agricultural products. Government policies on agricultural marketing have covered regulation of marketing practices, creation of infrastructure, price support, promotion of cooperative organisations, technology transfer, input supply and credit delivery. The following trends are evident:

a) Marketed surplus per farm rose.

b) Marketing charges payable by farmers were dropped, standardised, or liability shifted to buyers.

c) More and more reliable information was made available to farmers.

d) Better storage facilities and quicker means of transportation became available.

e) The trend of village sales declined and market sales increased.

f) Price-support programmes helped farmers reduce price risks.

g) In some areas, there was an increase in cooperative or group marketing by farmers.

There are 7 161 regulated markets, mostly primary wholesale markets and the number has increased overtime. Through APMC Acts, marketing boards have often accumulated substantial resources, based on fees that range between 0.5 and 2 percent of the value of produce transacted. These market development funds were supposed to be ploughed back into the development of infrastructure and other services. Whether this has indeed happened, is doubtful. In many states, state governments alone are allowed to set up markets for agricultural commodities. Such monopolistic practices prevent the development of efficient and transparent agricultural marketing and the use of preharvest and postharvest technologies, or the setting up and enforcement of quality standards. The food chain in India has several intermediaries from the farmer to the consumer. It is estimated that 20 percent of the food produced in India is wasted Less than 2 percent of fruit and vegetables is processed and wastage is estimated at 25 percent of total production. While private traders are allowed to trade in foodgrain, the APMC regulations create much complication in trade in agricultural commodities.

DEVELOPMENT OF CROP TECHNOLOGY

India spends 0.26 percent of the agricultural GDP on research and 90 percent of expenditure on R&D comes from the public sector. However, R&D has paid little attention to crops grown under dryland conditions. Agricultural research and extension services need rehabilitation, with a focus on subsistence crops and dryland agriculture. Krishi Vigyan Kendras (KVKs) need to be further integrated with state and district level extension mechanisms and selected agricultural universities and research laboratories need to focus research on high-yielding hybrid coarse grains, pulses and items for mass consumption.

The Protection of Plant Varieties and Farmers' Rights Act seeks to protect the rights of farmers, breeders and researchers. Issues such as the compulsory licensing of rights and preventing the import of varieties by incorporating the Genetic Use Restriction Technology (GURT), which makes it obligatory for farmers to depend on companies for seeds, are some of its more discussed features. It provides rights to farmers to use the seeds from their own crops for planting the next crop. Further, there are provisions for benefit sharing with farmers and a penalty for marketing spurious propagation material. The act came about after almost a decade of debate on the relative rights of farmers, breeders and developers. However, the success of illegal varieties of BT cotton illustrates the poor capacity of the government to enforce intellectual property rights. The structure now exists, and an efficient enforcement mechanism needs to be created.

This supplements the issue of upgrading facilities and expertise in ICAR (Indian Council of Agricultural Research) and state agricultural universities with full attention to the needs of the industry, with adequate funding linked to outcomes. However, there is a difference between developing technology and disseminating it. In emphasizing the former, one should not lose sight of the latter. During the last decade, the agricultural extension mechanisms of state governments have been considerably weakened.

Access

The discussions in the previous sections also reveal an important lesson. Improvements in yield necessarily require better technologies. However proper mechanisms for making technology accessible to a wide spectrum of potential users are essential for benefits to be realised. In the case of India, the success of the Green Revolution was based on precisely this aspect. However, as further spread was limited due to the inability of the government to develop and disseminate newer technologies and complementary inputs over a larger geographical area, sets of commodities and classes of farmers, significant yield improvements have not continued.

Moreover, once a particular technology delivery regime is initiated, it takes steady and sustained efforts (especially in democratic and multi-tiered governments) to switch to a new delivery regime. Production scenarios are therefore difficult to predict and are to a large extent dependent upon the government's ability to facilitate appropriate technology, input and service delivery mechanisms.

The decentralised system, with many small and large private entities that are cropping up across India, is a good way of ensuring that appropriate technology is developed and disseminated. In this regard it is essential that minimal regulation is applied to the activities of these merging private entities, while at the same time ensuring fair and quick dispute resolution.

EFFORTS FOR AGRICULTURAL AND RURAL DEVELOPMENT

India's UPA government, at the union or central level, came to power in 2004 and the UPA's National Common Minimum Programme (NCMP) forms the foundation of economic policy. The NCMP is as much a political document, as it is an economic one. Many of the statements are echoed across the political spectrum and are not likely to be opposed. There has been a growing consensus on many of the economic aspects of India - decentralisation within the government, the growing role of the private sector and the importance of a social security net are some examples. The importance of water conservation and rainfall harvesting and other environmental issues is also echoed across the political spectrum. The importance of rural infrastructure is also universally agreed. Rural credit is becoming easier to provide via increased private sector involvement, the roles of NGOs and microcredit institutions.

Current Agenda

Several antipoverty and employment generation schemes already exist in rural areas often through centrally sponsored schemes (CSSs). Through the Planning Commission an attempt is underway to rationalise these CSSs. For example, under the Department of Agriculture and Cooperation, assorted schemes on agricultural statistics may be merged. So will technology missions on oilseed, pulses and maize with technology missions on cotton and horticulture in the northeast being retained as separate programmes. There will be independent CSSs on on-farm water management agriculture census and macromanagement. In addition, there are CSSs through the Department of Animal Husbandry and Dairying, Ministry of Environment and Forests, Ministry of Land Resources and Ministry of Rural Development.

The NCMP of the UPA Government states, "All centrally sponsored schemes except n national priority areas like family planning will be transferred to States." Otherwise, here is the concept of devolving CSSs downwards to panchayati raj institutions (PRIs) and linking CSS releases to states only when states transfer funds, functionaries and financial resources to PRIs.

A National Food for Work Programme (NFWP) was launched in 2004 in 150 less-advanced districts. Subsequently, 167 were identified under the Rashtriya Sama Vikas Yojana (RSVY), with terrorism-affected districts being added to districts identified on he basis of economic criteria. Both the NFWP and RSVY identifications have now been uperceded by identification of 200 less-advanced districts, undertaken by the Planning Commission, although the list is not available in the public domain yet.

This identification became necessary because of a Backward Regions Grant Fund BRGF) and the National Rural Employment Guarantee Act (NREGA). The NREGA, in place since 1 February 2006 in 200 districts, provides that state governments will offer 00 days of unskilled manual work every year to every poor household in rural areas whose adult members volunteer for such work, at minimum wage rates. The 100-day ceiling applies to households and not to individuals. Existing CSSs, including the NFWP, will be subsumed under the NREGA. If states fail to provide jobs, there is an nemployment insurance component. There are also legislative plans for introducing ocial security for unorganised sector workers, including agricultural workers, and minimum protective legislation for agricultural workers.

Agricultural issues cannot be delinked from broader rural development issues. As was mentioned earlier, centrally sponsored schemes exist for rural roads (Pradhan Mantri Gram Sadak Yojana), rural housing (Indira Awas Yojana) and rural electrification nd drinking water (Pradhan Mantri Gramodaya Yojana). Yet data show a lack of physical and social infrastructure in rural areas.

The first two phases of the National Highway Development Programme (NHDP) have been reasonably successful and have improved road connectivity. Phase I involves the Golden Quadrangle and links the four metropolises of Delhi, Mumbai, Chennai and Kolkata. Phase II involves a North-South and East-West corridor. Since January 2005,

Phases III and IV have been added to the NHDP, to improve connectivity in places not located on the upgraded networks of Phases I and II. More importantly, the PMGSY (Pradhan Mantri Gram Sadak Yojana) is being revamped; it has not performed very well in states like West Bengal, Uttar Pradesh, Bihar, Madhya Pradesh, Jharkhand, Assam and Chhattisgarh.

There is now an attempt to push rural infrastructure through clusters, although most clusters have an autonomous origin and are rarely induced, emerging around local skil or natural resource bases. At a conceptual level, there are three kinds of clusters tha one can visualise: relatively modern, small firm-dominated industrial clusters that ofter tend to be located in relatively urban areas; artisan and rural industry-based clusters and clusters that are based on the agricultural economy. The latter two, particularly th last, tend to be natural resource based.

Most policy interventions have focused on the first of the three, rather than the las two. Earlier attempts at cluster formation have not always been successful, as they ofte tended to be ad hoc and piecemeal. The present PURA (Provision of Urban Amenitie in Rural Areas) scheme is more integrated, at least in intention, and builds on differen kinds of connectivity: physical connectivity through roads and power, electroni connectivity through village Internet kiosks, knowledge connectivity through educatio and flowing on from them, economic or market connectivity through access to market PURA manifests itself through the Bharat Nirman idea, focusing on rural infrastructur It has six components of irrigation, roads, water supply, housing, rural electrificatio and rural telecommunications connectivity, with target dates for completion in 200 The government is also establishing a food processing SEZ (special economic zone) i Kakkancherry in Kerala.

The government has forwarded an Integrated Food Processing Act to parliament pass and the ECA has been amended repeatedly. Around 200 orders issued by the sta governments under the ECA are likely to expire in 2006, unless the states make cas for their continuation and the GOI approves/agrees to the proposals. This is a ve positive development. Contract farming, as it is equated with corporate farming, extremely controversial.

The Food Safety and Standards Bill has been also placed before parliament; consolidates eight laws governing the food sector and establishes the Food Safety ar Standards Authority (FSSA) to regulate the sector. The FSSA will be aided by sever scientific panels and a central advisory committee to lay down standards for food safety. These standards will include specifications for ingredients, contaminants, pesticide residue, biological hazards and labels.

Several states have amended their APMC Acts, to allow direct marketing, contract farming and establishment of private markets in the private or cooperative sectors.

There seems to be a conscious shift from an agricultural to a rural development policy, with a focus on rural infrastructure like roads, electricity and water (drinking and irrigation). For agriculture proper, the mid-term appraisal of the Tenth Plan flags certain issues:

1. Rejuvenation of support systems in extension, credit and delivery systems for inputs like seeds, fertilisers and veterinary services.
2. Investments in irrigation/water management.
3. Appropriate pricing of water.
4. Mega-irrigation projects.
5. Watershed development.
6. Agricultural research.
7. PDS (public distribution system) pricing.

The broad directions are quite clear - a host of measures related to water conservation, management, availability and accessibility are being addressed. Some support is related to pricing of output, charges for inputs, some legal reform as well as people's participation in environment and resource management. This, combined with the devolution of greater powers to rural local bodies and the powers accorded to the common man through the Right to Information Act, can help ensure that adequate accountability is imposed on the bureaucracy.

Constraints

The critical constraining aspects are three-fold and will require deft political manoeuvring to push through. First, many of these policy measures will require large sums to be generated through other (non-agricultural) revenue means. It is not clear how this will be possible. Second, poor control and monitoring abilities of many state governments limit their ability to implement such large-scale programmes effectively.

These abilities will take some time to build up. An example is the collapse of the extension programmes and delivery systems. Without significant administrative reforms within the government, neither is their privatisation feasible in the next few years, nor is their rejuvenation likely across many if not most states of the country. Third, if reforms have not matched up to actual expectations, there is a reason. Constitutionally, most agriculture-related areas are state government subjects and it is difficult for the centre to provide incentives for such reforms. The central government will need to put in a significant amount of effort if it wants a nationally coordinated approach.

Private Sector

Despite constraints, the policy regime is increasingly moving towards the greater role of private entities; economic policies in general have been moving towards the larger role of the private sector, private-public partnerships, the increasing role of communities, self-help groups and non-profit organisations for delivering critical inputs and services. Following constitutional amendments in the early 1990s, Panchayats have been given greater powers. State-level finance commissions have been established to oversee the devolution of funds to these bodies. It is now generally agreed that the Agricultural Produce Marketing Act that limited transactions between buyers and sellers except through mandis will be changed to allow for contract farming. Over the next few years, it is also expected that the laws and rules limiting internal trade in agricultural commodities will be changed to enable greater choice for the farmer.

The government is finally tackling the building of rural roads, the objective being to connect all of the roughly 600 000 villages to each other and with cities via all-weather surface roads.

At the same time developments on the IT front have enabled the setting up of Internet kiosks on a mass scale throughout rural India. Basic primary education for everyone has become a thrust area and it is only a matter of a few years when literacy will be universal, at least among young adults. India already boasts some of the cheapest cellular phone tariffs, and services for Internet and cellular communication are emerging. Private entities are marrying the advantages of Internet infrastructure and distributional networks to provide purchase, retail, information and knowledge delivery under one umbrella.

Meanwhile private sector banks are increasingly becoming aggressive lenders in the rural domain as well. Insurance firms are targeting agricultural markets more and more.

Private seed companies have sprung up across the country and with some changes in the IPR regime are expected to become significant players. Moreover, after many years, the private sector has emerged as an investor in building and running quality agricultural warehousing facilities across India. But all of the above components are recent changes; they hold great promise but currently are at a nascent stage of development. More importantly carry-overs from the past continue, and currently it is difficult to see reforms in the following areas:

- Electricity for farmers is quite cheap or free, a result of competition between various political parties;
- Investments in irrigation facilities, tanks and ponds are not rising to the levels required. State government funding is constrained;
- Water tables are falling rapidly throughout western and northwestern India, primarily due to extraction of water for agriculture;
- Central and state governments are increasingly unable to find the resources to maintain the existing infrastructure (canals for example) or to provide traditional services (such as extension services).

In other words, the private sector will be constrained as well owing to externalities, expectations due to past policies and lack of public funds for critical public infrastructure.

REFERENCES

Arndt, C., Hertel, T., Dimaranam, B., Huff, K. & McDougall, R. 1997. China in 2005: implications for the rest of world. *Journal of Economic Integration,* 505-547.

Francois, J.F., MacDonald, B.J. & Nordström, H. 1996. *Trade liberalisation and capital accumulation in the GTAP model.* GTAP Technical Paper No. 7.

Hertel, T.W. & Martin, W. 1999. *Would developing countries gain from inclusion of manufactures in the WTO negotiations?* GTAP working paper, Purdue University.

Hertel, T.W. (ed). 1997. *Global trade analysis: modelling and applications.* Cambridge University Press.

Huang, J., Rozelle, S. & Chang, M. 2004. Tracking distortions in agriculture: China and its accession to the World Trade Organization. *World Bank Economic Review,* 18(1): 59-84.

Van Tongeren, F. & Huang, J. 2004. *China's food economy in the early 21st century.* Report #6.04.04. The Hague, Agricultural Economics Research Institute (LEI).

Walmsley, T.L., Betina, V.D. & Robert, A.M. 2000. *A base case scenario for the dynamic GTAP model.* West Lafayette, Center for Global Trade Analysis, Purdue University.

Walmsley, T.L. 1998. *Long-run simulations with GTAP: Illustrative results from APEC trade liberalization.* GTAP technical paper No. 9.

2

Oilseed

Oilseed crops have been the backbone of agricultural economy of India from time immemorial. Today these crops are cultivated on about 16.5 million hectares, with total production of 10 million tonnes. This area constitutes approximately one-tenth of the total cultivated area in India. On the oilseed map of the world, India occupies a prominent position, both in regard to acreage and production. The important oilseed crops grown in this country in order of importance are groundnut, rapeseed and mustard, sesame, linseed, safflower, castor, sunflower and niger.

India is one of the largest producers of oilseeds in the world. The oilseeds area and output is concentrated in Central and southern parts of India, mainly in Madhya Pradesh, Gujarat, Rajasthan, Andhra Pradesh and Karnataka. The nine major oilseeds cultivated in India are groundnut, mustard/rapeseed, sesame, safflower, linseed, niger seed, castorseed, soyabean and sunflower. Coconut is the most important source of edible oil amongst plantation crops, while in non-conventional oils, rice bran oil and cottonseed oil are the most important. Groundnut, soyabean and mustard together contribute about 85 percent of the country's oilseeds production.

While production of oilseeds has risen over the years, it has not been able to keep pace with the domestic demand for edible oils. Therefore, there has been a persistent gap between domestic demand and output of edible oils. The net domestic availability of edible oils was only 5.62 million tonnes during 2000-01 while the domestic requirement was close to 10 million tonnes. The Government, with a view to avoiding scarcity of this item and consequential rise in prices, has been allowing liberal import of edible oils, though with stiff tariff rates.

In pursuance of the policy of liberalisation, there have been progressive changes in the import policy in respect of edible oils during the past few years. Edible oil, which was in the negative list of imports, was first decanalised partially in April, 1994. This was followed by enlargement of the basket of oils under Open General License (OGL) import in March, 1995. Duty structures were further rationalised in subsequent years in order to harmonise the interests of domestic oilseeds growers, consumers and processors and to avoid large imports of edible oils to the extent possible.

Import duty on edible oils has been revised in March 2001. The custom duty on Crude Palm Oil (CPO) meant for vanaspati manufacture was raised from 25 percent to 75 percent except in the case of sick vanaspati units, where duty was kept at 55 percent. This concession was later done away with and a uniform rate of duty @ 75 percent on CPO for all the vanaspati units, whether sick or otherwise was imposed. The duty on refined oil was raised to 85 percent (basic) except in the cases of refined soyabean oil and refined mustard oil where the duties were 45 percent (basic) and 75 percent (basic) respectively. Special Additional Duty (SAD) was levied on import of refined oils at the rate of 4 percent.

In August 2001, the Government imposed a fixed tariff values for crude and refined palm oils as follows:

	$ per tonne
CPO	337
Crude Palmolein	357
RBD Palm oil	351
RBD Palmolein	372

This implies that imports would no longer be valued at the transaction value reflected in the price mentioned on the invoice, instead the basic duty will be calculated on the standard tariff value regardless of the actual invoice. This was done to avoid under invoicing by importers. The tariff values has since been revised twice, and the latest rates as revised on December 17, 2001 are as follows:

	$ per tonne
CPO	314
Crude Palmolein	334
RBD Palm oil	341
RBD Palmolein	349

On October 30, 2001, customs duty on CPO meant for vanaspati manufacture was reduced from 75 percent to 65 percent. Duty on crude sunflower/safflower was reduced from 75 percent to 50 percent w.e.f. November 16, 2001 upto an aggregate of 1,50,000 tonnes in a financial year.

GROUNDNUT

Groundnut (*Arachis hypogaea* L.)is believed to be the native of Brazil to Peru, Argentina and Ghana, from where it was introduced into Jamaica, Cuba and other West Indies islands. The plant was introduced by Portuguese into Africa from where it was introduced into North America. It was introduced into India during the first half of the sixteenth century from one of the Pacific islands of China, where it was introduced earlier from either central America or South America.

Economic Importance

The oil content of the seed varies from 44 to 50 per cent, depending on the varieties and agronomic conditions. Groundnut oil is an edible oil. It finds extensive use as a cooking medium both as refined oil and Vanaspati Ghee. It is also used in soap making, and manufacturing cosmetics and lubricants, olein stearin and their salts. Kernels are also eaten raw, roasted or sweetened. They are rich in protein and vitamins A, B and some members of B_2 group. Their calorific value is 349 per 100 grammes. The H.P.S. type of groundnut kernels are exported to foreign countries. The residual oilcake contains 7 to 8 per cent of N, 1.5 per cent of P_2O_5 and 1.2 per cent of K_2O and is used as a fertiliser. It is an important protein supplement in cattle and poultry rations. It is also consumed as confectionary product. The cake can be used for manufacturing artificial fibre. The haulms (plant stalks) are fed (green, dried or silaged) to livestock. Groundnut shell is used as fuel for manufacturing coarse boards, corksubstitutes etc. Groundnut is also of value as rotation crop. Being a legume with root nodules, it can synthesise atmospheric nitrogen and therefore improve soil fertility.

Botanical Description

Groundnut (Archis hypogaea L.) is a member of subfamily, Papilionaceae of the family Leguminosae. Archis hypogaea L. consists of two subspecies each containing two botanical varieties.

Plants of the botanical variety hypogaea are spreading (runner) to upright (erect bunch) in growth habit, have alternate branching, lack inflorescences on the main stem, possess appreciable fresh seed dormancy, flowers are longer and mature later than those of subspecies fasligiata. Variety hirsuta has been used only to a little extent.

Plants of this subspecies fasligiata are upright, have sequential branching and inflorescences in the main-stem leaf axils, possess little fresh seed dormancy, and are of shorter duration than those of the subspecies hypogaea. Subspecies fasligiata includes both the Spanish and Valencia types

Groundnut, in general, has a short-statured plant, with the main axis being upright (15 to 40 cm long) but the major part of the plant consists of the primary branches. Secondary and tertiary branches are found in the semi-spreading and spreading (Virginia) types, giving them a prostrate stature. The leaves are alternate, stipulate and quadri-foliate. The flowers are orange yellow, typically papilionaceous, with a long calyx tube within which is held the style borne on a superior ovary. The calyx and corolla lobes are borne in the axils of the leaves on the fruiting branches. They are bisexual, Zygomorphic, complete and sessile. Petals are five, with one large standard, two wings and two fused keel petals. There are two steering and eight fertile anthers, four of which are globose and four oblonge (dimorphic).

Archis hypogaea

Groundnut is predominantly a self-pollinated crop and pollination takes place early in the morning. As soon as the fertilisation is complete, the flowers fade. After fertilisation, an intercalary meristem becomes active at the base of the ovary above the point of attachment of the hypanthium, producing new tissue below itself and resulting in an elongated stalk in the peg. The pegs are positively geotropic, enter the soil and bend in a horizontal plane. Generally two, and occassionally one, three or four, fertilised ovules are borne at the tip of the peg which later swells to become the pod. The testa is generally pink, but varieties with red, white, purple and blotched testa, with various gradations of colours are available.

Distribution, Area and Production

The major groundnut-producing countries of the world are India, China, Nigeria, Senegal, Sudan, Burma and the USA. Out of the total area of 18.9 million hectares and the total production of 17.8 million tonnes in the world, these countries account for 69% of the area and 70% of the production. India occupies the position, both in regard to the area and the production, in the world. About 7.5 million hectares is put under it annually and the production is about 6 million tonnes. 70% of the area and 75% of the production are concentrated in the four states of Gujarat, Andhra Pradesh, Tamil Nadu and Karnataka. Andhra Pradesh, Tamil Nadu, Karnataka and Orissa have irrigated area forms about 6% of the total groundnut area in India.

Climate and soil

groundnut is grown throughout the tropics and its cultivation is extended to the subtropical countries lying between 45 degrees N and 35 degrees S and upto an altitude of 1000 metres. The crop can be grown successfully in places receiving a minimum rainfall of 1,250 mm. The rainfall should be well distributed well during the flowering and pegging of the crop. The total amount required for presowing operations (preparatory cultivation) is 100 mm; for sowing, it is 150 mm and for flowering and pod development an evenly distributed rainfall of 400-500 mm is required. The groundnut crop, however, cannot stand frost, long and severe draught or water stagnation.

Groundnut is grown on wide variety of soil types. However, the crop does best on sandy loam and loamy soils and in the black soils with good drainage. Heavy and stiff clays are unsuitable for groundnut cultivation as the pod development is hampered in these soils.

Rotation and mixed cropping

Generally, as a kharif crop, groundnut is grown year after year. In certain places, it is rotated with wheat, jowar, bajra, garden crops, such as potatoes, onions, chillies, garlic, ginger and turmeric. The yields of the cereal crops following groundnut are usually increased by about 25%. Pulses, e.g. red gram (arhar), mash and moong are grown mixed with groundnut. In certain places, millets e.g. jowar and bajra, and castor are grown mixed with groundnut.

Season

Groundnut is raised mostly as a rainfed kharif crop, being sown from May to June, Depending on the monsoon rains. In some areas or where the mansoon is delayed, it is sown as late as August or early September. As an irrigated crop it is grown to limited between January and March and between May and July.

Cultivation

For a kharif crop, with one set of rains in May-June, the field is given two ploughings and the soil is pulvarised well to obtain a good tilth. The third ploughing may be given just before sowing. Harrows or tillers can be used for cultivation. When the soils are heavily infested with perennial weeds, e.g. Cynodon or Cyperus, very deep ploughing is needed. If a field is infested with white grubs, chemicals, such as Heptachl or chlordane, are drilled at the rate of 25 kg per ha before final borrowing. For the irrigated crop, beds of convenient size may be made, depending on the tropography of the land, the nature of the irrigation source and the mode of lifting water. Since groundnut is a deep rooted plant it uses up both moisture and nutrients in the deeper layers of the soil. Generally, the following fertiliser schedules are recommended, depending upon the soil fertility, the variety grown, and the quantum and distribution of rainfall.

Farmyard manure	*N*	*P*	*K*	
Rainfed crop	6.25 Tonnes	10-20	20-40	20-40 kg/ha
Irrigated crop	12.5 Tonnes	20-40	40-90	20-40 kg/ha

Usually, the application of only N and P is resorted to, since most of the Indian soils are rich in K. The application of N in two equal split doses, one before sowing and the other 30 days after sowing, is found to increase the yield. In virgin lands, when groundnut is newly introduced, the application of a culture of *Rhizobium* as seed treatment is beneficial in increasing nodulation and nitrogen fixation. The application of gypsom at 500 kg per ha at the pegging stage will enhance pod formation.

Healthy and well developed pods should be hand shelledor shelled with a suitable groundnut-sheer about fornight before sowing. Well defilled kernels should be selected and treated with 5 gm of thiram or 3 g of Captanper kg of kernels. The kharif crop is sown with a speed-drill or with a suitable planter at a depth of 8 to 10 cm. specings adopted differ from place to place. For the rainfed bunch groundnut the inter-raw specings vary between the adjuscent rows varies from 20 and 30 and intra-raw specings between 10 and 20 cm. The quantity of well developed seeds required per hectare with the above specings will be about 110 kg for semi-spreading and spreading varieties and 120 kg for the bunch varieties.

Weeds cause considerable reduction in yield. A reduction of 20 to 45% in yield due to weeds has been recorded. For controlling weeds, and also to keep the soil in a friable condition, the crop should generally receive a hand-weeding and one or two hoeings, with bulllock-drawn implements, the first about three weeks after sowing and the second and the third about a fortnight and a month later. No interculture would be done after the pegs have commmenced going underground. Earthing up can be done in the case of bunch and semi-spreading types to fascilitate the maximum penetration of the pegs into the soil. Weeding can also be done quickly and efficiently with a wheel hand-hoe, if the soil is light. It should be run before weeds make excessive growth. Weeds can also be controlled effectively with Lasso or Tok-E-25 weedicides at the rate of 5 litres in 500 litres of water per hectare as a pre-emergence soil spray within two days of sowing groundnut. If the Kharif crop is cought in a long spell of draught, especially at the pod formation stage, supplemental irrigation is given, if water is available. For the irrigated groundnut, the frequency of irrigation depends on soil texture, and interval between irrigation ranges from 8 to 12 days. Studies on the water requirement of indicate that the period between the fifty first and fifty seventh day (i.e. at peg formation stage) the crop depends on timely irrigation to it during this period.

The prominent symptoms of maturity are the yellowing of leaves, the shedding of older leaves, the development of oroper colour of the testa and the dark tint inside the shell. The bunch and semispredading varieties are usually harvested by hand pulling when there is adequate moisture in a soil. The spreading types, on the other hand, are harvested by digging with a mammity or by ploughing or working a blade harrow. The pulled out plants are stacked for a few days for drying and stripped afterwards.

The stripped pods are cleaned and dried to a safe moisture content of not more than 5%. Damp nuts, if stored, will ferment and allow the development of poisonus moulds, e.g. aspergillus flavus in the Kernels, leading to contamination with aflatoxin—a health hazard both for human beings and livestock. The oil express from such produced will be rancid and the cake. When fed to poultry results in a heavy mortality of the birds. It is desirable to store groundnut in gunnt bags are stacked in a storeroom in tires comprising not more in each in such a way that the air keeps circulating planks to avoid damage from dampness, rats, etc. The store rooms should be periodically inspected to ensure that there is no storage pest or to decompose of the stock immediately after proper cleaning.

Pest and diseases

Aphids (Aphis craccivora Koch-), Leaf miner (Stomopteryx subseciva Zell)., the red hairy cater piller (Amsacta albistriga Walk., and A moorei Bltr.) and the white grub (Lachnosterna consanguica) are the serious pests of the groundnut crop. Dusting B.H.C 10% on the young larve or spraying Metasystox) 0.05% on grown up caterpillars is recommended for controlling the red hairy caterpillar. Spot-treatment with Phorate or Mephospholan granules at the rate of 1.5 kg a.i. per ha into the soil before planting is recommended for controlling the white grub.

The tikka leaf-spot (Cercospora arachidicola and C. personata) and the coller- rot (Aspergillus niger and A. pulverulentum) are important diseases of groundnut. Rust (Puccinia arachidis) has been occurring in a serious form in recent years in certain groundnut growing areas. It is desirable to store groundnut in gunnt bags are stacked in a storeroom in tires comprising not more in each in such a way that the air keeps circulating planks. Dry seed dressing with Thiram or captan @ 4g/kg of seed is recommended for controlling collar-rot. Yet spraying with Hinosan at the rate of 0.02% is recommended to check the spread of the Pathogen.

Yield

Under rainfed conditions, the average yield of semi-spreading and spreading varieties is 1200-1400 kg of unshelled pods per hectare and that of bunch types is 800-1,000 kg. The crops grown with supplemental irritations produce 3,000 kg more of pods per hectare. The pods yield 70 to 75 per cent of kernels by weight. The yield of haulms is usually two to two-and-half times that of pods.

Varieties

The varieties under cultivation fall into three groups in respect of the habit of growth, namely bunch (Spanish), semi-spreading (Virginia bunch) and spreading (Virginia runner). In the bunch group, the plants grow erect, possess light-green foliage, produce pods in clusters at the base of the plant and have round, plump non-dormant seeds, with light-rose testa. In the case of the semi-spreading and spreading varieties, the branches trail either partially or completely on the surface of the soil, produce pods all along them, possess dark-green foliage and have oblong, dormant brownish seeds. The semi-spreading and spreading types are usually heavier yielding and later-maturing than the bunch varieties.

RAPESEED AND MUSTARD

Under the names rapeseed and mustard, several oilseeds belonging to the crucieferae are grown in India.

1. Brown mustard, commonly called rai (raya or laha)—Brassica juncea (L.) Czern. & Coss
2. Sarson
   - Yellow sarson—B. campestris L. var. sarson Prain
   - Brown sarson—B. campestris L. var. dichotoma Watt
3. Toria (lahi or Maghi Labi)—B. campestris L. var. toria Duth.
4. Taramira or tara (eruca sativa Mill.)

In trade, sarson, toria and Taramira are known as rapeseed, and rai as mustard.

Banarsi rai (B. nigra Koch.) which does not fall under any of the four groups is a garden crop used as spice. The cultivation of white mustard (Brassica alba syn. Sinapis alba) is no longer found in India. Rai and Yellow sarson are self-fertile and the rest of the cruciferous oilseeds, viz. brown sarson, toria, taramira, Banarasi rai and white mustard, are self-incompatible.

Brassica juncea

The acreage under yellow sarson is scantily (mainly in Bihar) and constantly on the decrease. In the recent past, the acreage under brown mustard is steadily on the increase (over 65%) at the expense of other Brassica due to its higher production, greater to pests and diseases and moisture stress.

Brassica juncea L. (rai) was originally introduced from China into north-eastern India. From where it has extended into Afghanistan via the Punjab. Eastern Afghanistan, together with the adjoining north-western India is one of the independent centres of brown sarson (Brassica campestris var. brown sarson). Yellow sarson (B. campestris var. yellow sarson) is commonly grown in the eastern parts of India where it shows much diversity of forms. Taramira is a relatively recent introduction into India. It is believed to be a native of southern Europe and North Africa.

Economic Importance

Rapeseed and mustard yield the most important edible oil content of the seeds of different ranges from 30 to 48 per cent. In the case of white mustard, the oil content ranges from 25 to 33 per cent. The oil obtained is the main cooking medium in Northern India and can not be replaced by any other edible oil. The seed and oil are used as a condiment in the preparation of pickles and for flavouring curries and vegetables. The oilcake is mostly used as cattle feed. The leaves of young plants are used as a green vegetable. The use of mustard oil for industrial purposes is rather limited on account of its high cost.

Botanical Description

Rapeseed and mustard include annual herbs ranging in height from 0.45 to 1.75 m. Roots, in general, are long and tapering. Toria is more or less a surface-feeder and brown sarson has long roots, with a limited lateral spread, enabling its successful cultivation under drier conditions. Yellow sarson has both extensive and lateral spread. The height of the stem varies from 45 cm (in some varieties of toria) to 1.75 m (in yellow sarson). In toria and brown sarson, the branches arise at an angle of 30^0 to 40^0.In yellow sarson, branches arise laterally at an angle about 10^0 to 20^0 and give the plant a narrow and pyramidal shape. The inflorence is coroymbose raceme. In the case of sarson the four petals are spread apart, whereas in brown sarson and toria the petals overlap or may be placed apart, depending upon the variety. The flowers bear a hypogynous syncarpous ovary. In brown sarson and toria, the ovary is bicarpellary, whereas in the case of yellow sarson, it may also be tri or tetra-carpellary. The fruit is a siliqua. The pods are two-valved, three-valved or four-valved, depending upon the number of carpels in the ovary. The flowers begin to open from 8 a.m. and continue up to 12 noon.

Distribution, Area and Production

The crop is grown both in subtropical and tropical countries. In Asia, it is chiefly grown in China, India and Pakistan. It is also grown in Europe, Canada and the USSR, but the forms of rapeseed and mustard grown there are different from those grown in India. India occupies the first position, both with regard to acreage and production of rapeseed and mustard in the world. In India the Braddica crops occupy the second largest position after groundnut, with 3-5 million hectares, producing about 2 million tonnes of seed annually. The chief states producing them are Uttar Pradesh, Punjab, Haryana, Assam, Bihar, Madhya Pradesh, Rajasthan, West Bengal and Orissa.

Climate and Soil

The rapeseed and mustard crops are of the tropical as well as of the temperate zone and require relatively cool temperatures for satisfactory growth. In India, they are grown in the rabi season from September-October to February-March. The rapeseed and mustard crops grow well in areas having 25 to 40 cm of rainfall. Sarson and toria are preferred in low-rainfall areas, whereas raya and toria are grown in medium and high rainfall areas respectively.

The rapesed and mustard thrive best in light to heavy loams. Raya may be grown on all types of soil, but toria does best in loam to heavy loam. Sarson is suited to light-loam soil and taramira is mostly grown on very light soils.

Cultivation

A fine seed-bed is required to ensure good germination. In irrigated areas, the first ploughing is done with a medium sized soil-turning plough, followed by two to four ploughing with a desi plough or a cultivator. Sohaga (Planking) is given after every ploughing. In rainfed areas, one to two ploughings with a desi plough or a cultivator, each followed by planking, may be given. Toria, in particular, requires a fairly moist seed-bed for good germination, but excessive moisture should be avoided. The rate in the case of mixed cropping depends on the proportion of the rapeseed to the main crop.

When sown pure, 5 kg of seed per ha is used for all rapeseeds and mustard. When sown mixed with other crops, the time of sowing rapeseed and mustard is governed by the sowing of the main crop. The first half of September is best for sowing toria (if wheat is to follow, it should be sown by the end of August), 25th September to 15th October for sarson, 30th September to 15th October for raya, and tarantira is sown throughout October. The seed from healthy and desirable plants, grown in isolation in the case of self-sterile form, should be used. Whenever moisture in the field is inadequate, the seed is mixed with moist soil and kept overnight. For distributing evenly, the seed is usually mixed with sand before sowing.

In mixed cropping rapeseed and mustard are sown in rows 1.8 to 2.4 metres apart across the main crop. The pure crop of rapeseed or mustard is sown at a depth of 4 to 5 cm in lines, 30 cm apart, with a drill, or with a (tube) attached to the plough. Thinning is done three weeks after sowing to maintain a plant-to-plant distance of 10 to 15 cm.

One hoeing in the case of toria in the third week after sowing and 1-2 hoeings in the case of sarson and raya are adequate. Forty kg of W per ha is optimum for all rapeseed and mustard crops in rainfed areas. Under irrigated conditions doses of 40 and 80 kg of W per ha are considered optimum for toria, sarson and rapeseed respectively. For tarantira, 20 kg of W per ha is ancient. All the fertilisers should be drilled before sowing. Among the Brassicae, raya is most responsive to irrigation, followed by yellow sarson. Two irrigations, one at flowering and the other at pod formation, result in the maximum yield in the case of toria, sarson and raya. Its flowering is most economical, if there is water scarcity. Tarantira is not irrigated.

Harvesting is done as soon as the crop begins to turn yellow. Tma, which takes 75 to 90 days to mature, is the earliest crop to be harvested. Harvesting is done with hand-sickler. Threshing is done by beating with a wooden stick the seed bearing part of the plants, taken in convenient bundles or by trampling them under the feet. Winnowing is done with by slowly dropping the threshed produce from a basket held shoulder-high. The seed after being dried in the sun is stored in gunny bags or bins.

Pest and Diseases

The most serious pests of rapeseed and mustard are the mustard aphid (Lipaphis erysinbi (Holt) syn. Rhopalosiphun pseudobrwsicae (Davis), mustard sawfly (Athalia proximc King) and cutworm (Agrotis ypsi" Poott.). To control aphidy the spraying of Methy Demeton 0.02% or limetboate 0.03% or Phosphomidon 0.0V/ two or three times depending on the intensity is recommended. The mustard sawfly is controlled by dusting 10% BHC at the rate of 25 kg per ha. Cutworms are controlled by drilling into the soi 5% dust of Aldrin or Heptachlor or Chlordane at the rate of 25 kg per ha.

Alternaria blight (Alternaria brassicae Bork.) is the most widespread and destructive disease of rapeseed and mustard, whereas rust (Cystopus candidus) is serious in certair areas. Spraying with Dithane M-45 or Difolaton at the rate of 11 kg per ha is recommended for controlling these diseases. Orobanche, a phanerogamic parasite, has also been reported to be serious in Punjab and in certain parts of Uttar Flradesh. Hand pulling of the plants of the parasite and a long rotation are recomnended. Yield. The average yield of rapeseed and mustard is about 500 kg per ha. Under good crop management, toria, which is a short-duration crop, gives a yield of about 450 to 650 kg per ha, sarson 800 to 1,000 kg and rai 1,000 to 1,200 kg per hectare.

Oil and Protein Quality in the brassicae

The oil of rapeseede and mustard possesses a Sizable amount of crucie acid (38 to 57%) together with linolenic acid up to 4.7 to 13.0 % . The oleic and linolenic acid, which have nutritive value, together constitute only about 27 per cent. It is desirable to increase the quantity of oleic and linolic acid and erucic acid. A lower proportion of erucic acid will make the oil more palatable, nutritive besides reducing metabolic disorder. The Protein contained in rapeseed and mustard normally ranges between 24 and 30%. On the whole seed basis and between 35 and 40% on the meal basis. But the presence of toxi glucosinolates in the mustard cake is unsuitable as source of human protein and is a present usé as a manual and as a cattle feed.

Varieties

The varieties of different forms of rapeseed differ greatly in branching habit, maturity shape and colour of foliage, colour and size of flower, seed, etc.

Hints for enhancing the yields of rapeseed and mustard

1. Select the right crop (toria, sarson, raya and taramira) according to the soil type

2. Sow the seed of an approved variety only.
3. Prepare a fine seed-bed.
4. Sow the crop during the optimum sowing period, using the recommended spacing and seed-rate.
5. Apply the recommended doses of fertilisers.
6. Protect the crop against insect pests, particularly aphids, by frequent spraying, using proper insecticides.
7. Harvest the crop in time to avoid loss owing to shattering, especially in the case of Raya.

SESAMUM

Sesamum (*Sesamum indicicum* L.), also known as sesame, til and gingerly, is an important and ancient oil-yielding crop. Africa is considered to be the primary centre of origin of this crop because of the presence of its diverse wild species in that continent. India is the secondary centre of origin and another secondary centre is Japan.

Economic Importance

The sesamum seed is a rich source of edible oil. Its oil content generally varies from 46 to 52 per cent. Its grains may be eaten fried, mixed with sugar or in the form of sweetmeats. Sesamum oil is used as a cooking-oil in southern India. It is also used for anointing the body, for manufacturing perfumed oils and for medicinal purposes. Sesamum-cake is a rich source of protein, carbohydrates and mineral nutrients, such as calcium and phosphorus. The cake is edible and is eaten avidly by working classes. It is also a valuable and nutritious feed for milch cattle.

Botanical Description

The genus Sesamum belongs to the family Pedaliaceae which has a wide distribution, covering tropical Africa, Madagascar, Arabia, India, Sri Lanka, tropical Australia and a few of the eastern islands of the Malayan Archipelago.

Sesamum is an erect and annual herb. The plant is simple or branched. The leaves are opposite below and alternate above. The flowers are zygomorphic, solitary, occasionally two or three together, auxiliary, short- pedicelled, borne on the upper part of the stem or branches geniculate. The tube, bent downwords and is dialated above the oblique base. The calyx is small, and five sectioned, 0.5-0.6 cm long. The corolla is tubular, campanulate, approximately 3 cm long, with an applicate margin and slightly bilabiate, the upper lip is entire and the lower one is tripartite. There are four functional stamens and often one is sterile, didynamous. The antliers are sagittate. The ovary is superior, bilocular, but may be subdivided by false internal walls. The fruit is a capsule, erect and oblong. The capsule contains numerous small ovate seeds. The testa may be smooth or reticulate and may be white, yellow, reddish-brown or black. The seed is dicotyledonous, albuminous and oleaginous.

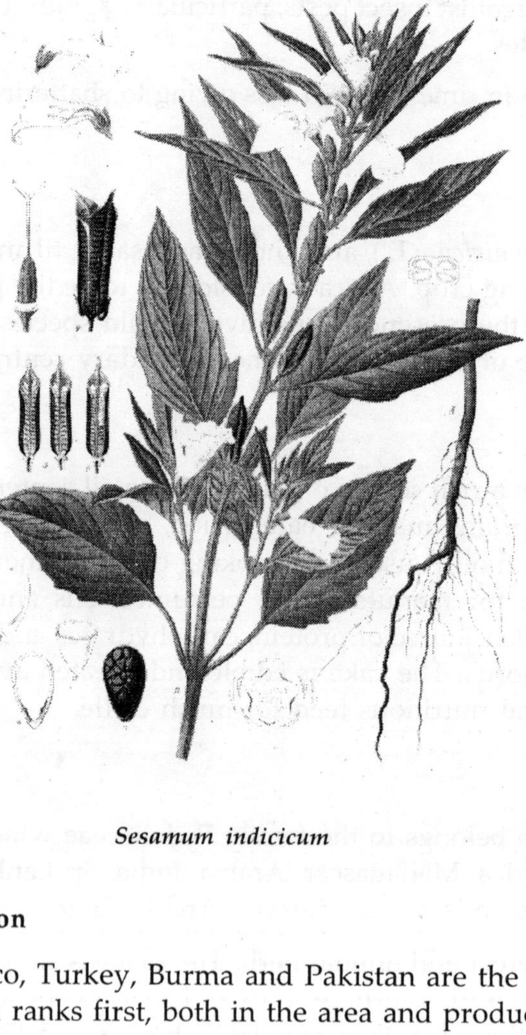

Sesamum indicicum

Distribution and Production

India, China Sudan, Mexico, Turkey, Burma and Pakistan are the important sesamum producing countries. India ranks first, both in the area and production of sesamum in the world. The annual area put under it in India is about 2-5 million hectares (45 per cent of the world hectarage) and the total production is nearly 52 thousand tonnes. Sesamum is grown on 21 lakh hectares in only eight states, viz. Uttar Pradesh (673,000), Rajasthan (562,000), Madhya Pradesh (345,000), Andhra Pradesh (237,000) Maharashtra (139,000), Gujarat (118,000), Tamil Nadu (117,000) and Orissa (103,000). Among other states only Karnataka has a sizable area (68,000) under sesamum. In the remaining states it is grown only on a small area and hence is a very minor crop there.

Climate and Soil

Sesamum grows in the plains and at elevations up to 1,200 m. It cannot stand frost, continuous heavy rain or prolonged drought. It is grown in sandy-loam to heavy black soils. Most of the crop is confined to lighter soils and its cultivation in heavy soils is limited to certain regions. The pH of sesamum growing soils ranges from 5.5 to 8.2. The sesamum growing soils are shallow to medium in Andhra Pradesh, Orissa, Andhra Pradesh and Tamil Nadu and are deep in Rajasthan, Gujarat and Maharashtra.

Mixed Cropping

Sesamum is grown both as pure and mixed with other crops, whereas the semi-rabi and summer crops are taken as pure. The common component crops are red-gram (arhar) in Andhra Pradesh, bajra and red-gram in Gujarat, red-gram and jowar in Madhya Pradesh cotton, groundnut and red-gram in Maharashtra, and black-gram, green-gram and horse-gram in Orissa.

Cultivation

In India, seasmum is grown in three seasons, viz. kharif, semi-rabi and summer. The kharif crop occupies over 70 per cent of the area, whereas the semi-rabi and summer crops occupy 20 per cent and 10 per cent area respectively. The kharif sesamum is sown in June-July with the onset of the monsoon and is harvested in December-January. The kharif and semi-rabi crops are entirely rainfed, whereas summer crop is grown under irrigation. The yield of the kharif crop is poor, whereas those of the semi-rabi and summer crops are high, as they are grown in rich soils and under better management. The preparatory cultivation for the kharif crop is usually not thorough. For the semi-rabi and summer crops, the land is ploughed and harrowed repeatedly to secure a clean and fine seedbed. Except in Maharashtra and Gujarat, where line-sowing is practised, in all other states it is sown broadcast. The recommended spacings for the line-sown crop in different regions, however, range from 25 to 35 cm between the rows and 10 to 20 cm between the plants in the row. The seed being small, it is often mixed with sand to ensure its even distribution and the drill is operated rather shallow to avoid deep sowing. After sowing, the seed is covered lightly with a brush harrow. The seed-rate varies from 3 to 5 kg per hectare. The seed should be treated with Captan or Thiram at the rate of 3 g per kg of seed to control seed-borne diseases.

The rainfed crop is weeded and hoed once or twice and the irrigated crop is weeded often. Wherever the weed control with mechanical means is a problem, chemical weed control with pre-emergence application of Lasso at the rate of 3 litres/ha is recommended.

The crop is generally recommended for different regions as 25-30 kg of W, 20-30 kg of P_2O_5 and 0-20 kg of K_2O per hectare. The crop is harvested when the leaves, stems and capsules begin to turn yellow and the lower leaves start shedding. To prevent the shedding of grains, the crop should not be allowed to become a drip in the field. The ripe plants are cut, carried to the threshing-yard, stacked for a week in the sun with the cut-ends downwards and well shaken or beaten to take out the grains from the dry capsules. Winnowing and cleaning completes the process.

Pest and Diseases

The leaf-eating caterpillar (Antigastra Catalaulmlis Dup.) and the gall-fly (Asphon-dylia sesalili Felt) are the serious pests of sesamum. Two disulphur (0.5 kg/ha) sprays of Carbaryl 0.2%, or N on the 40th day and the 60th day, however give an elective control of both the pests.

Stem and root-rot (Macrophominia phaseoli Maubl.) Ashby, phyllody (virus, mycoplasina), bacterial leafspot (Pseudomonas sesami Matkoff) and leaf-curl are the important diseases of sesamum. Seed treatment with Ciptan at the rate of 3-9 Per kg of seed for controlling Macfphomila rot, 3 sprays of Agrimyein at the rate 01-100 ppm at 15-day intervals for controlling bacterial leaf spot are recommended. Both phyllody and leaf-curl can be controlled by spraying, the insecticides, such as Phosphomidon 0.05% or Dimethoate 0.05% or Dichlorovas 0.05%, which reduce the population of the insect vectors. The average grain yield of the kharif crop grown unmixed is 200 to 500 kg per hectare and the average yields of the semi-rabi crop and of the irrigated summer crop are 300 to 600 kg per hectare.

Varieties

The varieties differ not only in respect of the season of growth and the time of maturity but also in the number of capsules formed in the leaf axils, the habit of growth (Single-stemmed or branched) and the size, colour and oil content of the seeds. The seeds may be white, dull white, light brown dark brown, red or black and their oil content varies from 48-52%. The Table 2 gives a list of improved varictica recommended for cultivation in different regions.

Important hints for obtaining high yields

1. A suitable variety must he chosen for the tract and the season.
2. The crop should be sown in lines.
3. The application of fertilisers, inter-cultivation and plant-protection measures must be taken up at the appropriate time.

4. Harvesting should be done after the crop matures but must not be delayed up to the stage when the capsules shed seeds.

LINSEED

Geographic Origin

Linseed (*Linum uaitdissimum* L.) belongs to the family Linaceae and the genus Linum which has 100 species. It has been grown from ancient times for fibre (flax) and for its seed which is rich in oil The oldest regions of cultivated flax are reported to be in Asia and on the Mediterranean coast.

Economic Importance

Linseed occupies a greater importance among oilseeds; owing to its various uses and special qualities. In India, it is grown mainly for seeds, used for extracting oil. The oil content of the seed varies from 33 to 47 per cent. Linseed oil is an excellent drying oil used in manufacturing paint and varnishes, oilcloth, waterproof fabrics and linoleum and as an edible oil in some areas. Linseed-cake is a very good manure and animal feed. Linseed straw produces fibre of good quality. Linseed is used in making paper and plastics.

Area and Distribution

Linseed is extensively grown in the countries of the temperate zone as well as in those of the tropical zone. The major linseed-growing countries are Argentina, the USSR, India, the USA, Canada, Pakistan and Australia. India accounts for about 1.9 million hectares, with a seed production of 4.98 lakhs of tonnes and occupies the third rank among the linseed-producing countries. Australia and Canada have the highest productivity of about 7 quintals per hectare, whereas India averages 255 kg per ha. This yield is the lowest in the world.

In India, Madhya Pradesh leads in yield and acreage, followed by Uttar Pradesh and Maharashtra, Bihar, Rajasthan, Karnataka and West Bengal also grow linseed in large areas. Madhya Pradesh and Uttar Pradesh together contribute to the national linseed production to the extent of about 70 per cent.

Climate and Soil

Linseed is grown in the range of latitudes between the 10th and 65th parallels, both north and south. Its cultivation is confined to low elevations, but it can he successfully grown up to 770 metres. Areas with the annual rainfall ranging from 45-75 cm are best suited for its cultivation. The seed crop does well under moderate cold, but the fibre crop grows best in cool moist climates. In India, the crop is grown in the rabi season from September-October to February-March.

Linseed

Linseed can be grown on different kinds of soils, except the sandy and badly drained heavy clays or clay loams. It does well on the deep clayey black soils of central and peninsular India and on the alluvium loams of the Indo-Gangetic plains.

Cultivation

Fields of linseed are prepared as for other crops. Frequent ploughings are given to retain moisture and to check weeds in the field during the rainy season and a fine weed-free seedbed is prepared for sowing. BHC 10% or Aldrin or Chlordane 5% dust at the rate of 26-30 kg per ha is mixed within soil during the last ploughing before sowing to protect the seedlings against white-ants and the early damage from cutworms.

The main season for sowing linseed is October to November, depending upon the cessation of the rains. The crop is grown either broadcast or is drilled in lines 20 to 30 cm apart: In the case of broadcasting, which includes uterasowing also, the seed-rate is kept at 40 kg per ha. For line-sowing, the seed-rate varies from 20 to 510 kg per ha, depending upon the seed size. Thin sowing gives a good number of tillers, but for fibre and weed, thick sowing is desirable.

The application of fertilisers and irrigation increases the yield of the crop. Under rainfed conditions, the fertiliser dose including 30 kg of N and 15 kg of P per ha is given. If irrigation is available, two to three irrigations are sufficient. Under irrigated conditions, the linseed crop requires 60 kg of K and 30 kg of P per ha.

The deep placement of fertiliser at sowing in the case of the rainfed crop gives better results. Under irrigated conditions, nitrogen is applied in two splits, half the dose as basal and the other half at the first irrigation, 40 to 50 days after sowing. The utera crop is fertilised at the rate of 10 kg of N per ha applied at the time of sowing. The rainfed crop generally requires no interculture. When the crop is irrigated, one weeding becomes essential.

Pest and Diseases

Linseed gall midge (Dasyneura lini Barnes) is the most wide spread and serious pest of linseed. The growing of early varieties and mixed cropping are reported to reduce the incidence of the pest. Linseed rust and wilt are the two most serious diseases. The growing of resistant varieties is recommended to control these diseases.

The linseed crop starts maturing by the mid of February, depending on winter spread and sowing time. Plants turn golden yellow when the crop is mature and ready for harvesting. Harvesting is done when the crop is dead ripe with a sickle or by uprooting the plants. When the fibre is also desired along with seed, the harvesting of the crop is done at the stage of capsule maturity even when the crop is slight green. Linseed is threshed after it is completely dry. The crop is threshed either by beating it with a stick or by trampling it under the feet of the bullocks. Hand-beating is better to ge lien fibre, along with seed. Seed is separated from the chaff by winnowing and is stored in a dry place.

Yield

The average yield of a pure crop varies from 210 to 450 kg per ha of seed under rainfed cultivation. The crop in northern India generally gives higher yield than in central and peninsular India. The irrigated crop may yield 1,200 to 1,500 kg per ha. The percentage of oil in the seeds varies from 37 to 43, but the commercial extraction in expellers of rotaries gives about 33 per cent from small seeds and 34 to 36 per cent from hold seeds. Crushed in village ghanis, the seeds yield only 25 to 30 per cent of oil.

Varieties

In India, the linseed varieties are of two types

(i) Peninsular types which have deep root-system.

(ii) Alluvial types which are shallow-rooted and are characterised by profuse tillering. Varieties differ in maturity, growth habit, and seed size and colour. The improved varieties recommended for cultivation in different states are given in Table 3.

Important hints for getting maximum yields

(a) Clean, fine, weed-free seed-bed should be prepared by ploughing during the rainy season.

(b) 10% B.E.C. or 5% Aldrin or Chlordane at the rate of 25 to 30 kg/ha be mixed with the soil at the last ploughing before sowing.

(c) Early sowing escapes the attack of many diseases and pests. Under utera, sowing be done at the dough stage of paddy.

(d) The application of 30 N+15 1 for the rainfed and 60 N+30 P for the irrigated crop and 10 N kg per ha for the utera sowing is recommended.

(e) Use the optimum seed rate of 30 to 40 kg per ha and a closer row spacing between 20 and 30 cm and 40 kg seed per ha for utera sowing.

(f) Cultivate the recommended varieties.

(g) Carry out plant-protection measures, if needed.

CASTOR

Geographic Origin

According to the available literature, castor (*Ricinus comrnunis* L.) is indigenous to Eastern Africa and most probably originated in Ethopia.

Economic Importance

It is generally grown for its, oil-yielding seeds. The oil content of the seeds varies from 35-58 per cent in different varieties, the average being about 47 per cent. Castor-oil is being used widely for various purposes. It is used as a lubricant in high-speed engines and aeroplanes, in the manufacture of soaps, transparent paper, printing-inks, varnishes, linoleum and plasticisers. It is also used for medicinal and lighting purposes. The cake is used as a manure and plant stalks as fuel or as thatching material or for preparing paper-pulp. In the silk-producing areas, leaves are fed to the silkworms.

Botanical Description

Castor belongs to the gellus Ricinus, its a member of the Euphorbiaccae which consists a vast number of plants mostly native to the tropics. The geiut; Ricinus is considered to be momotypic and *R. cwnmunis* as the only species which includes many polymorphic typco. Thc cultivated types are dwarf annuals

The stem is erect, circular in section, partially, hollow, smooth, glabrous, with good branching. The stem is marked by a number of well-defined nodes, from each of which a leaf arises. The lower internodes are shorter and their length increases with the height. Leaves are alternate, large, palniate with 5-11 lobes, acuminate, margins notchedy serrate or indented. They are carried on long stout petioles.

Ricinus comrnunis

The inflorescences are borne terminally on the main and lateral branches. Flowers are large, in terminal sub-panelled racemes, monoccious, apetalous, the upper portion of the raceme being occupied by the female flowers and the lower by the male flowers. The fruit is a roundish glaucous capsule, with three projecting sides covered with tough spines or smooth, three-loculed and three-seeded. Seeds are albuminous, anatropous, broad, oval, compressed with a marked carnnele and longitudinal raphc. The testa is thin, brittle, varying in colour and mottling. Below the testa is the thin tegmen, covering the whitish oily endospeim containing the embryo.

Distribution, Area and Production

The main castor-growing countries are Brazil, India, USSR and Argentina. India ranks first in area (44,8000 hectares) but second in production (140,000 tonnes), Brazil being the, first. Andhra Pradesh (67.2%), Gujarat (12.7%) Karnataka (7.1%) and Orissa (5.8%) account for over 90% of the area and also production.

In lighter soils, the land is repeatedly ploughed in summer, as and when the rains are received, and with the onset of monsoon rains the clods are crushed by working a country plough or harrow. In clay loams, only harrowing is done with a blade-harrow two to three times to bring the soil into proper tilth.

The seed is sown either in the plough furrow or with a seed-drill or by hand-dibbling. The seed-rate used for sowing behind the plough is 12 kg per hectare. For hand-dibbling, 8 to 10 kg of seed will be sufficient per hectare. Before sowing, castor seed should be treated with Thiram 3 g/kg of seed to prevent the attack of root-rot and Alternaria blight. The spacing adopted generally ranges from 90-120 cm between rows and 45-60 cm between plants in the row. One or two weedings and hoeings with bullock-drawn implements are given and sometimes, the plants are earthed up. The fertiliser dose recommended E for castor is 40 N-40 P-20 K kg/ha. The entire dose of P and K and half of W is applied basally and the remaining half of W about 30 days after sowing, as and when the soil moisture is adequate.

The improved varieties cultivated at present are early maturing and take about 150 to 180 days. When one or two capsules in a bunch show signs of drying, the whole cluster is generally removed and stacked, covered and weighed in a corner of the field or in a pit. The harvesting of unripe capsules in this manner has an adverse effect on the oil content of the seed. It is preferable to collect the fruits, as and when they ripen. When the whole crop is gathered, it is dried in the sun for a few days and the threshiug is done either by beating the dried capsules with a stick or by treading them under the feet of bullocks. Winnowing is done in the usual manner.

Pests and Diseases

The castor semi-looper (*Achow janaga* L.) and the capsule-borer (Dicitocrocis punctiferalis) are the two most serious pests of castor. Dusting BHC 10% in early stages or spraying 0.44% Ekalax or 0.1% Carbaryl or 0.05% 1 arathion on the crop will give an effective control on these pests.The seedling blight (Phytopiathora colocasic Racib. emend. Thomas & T.S. Ramkar) and Alternaria blight Alternaria ricini (Yoshii) Hansford cause serious losses to this crop. The Sowing of the crop in low-lying and water-logged areas should be avoided to prevent the seedling blight from appearing. For alternaria blight, seed treatment with Thiram at the rate of 3 g/kg of seed is recommended.

Yield

The average yield of rainfed castor varies from 200 to 500 kg per hectare, that of the mixed crop from 100 to 200 kg and of the irrigated crop from 500 to 800 kg. In eleven years (from 1964-65 to 1974-75), the average total yield of castor in India has risen to 350 kg per ha.

Varieties

The varieties differ in the branching habit of the plants, colour of the stem and branches (red and green), the nature of capsules (smooth or shiny), duration (early or late) and the size of seed. The following are the principal improved varieties of castor recommended for cultivation in different states. As mentioned in (Table 4)

Important hints for getting Maximum yields

1. Recommended variety for a particular area should he chosen.
2. Seed treatment with Thiram (3 g/kg of seed) should he done before sowing to prevent the attack of root rot and alternaria blight.
3. Sowing should be done at the right time as recommended for different agro-climatic conditions. Sowing should he done with proper spacing as recommended for different regions.
4. The recommended fertiliser schedule for the region should he adopted.
5. Proper, timely and adequate plant-protection manure should be adopted to save the crop from insect pests and diseases.
6. Two to three intercultures should be done to provide aeration for the root-system and to keep the crop free from weeds.
7. The nipping of branches increases the size of the main spike and induces early maturity.
8. Harvesting must be done at the right time, as too early or too late harvesting results in reduced yield, oil and shattering of seed respectively.

SAFFLOWER

The three centres of origin, as suggested by Vavilov for safflower (*Carthamus tinctorius* L.), are India, Afghanistan and Ethopia. Decandole was of the opinion that Arabia was the most probable centre of origin.

Economic Importance

The crop is now cultivated, primarily for its seeds which yield oil, though at one time it used to be grown for the extraction of a dye also. The seeds are edible and are eaten after roasting. Their oil content varies from 24 to 36 per cent, depending on the variety, soil, climate and other conditions. The cold-pressed oil is golden yellow and is used for culinary purposes, or for making soap. The oil obtained by dry hot distillation is black and sticky and is used only for greasing well ropes and leather goods exposed to water. Safflower oil has also good drying properties and is, therefore, used in the manufacture of paints, varnishes and linoleum. It can be mixed with white paint without any after-yellowing effects. The cake, particularly from decorticated seed, is used as a concentrated cattle feed, and that from undecorticated seed is sometimes used as a manure.

Botanical Description

The safflower plant is highly branched, herbaceous, thistle-like annual, varying in height from 30 to 160 cm. It has many branches each terminating in a flower. The inflorescence is broad, flat or slightly curved and densely bristled owing to the presence of numerous floral tracts. The flowers are yellow to red, containing 20 individual ficrets, each of which produces a seed.

Carthamus tinctorius

Distribution, area and Production

The important safflower growing countries, besides India, are the USA, Mexico, Ethopia, Spain, the USSR. and Australia. In India, it occupies 590,000 hectares with a production of nearly 130,000 tonnes. Over 98% of the area is concentrated in the states of Maharashtra (04.4%), Karnataka (26.0%) and Andhra Pradesh (8.0%).

Climate and Soil

The crop is grown during rabi primarily as a rainfed crop, but in some areas it is raised under irrigation. At all stages of growth of the crop, excessive rainfall or humidity increases damage from fungal diseases. Water-logging due to poor drainage or prolonged rains even for relatively short periods can cause substantial fall in seed-yield. Being drought-resistant, it is cultivated on all types of soil, including sandy soil, but it thrives best on water-retentive black soils and alluvial loams. The crop is fairly resistant to saline condition.

Cultivation

Saffiower is grown mostly as a mixture with wheat, barley, gram and rabi jowar, three rows of safflower being planted after every nine, 12 or more rows of the main crop. Sometimes, it is sown as border rows surrounding the crops of wheat, rabi, jowar, etc., because being spiny, it protects the main crop against cattle trespass. As a pure crop, it is rotated with wheat, gram and rabi jowar. Occasionally, it is grown as a second crop after any quick-maturing kharif crop, e.g. green-gram (moong), black-gram (mash), groundnut, coriander, etc.

The number of ploughings should be as few as possible and be aimed at preserving the maximum amount of soil moisture. One or two ploughings with a country plough will be sufficient to remove the weeds and break the clods. Sowing is generally done in September and October. The seed-rate adopted varies from 5 to 12 kg per hectare, depending upon the soil fertility and the nature of the crop (pure or mixed). When grown in strips, a row spacing of 45 cm is adopted. Plants too close tend to have thinner stems or wear a superficial root-system, with a reduced number of flowers. In general, the pure crop is given one or two weedings, combined with hoeing on the 20th and the 45th day after sowing. The application of 20-40 kg of N per hectare results in a substantial increase in yield. The mixed crop shares the preparatory tillage, manuring and cultivation given to the main crop. When the plants have developed the central flowering head, they are topped to promote branching, flowering and seed production.

Harvesting and Yield

The crop matures in 4 or 5 months after sowing. The ripe plants are either cut or pulled out, stacked for a few days to dry, threshed by cutting with sticks, and the resulting material is winnowed to obtain clean seeds. The average yield of the pure crop is 400-500 kg of seed per hectare, whereas that of a mixed crop is about 100 kg.

Pests and Diseases

The safflower aphid (Dactynotus carthami), thrips (FranklinieUa sulphurea) and the flower bud fly (Acanihiphilus hclianthi Rossi) are the serious pests of the crop. Spraying the crop with 0.1% Fenithion, 0.03% Dimethoate and 0.07% Endosulfan is recommended for controlling aphids, thrips and budfly, respectively.

Rust (Puceinia carthami), alternaria leaf-spot (Alternai carthami) and root-rot (Phytophthora drechskri) are the important diseases of safflower. The growing of resistant varieties is recommended to avoid damage from these diseases.

Rust (Puceinia carthami), alternaria leaf-spot (Alternai carthami) and root-rot (Phytophthora drechskri) are the important diseases of safflower. The growing of resistant varieties is recommended to avoid damage from these diseases.

Varieties

The varieties under cultivation in different regions of the country are given in Table 5.

Table 5.

State	*Name of the variety*	*Duration in days*	*Special characteristics*
Karnataka	'A.I'	125	
		125	Yields 800-850kg/ha, oil content 30.8%
	'A-300'	125	
		125	Yields 750-800kg/ha, oil content 30.9%
Madhya Pradesh	'NO-7'	140	Yields 700-850kg/ha, oil content 30.3%
Maharashtra	'NO-62.8'	140	Yields 1,000-1,200 kg/ha,oil content 30.%
	'NAG-7'	140	Yields 1,000-1,200 kg/ha, oil content 30.%
	Tara	130	Yields 1,000-1,400 kg/ha, oil content 32.5%
Tamil Nadu	'K-1'	120	Yields 600-800 kg/ha, oil content 30.5%
Andhra Pradesh	'Manjira' C 438	110	Yields 1200 kg/ha, oil content 32%

Important hints for Obtaining the maximum yields

1. The sowing time should be adjusted in such a way that there will be no rainfal after flowering.
2. Nitrogen amounting to 20-45 kg per hectare has to be applied, preferably in two split doses, half at the time of sowing and the remaining half at flowering.
3. Timely inter-cultivation for the eradication of weeds and control measures agains pests and diseases should be adopted.

SUNFLOWER

Origin and Distribution

The cultivated sunflower (*Helianthus annus* L.) is a native of southern United States and Mexico, from where it was taken to Spain before the, middle of the sixteenth century In the nineteenth century, the cultivation of sunflower as an oilseed crop began in th Soviet Union and the majority of the present day varieties grown all-over the world trac back their origin to the USSR. The major sunflower-producing countries of the worl are the Soviet Union, Argentina, Bulgaria, Pumania, Turkey and South Americ Sunflower, as an oilseed crop, was introduced into India in 1969 and, according to th estimates of the trade and the Vanaspati Manufacturers Association, 3,88,000 hectare of sunflower was grown in India during 1975-76.

Economic Importance

Sunflower is mainly grown for its oil. The oil is used for culinary purposes, in preparation of vanaspati and in the manufacture of soaps and cosmetics. It is especially recommended for heart patients. Its cake is rich in protein and is used as a cattle and poultry feed.

Botanical Description

Sunflower belongs to the family Compositac and the genus Helianthus. The cultivated genotypes are characterised by a single stem terminating in a capitulum. Sunflower is protandrous, in which the male and female elements mature at different times. There appears to be a time-lag of 18-24 hours in the maturity of the male and female elements. So it is essentially a cross-pollinated plant, besides showing varying degrees of self-incompatibility.

Helianthus annus

Climatic and Soil Requirements

Sunflower is a day nuteral plant can be grown successgully in different seasons under varying conditions of day length, provided the temperature is favourable. Sunflower, basically a temperature plant, can also be grown commercially in tropical or sub-tropical conditions. It grows best with clear sky and occasionally rain shower during early stages. This corps requires a cool climate during germination and seedling growth and warm non-cloudy weather and high temperature during flowering to maturity. Sunflower can tolerate temperature range of 8 to 30⁰C but the temperature below 16⁰C and above 40⁰C reduces yield and oil content. High temperature affects pollination and fertilisation, thus care should be taken to adjust the sowing date in such a way that flowering does not coincide with high temperature. The optimum temperature for best performance under controlled conditions is 27-28 ⁰C.

Sunflower can be planted successfully when soil temperature exceeds 10⁰C, though germination begins at 3-5⁰C. Young plants (1 or 2 pairs of leaves) can withstand frost to -6⁰C. However, plants are frost sensetive from 6-8 leaf stage until flowering finishes. Late frost does not destroy the whole plant but destroys the fruiting head, leading to increased branching, resulting in small heads with many empty fruits. During flowering temperature should be between 18-20⁰C. Goods yields are obtained where the ruling temperature is 18-22⁰C showing that generally speaking the crop needs a warm climate. Average minimum temperature below 17⁰C between flowering and maturity is necessary to ensure best oil quality. The crop is grown from 45⁰S to 55⁰N but greater production is between 20-50⁰S. The crop can be cultivated upto an altitude of 2500m but highest yield of oil is obtained below 1500m. Sunflower cannot tolerate frost and requires a frost free period of 120 days. It is resistant to drought but requires continuous availability of soil moisture for the best performance.

Sunflower can thrive on a wide range of soils such as sandy loam, black soil and alluvial soil, but it performs better on sandy loam soil than on clay soil under similar management. However, good drainage is more important than soil type. Thus, fertile well drained, medium textured soils having heavy water holding capacity are best suited for the cultivation of sunflower. Sunflower is highly susceptible to water logging. Avoid cultivation of sunflower on soils affected by salts and with poor drainage the optimum pH requirements 6.0 to 8.5.

Preparatory Tillage

Sunflower requires a fine seedbed, free of weeds and clods, for proper germination and good crop stand to bury the residues or previous crop, ploughing with disc harrow followed by 2-3 cultivations and plankings may be required fields vacated by potato may not require much seedbed preparation. However, adequate soil moisture is essential forgetting good germination.

Variety

MSFH8, PSFH-67,GKSFH-2002, Jwalamukhi, Mega-363 and NSFH-592 have been recommended. These hybrids are medium tall with dark green leaves their seeds are black in colour and mature in about 110-125 days mega-363 is a long duration hybrid and hence should not be selected for cotton areas. These hybrids have an average yield about 20 quintals per hectare with an oil content of 40.5 to 43.0 percent. In hybrids fresh seed is to be sown every time which adds to its cost of cultivation.

Sowing Time

Spring season is the best for getting higher and stable seed yield sunflower. Sunflower being a cross pollinated crop depends on honey bees for pollination and seed setting. Availability of honey bees is abundant during spring season due to favourable. Planting during spring should be made after the soil temperature exceeds 10^{0}C at 10cm depth at 8 a.m. Earlier planting at lower temperature increases time of flowering without much earliness. Best results are obtained when crop is sown in the first fortnight of January which produces higher seed yield than later planting. Sowing of short duration hybrids like GKSFH-2002,PSFH-67 and nsfh-592, however, can be extended upto second fortnight of January. Besides low yields, the late sowing may also face early mansoon rain at harvesting and threshing causing difficulty in its management.

Seed Rate and Seed Treatment

Use 5kg seed per hectare for hybrids to achieve the required plant stand. Low seed rate recommendation for hybrids is mainly because of smaller seed size and wider spacing for hybrids. Seed should be treated with Brassicol or Thiram @ 2 g per kg of seed to control seed and soil borne fungal disease.

Method of Sowing

A spacing of 60×30cm for hybrids has been found to be an optimum. These spacings result in better development of head, more seed filling and less stem breakage.

Crust Breaking

Rain after sowing, results in crust formation, which hinders the proper emergence of seedlings. Thus it may be necessary to break the crust, which can be done with harrowing. Harrow as soon as possible after the crust has performed and if possible, before seedlings have emerged.

Fertiliser Management

Fields studies have shown that hybrid sunflower responds well to nitrogen application upto 60 kg/ha. further increase in nitrogen results in lower seed and oil yield. Application phosphatic ferticisers @ 30kg P_2O_5/ha increase sunflower yield sunflower responds to K application only on soils deficient in available potassium and it is recommended to be applied @ 30 kg K_2O/ha on such oils.

Regarding application method, drilling the fertilisers besides the seed row has been found to be more beneficial for getting higher seed yield. Application of fertilisers below the seed was found to be injurious to germinating seeds and reduced seed emergence count and seed yield.

Entire dose of N, P and K should be applied at the same time of sowing. Prefer single superphosphate for P application as it also contains sulphur to which sunflower responds favourably. If single superphosphate is not available apply gypsum @ 25-30 kg/ha at sowing time. A supplementary dose of nitrogen @ 25-30 kg/ha may be top dressed a month after sowing sunflower if the crop exhibits symptoms of nitrogen deficiency.

Nitrogen deficiency appears as yellowing of the older leaves, which die prematurely. Leaves are also fewer and smaller. Plants with a severe deficiency are yellow and stunned, while young plants are of purplish colour. Plants are slender with longer internodes. The flower head is small with many unfilled seeds and the oil content is reduced.

Phosphorous deficiency results is delayed flowering, stunned growth and small leaves which prematurely die back from the leaf tip. Phosphorous deficiency is shown in small dark green leaves later turning pale with necrosis. Growth is restricted with short stiff stems, flowers are small, stuned malformed. Ripening is delayed, oil content and yield are reduced.

Potassium deficient plants are squat with a proportionately greater reduction in stem length then in leaf size. Initially leaves are dark green but turn yellow at the edges and die except the lower leaves. Potassium deficient plants have restricted growth with a rosette arrangement of the younger leaves later the leaves develop dark brown necrotic patches.

Water Management

Sunflower can withstand moderate drought because of its deep root system. However, it responds favourably to irrigation and its seed yield can be increased substantially by substantially by sustaining adequate moisture in soil by timely irrigations besides improvements in oil content sunflower is highly moisture sensitive plant and its broad leaves exhibit wilting with slight moisture deficiency.

First irrigation to sunflower should be applied a month after sowing and interval during the subsequent months be reduced to 2-3 weeks during March and 8-10 days during the hots months of April and May. Irrigation should be stopped about 12-14 days before harvest. Sunflower is a tall growing plant with heavy head, therefore , irrigation at flowering and thereafter should be given on clam days to avoid lodging. The incidence of Sclerotinia stem-rot is more under irrigated conditions especially where the soil drainage is poor. Excessive and frequent irrigations on fine textured soils should be avoided.

Weed Management

Mechanical methods such as hoeing at 3-4 weeks after sowing followed by earthing up before flowering has been found to be effective in checking weeds. Use of tractor drawn implements before the crop attains 60-70 cm height has also been found effective for better efficiency of manual weeding an improved 3 tine wheel hand-hoe could also be used. Removal of weeds through mechanical/manual means quite effective but expensive. Moreover, often timely labour is not available. The studies initiated to find effective herbicides for weed control in sunflower, revealed that pendimethalin (STOMP @2.51/ha) as pre emergence application gave effective weed kill.

Harvesting and Threshing

The crop is ready for harvesting when the back of flower heads turn lemon yellow or yellowish brown near the stalk and the discs start drying. At this stage the seeds give a blakish look and are fooly ripe the shading of disc-flower residues from undamaged heads also indicate psysiological maturity. Promptly harvesting seed after psysiological maturity has many advantages; seed is least exposed to bird and rodent attack; the crop cycle is shorter resulting in more time to prepare for the next crop; reduces harvest seed losses; provides high quality seed and reduces exposures to bad weather.

The harvested heads are mildly dried and threshed by beating or trampling with tractors or sunflower threshers. The sunflower threshers are more efficient and economical. After threshing the produce should be thoroughly dried before storing otherwise in storage it gets affected by fungus and decomposes.

Allelopathic Effects of Sunflower

Researches show that sunflower depresses the growth and productivity of succeeding crops. The reason postulated are the allelochemicals being secreted by sunflower crop. Decomposing residues of sunflower could also bring about allelopathic effects. Phytotoxic compounds released into the soil rhizosphere have been observed to cause varying degree of allelopathy/autotoxicity in crop. Field studies conducted in the department of Agronomy, Punjab Agricultural University, Ludhiana on the allelopathic effect of spring sunflower on succeeding Kharif crops revealed that maize, sorghum, cowpea, pearlmillet, pigeonpea, mungbean and sesamum suffered 41.7, 36.7, 27.6, 28.8, 23.0, 10.8 and 9 percent reduction in yield, respectively, as compared to these when grown after wheat. Studies in allelopathic effects of different plant parts of sunflower indicated that sunflower roots reduced pigeonpea, cotton and mungbean yields the most followed by roots plus shoots treatment. Sunflower shoots did not have any harmful effect on growth of these crops.

Bioassay studies involving use of aqueous sunflower extract, decomposing plant residue and soil from area in proximity of sunflower crop revealed inhibition of gemination and growth of cotton, pigeonpea, sunflower and mungbean. These studies further showed that effect of leaf extract/residue was more than roots. The allelopathic effect of sunflower plant residue persisted upto 24 days after incorporation and beyond this period the effect was rather meagre.

Chromatograhic and physico-chemical analysis of allelochemicals in sunflower parts showed the presence of chlorogenic and isochlorogenic acids in both stems and leaves besides these some unidentified phytotoxins were also observed. The concentration of these was more in leaves than in stems. Further studies revealed that total phenolics was more in leaves followed by in roots and shoots. Sunflower raised under moisture stress conditions had higher phenolics as compared to the raised without any moisture stress.

Cut worm (Agrotis spp.)

It is polyphagous pest and cuts off the tender stalks, near ground, of young crop plants during night. During day time the caterpillars remain hidden in the soil. It is more serious during March-April when sunflower follows potato. When flat sowing is practised, apply 5 litres of Dursban/Radar 20 EC (Chlorpyriphos) per hectare before sowing. These insecticides should be mixed in 25 kg fine soil and broadcast uniformly after last ploughing but before planking. Sow crop on ridges to avoid cut worm damage in the early stages.

Tobacco caterpillar (Spodoptera litura)

The polyphagus pest remain active throughout the year feeding on one or the other crop. The newly hatched caterpillar is pale-greenish with a black head and feeds on the outer epidermis of leaves. It poses a serous threat during April-May and defoliates the plants.

Cabbage Semi-looper (Plusia orichalcea)

This insects feeds on many crops including the sunflower. The newly hatched caterpillars defoliate the crop and cause severe damage to it during April-May.

Bihar hairy caterpillar (Spilosoma obliqua)

It is also called Bhabbu Kuta. This is highly polyphagous and voracious feeder. The nefarious pest causes severe damage and plants looke denuded. Newly hatched larva is bright brown and its body is covered with hair and legs are brown. At the older stage it is much darker and with dark brown hair of almost uniform length these insects feed gregariously and should be collected along with the plant and buried deep in soil. For the control of above mentioned pests, spray 1.125 L. Folithion/Sumithion 50 EC (fenotrothin) or 1.225 L. Thiodan 35 EC (endosulfan) or 500 ml Nuvan 100 EC (dichlorvos) in 250 litres of water per hectare. Repeat after two weeks, if necessary.

Head borer (Helicoverpa armigera)

Head borer or American boll worm has become the noxious polyphagous pest in India and also causes heavy losses in sunflower. Newly hatched larva is green yellow which changes its colour with every instar.

The full grown caterpillar is brownish green having grey longitudinal streaks on it. The adults lay eggs on the flower buds. After hatching the larvae feed on developing seeds and make tunnels through them.

Head borer can be controlled by spraying 2.5 Kg of Sevin 50 WP (Carbaryl) or 2.0 litres of Asataf 75 SP (Acephate) or 2.125 litre Folothion 50 EC (Fenitrothion) or 2.5 litres of Dursban 20 EC (Chlorpyriphos) or Thiodan 35 EC (Endosulfan) or 1.25 litre of Nuvacron 30 SI (Monocrotophous) in 375 litres of water per hectare. Repeat after two weeks if necessary. Endosulfan is comparatively safe for honey bees. Spray when honey bee activity is minimum i.e. early morning or late evening.

Precautions for spraying insecticides

- Spraying during flowering should be avoided to protect the pollinators.
- Spray in the afternoon when pollinators are less so that there is a time gap till morning or mid day when their activity is more.
- Spray only when pest incidence is observed in abundance.
- Spray on calm day to avoid drift.
- Precautionary measures are advisable if pest outbreak is observed every year.

Diseases and their Control

Steam rot / cottony rot

It is soil and borne disease. Infestation of basal portion of stem takes place via mycelium that grows with the roots and the stem becomes soft and cotyledon leaves flatten on the ground. At bud formation soft decaying area appears on its stem and base and white mycelia appear later on. The stem becomes weak, leaves dry up and drop off with strong wind.

Infestation takes place on heads when fully developed and seed formation has taken place. The back of head becomes soft and turns brown. Fungus like cotton appears and seeds are also infested. The sclerotia of this pathogen, almost of the size of sunflower seed are produced in abundance which are visible in pith of rotton stem and also in the infested head. The affected tissue turns black and shredded. This disease can be effectively controlled by treating seed with thiram @ 2 g/kg seed before sowing.

Root rot or charcoal rot

It is soil borne disease that infects the roots. Roots turn brown and rot. Infection at early stages kills the seedlings rapidaly. This also causes neck rot i.e. stem rotting immediately behind the head where infected stem turns brownish and darker towards the base. Roots turn grey and stunted. At later stages, it causes premature ripening, smaller and poor filled heads and a lower yield. For its control, treat the seed with Thiram @ 2 g / Kg seed and avoid moisture stress.

Head rot

High humidity and injury to plants favour infection at lower surface of head where it is attached to the stem irregular water soaked spots appear, which gradually enlarge and become soft. Shredding of back of heads takes place. Cracking of front side is also common. Seed filling is affected severely and seeds remain small and shrivelled. The yield losses range from 20 to 60 percent and oil reduction to the tune of 25-53 percent and increase in free fatty acids, which is undesirable. To avoid this disease plough the soil with furrow turning plough to bury the debris of previous crop and residues of sunflower; treat the seed with Thiram @ 2g/kg seed; avoid excessive stress of irrigation; grow recommended hybrid and don't grow sunflower repeatedly in the same field; Spray with Indofil M-45 @ 625 g per hectare in 250 litre of water at 15 days interval starting at about 60 days stage.

Utilisation

Sunflower crop has many uses :

- Edible oil of sunflower is of superior quality because of anticholestrol and non cholestrol contents and is thus preferred over other oils for cooking food.
- Its oil is used for making vanaspati, soap and other allied products.
- Its oil is used in the manufacturing margarine.
- Sunflower meal in the form of cake is used for animal feed or manuring field.
- Hulls and stalks are used for the preparation of furfural.
- Green plants are used as forage or silage.
- Dry stalks are used as fuel.

- Pulp is also extract from plants.
- Sunflower husk finds use in the preparation of ethyl alcohol and yeast.
- Salted and fried kernels of sunflower are eaten as such and its meal is used for confectionery.
- The concentrates and isolates of defatted meals are used for protein additives, meat analogues, meat extenders, emulsifying agents, bakery products, beverages, snak products, etc.
- Seed is included in many commercial bird feeds.

Exploiting male Sterility

Sunflower being a Cross-pollinated crop can be exploited for increasing the production of edible oil by developing hybrid sunflowers. The development of commercial hybrids is only possible by using male sterility, and fortunately such genetic systems are available in sunflower. Two types of male sterility occur in sunflower: (1) genetic male sterility linked with certain marker genes. This sterility is due to a recessive nuclear gene and linked to the green colouring matter in the seedling; (2) the cytoplasmic male sterility was first obtained from the interspecific cross H. petiolaria x H. a.n.nuue by Leetereq in France in 1965. This type is being widely used in developing hybrid sunflowers all over the world. Work on the development of hybrid sunflower is in progress at the various centres of the A.I.C. O.P..P. 0. (Sunflower).

NIGER

Geographic Origin

The Niger (*Guizotia abyssinica)* L.F. Cass. crop is said to be indigenous to tropical Africa and more specifically to Ethiopia.

Economic Importance

The crop is grown for its seed, used for extracting oil which is about 37 to 43 per cent of the seed weight. The oil is used for culinary purposes, for anointing the body, for manufacturing paints and soft soaps, for lighting and lubrication and for manufacturing cosmetics. The oilcake is a well-known cattle feed. The crop sheds a large quantity of dry leaves in the field and thereby adds organic matter to the soil.

Botanical Description

Niger belongs to the family Compositae. The plant is a stout, erect, smooth orscabrous leafy annual herb, about one metre tall. The leaves are opposite or the upper ones are alternate. The capitulum is peduncled, axillary and terminal heterogamous. Bay-florets are pistillate, scriate, fertile; ligule 2-3 toothed and yellow. Disc florets are hermaphrodite, fertile; limb campanulate and 5-fid. Achenes are glabrous, dorsally compressed, tip rounded, pappus absent.

Guizotia abyssinica

Distribution, Area and Production

India is considered to be the chief Niger-producing country in the world, with an area of 4.8 lakh hectares, and with an annual estimated production of 1.2 lakh tonnes of seed. It is mainly grown in the states of Madhya Pradesh, Bihar, Maharashtra, Orissa and Tamil Nadu, of which the largest area of 2.38 lakh hectares is in Madhya Pradesh.

Climate and Soil

Niger is grown as a kharif crop under rainfed conditions. A moderate rainfall of 1,000-1,250 mm suits this crop and, as such, it is not cultivated in regions of heavy rainfall.

Light red soils and brownish loams with sufficient depth and good texture are considered suitable for this crop. It is often grown on poor soils of coarse texture. It can be grown on well-drained heavy soils.

Cultivation

Niger is grown either as a pure crop or is grown mixed with minor millets. Two to three ploughings are enough to prepare the land for sowing. The field is made weed-free and levelled by planking before sowing. The crop is sown from June to August, either behind a country-plough or with a four-coultered drill. A spacing of 30 cm between lines and that of 10-15 cm between plants in the line are generally adopted. A seed-rate of 7-8 kg per hectare is adequate to maintain a good crop-stand and a pure crop. About a fortnight after sowing, thinning is done to regulate spacing between plants. One or two hoeings are required for controlling weeds. In general, the Niger crop is not manured. But the application of 20 kg of N and 20 kg of P_2O_5 per hectare is recommended for obtaining higher yields of the crop. When sown as a mixed crop, Niger shares the manuring given to the main crop. The crop matures in November and December. The mature plants are cut with sickles at the base and heaped in the threshing-yard for a week. They are then spread in the sun to dry for two to three days, threshed by beating with sticks, and winnowed.

Yield

The average yield of Niger is about 200 kg/ ha, but under favourable conditions and with best management practices it yields up to 600 kg.

TRADITIONAL OIL PROCESSING METHOD

For thousands of years, fats and oils have been important in food preparation in India. Metal frying pans that are remarkably similar in design to those used today have been found in archaeological excavations of the Harappan civilisation of circa 2000 BC. A number of oleaginous materials such as sesame, rape and mustard seeds and coconut were known sources of oil. In addition, a variety of animal fats were used. However, the exact way that oil was obtained from oilseeds is uncertain.

In Sanskrit literature of about 500 BC there is a specific reference to an oil-press, although it was never described. Juices were extracted from vegetable materials as early as 1500 BC using either a mortar and pestle or a grinding stone working on a flat stone. Linguistic evidence suggests that it is from these two crushing systems that presses for both oilseeds and sugar cane developed in the form of a mortar-and-pestle arrangement powered by animals. This system is commonly called the ghani, or the kolhu or chekku.

Ghani operation has been noted in Afghanistan, Sri Lanka and Myanmar, which had cultural ties with India. The device is widely used in the Sudan to crush sesame seeds, though its antiquity there has not been documented. Despite subsidies by the Indian Government to support village industries, in recent times ghani operation has declined precipitously in the land of its origin.

Ghani Technology

The oilseeds and subsequently the expressed oil are held in a scooped circular pit in the exact centre of a circular mortar made of stone or wood. In it works a stout, upright pestle which descends from a top curved or angled piece, in which the pestle rests in a scooped-out hollow that permits the pestle to rotate, eased by some soapy or oily lubricant. Today the single angled piece takes the form of two shorter pieces pinioned or chained together. The bottom of the lower angled piece is attached to a load-beam; one end of the load-beam rides around the outside of the barrel, while the other is yoked to the animal. The load-beam is weighted down with either heavy stones or even the seated operator. As the animal moves in a circular ambit, the pestle rotates, exerting lateral pressure on the upper chest of the pit, first pulverising the oilseed and then crushing out its oil.

Within India there are regional variations in ghani design, which probably arose from the nature of the oilseeds that were regionally available for crushing. The large granite ghanis of southern India have a capacity of 35 to 40 kg, requiring two animals yoked side by side and two operators, one for the animals and the other near the mortar. The load-beam is very long and curved and rides on a strong outer groove on the mortar. These ghanis have a life of four to five years, after which the pit is too worn to be useful. The wooden ghani of western India has a capacity of 8 to 15 kg, has an oil outlet at the base of the pit (which is kept plugged during crushing) and frequently has the operator seated on the load-beam.

The Bengal ghani has a small capacity of only 5 to 10 kg per charge and is usually used to crush a mixture of rape and mustard seeds to yield a bouquet of flavours. The pit is small and the pestle is tall and has a stout base. The operation is prolonged so as to permit slow enzymatic liberation of several pungent alkyl isothiocyanates from the precursor glucosinolates in the prevailing warm, moist conditions. Punjab ghanis are of similarly small capacity but generally carry a short pestle.

For the mortar, the trunks of hard woods such as the tamarind tree (*Tamarindus indica*), neem (*Azadirachta indica*), *jack* (*Artocarpus heterophyllus*), baheda (*Terminalia bellirica*), shirish (*Albizia lebbeck*) and sal (*Shorea robusta*) have been utilized regionally, all these being very large trees. Pit designs also vary with region, and could even take the form of a wooden sleeve that sits snugly in the cavity and is less expensive to replace. Even wooden strips laid radially in the pit cavity are in use. The pestle is generally made of baheda, shirish or babul (*Acacia nilotica*) wood, with a bulbous tip sometimes clad with lengthwise metal strips. The shape and design of the pestle end must match that of the pit to avoid excessive dead space.

The curved or angled piece was once fashioned out of a single large piece of wood; a shortage of these has led to the use of two pieces, the top one angled or curved and the bottom one straight; these are tied, chained or pinioned together for easy detachment when the pestle has to be removed. The strong load-beam has to be designed so that lateral pressure on the animal does not force it to lean from a comfortable upright stance during ambulation. Trained male animals, cattle or buffalo, are generally used, usually blindfolded to avoid dizziness and distraction; however, on small ghanis in certain areas even human labour, both of men and women, is employed.

Crushing Oilseeds

In the crushing of 10 kg of sesame seed in a ghani, about three-fourths of the material is placed in the pit and the rest is evenly laid out all around the flat rim. The animal is prodded and allowed to perambulate for a few minutes until pulverised seed is found to climb the walls of the pit. The animal is halted, and 180 ml of water is sprinkled around the chest and 120 ml poured into the pit. A further 5 minutes of pestle rotation will cause about three-fourths of the seed to be pulverised, after which another 300 ml of water is poured evenly around the pithead. The material built up in the chest is raked using a crowbar, and the pieces are broken up by hand and cast into the pit. After the animal has resumed movement, the rest of the seed is evenly pushed in all around.

The operator now tests the solid material by balling it in his or her palm; if it crumbles too easily, more water is needed. The layer of built-up material is again broken up, and brisk ambulation is resumed. After about 45 minutes, a sudden release of frothy oil floods the surface. Another 100 ml of water is sprinkled over the oil, the animal is stopped and the oil is allowed to settle. A final quantity of about 20 ml of water is now brushed over the compacted cake surface using the edge of the palm, after which the animal makes a few more rounds. The operation is stopped, the two curved pieces are detached and the pestle is lifted out and laid aside. If the ghani has a drainpipe, it is unplugged and the oil is drawn into a vessel. Otherwise the oil released into the pit is mopped up with a piece of cloth and wrung out by hand into vessels. While the cake is still hot and before it has set really hard, it is prised out as thick slabs from the chest using a crowbar.

Rape and mustard seeds need more water during crushing than sesame, and copra rather less. The oilcake is not raked during linseed crushing, but only at the very end. Safflower seeds are always very carefully decorticated by passage between grinding stones, sieving and winnowing: only practically pure meats are subjected to ghani crushing. During crushing of groundnuts at least part of the shells are retained in the ghani so as to ensure formation of a granular and compact cake.

At the point of maximum contact, the pressure in a ghani is about 10 kg/cm^2, about one-third of that in a small screw-press and about one-tenth to one-hundredth of that in a large modern expeller. The pressure in the ghani is largely determined by the weight placed on the load-beam, usually 115 to 160 kg, which is transmitted by way of the curved piece to the top of the pestle.

The fit of the pestle within the pit is important. Experiments have shown that an inclination that exceeds 21° from the perpendicular causes so much lateral pressure that the mass will not climb the walls of the pit. Too much dead space in the pit will have the same effect.

The phased additions of about 7,5 percent water during ghani operation have a major role. The first addition provides the pestle with a grip on the dry oilseed, and the friction produces heat. The second portion, with the heat present, cooks the ground seed. This is analogous with what happens in the stack precooker in modern screw-press operation. Protein is denatured and coagulated, and as the moisture level reaches a critical point, oil is rather suddenly displaced from the cells. The cake at this stage turns granular and cohesive, and will not reabsorb the expelled oil. After the oil has appeared, the third addition of water serves to hydrate and coagulate gums and phospholipids. This phase is analogous to modern oil degumming. The last brushing with water serves to clear surface oil on the cake and give it a sheen.

Oil yield

An oil-rich seed such as sesame seed or groundnut yields about 5 percent less oil in a ghani than in a modern expeller, mainly because of insufficient pressure. Ghani oilcake carries about 15 percent residual fat, about twice that of screw-press oilcake. In fact, in modern commercial Indian practice oilcake produced by crushing rape and mustard seeds in the ghani is put through screw-presses to obtain about 2 percent oil; this is added to the pungent ghani oil already obtained, to raise total yield.

Recent Evolution of Oil Processing

At the beginning of the twentieth century, there were an estimated 500 000 ghanis in India which crushed 97 percent of all oilseeds. The remaining 3 percent, mostly rape and mustard seeds, coconut and groundnuts, were processed in rotary units which were no more than mechanised ghanis installed in factories. Virtually all the oil used in the country, about 800 000 tonnes, was processed by ghanis. These numbers dropped sharply as power-driven screw-presses, hydraulic presses and solvent-extraction units came into operation. A further switch to rotaries for rape and mustard seeds and coconut, and to expellers for groundnuts and castor beans, rapidly brought down ghani usage for all oilseeds to 40 percent in the 1930s and 28 percent in the 1940s.

In the 1930s, Mahatma Gandhi inspired the formation of an organisation devoted to the resuscitation of village industries, including oilseed crushing, which is now called the Khadi and Village Industries Commission (KVIC). This organisation has steadily evolved a succession of improved oilseed crushing devices. The first of these, the Wardha ghani, was still worked by animals, but it was followed by several units that employed electric power for traction: the rotating barrel power ghani, the overhead power ghani, the portable power ghani and the portable overhead power ghani. Assistance by way of loans for both capital and working expenditure is granted by the KVIC both to individuals and to cooperative organisations to modernise traditional ghanis with power-driven pestles or to install the new power ghanis.

By the mid-1950s the number of ghanis had fallen to 300 000, and in 1983, despite the efforts of the KVIC, the figure was placed at 100 000 to 150 000. About 60 000 units were beneficiaries of KVIC assistance, of which 37 percent were traditional ghanis, 50 percent earlier forms of the power ghani and 13 percent later designs.

At present, just 4 percent of all oilseeds are pressed in ghanis. The proportions of individual oilseeds crushed in ghanis have been estimated as follows: safflower seeds, 40 percent; sesame seeds, 24 percent; rape and mustard seeds, 6 percent; groundnuts, 2.5 percent; and copra, 1 percent. The major drawback is the low production capacity of ghanis. Even modern ghani units can press 100 kg of oilseeds per working day at best. As long as the present subsidies, such as exemption of sales tax on ghani oil, are retained as a measure of commitment to village industries, ghani operations can be expected to survive.

Advantages and Disadvantages of Ghani Crushing

When ghani crushing was widespread, fresh oil was in greater demand than it is today. Flavour, which was traditionally an important attribute of all oils, and particularly of rape and mustard, coconut, groundnut and safflower oils, was best in oils produced from mild ghani crushing. Both storage quality and nutritive value were perceived as being high, although this is not borne out by modern studies. Today homemakers, especially in urban areas, demand bland and refined packaged oils.

Since vegetable oils are naturally sterile, problems of hygiene in ghani oil are unlikely. Turnover of oil in the home is so rapid, and usage of oil in India so low, that oxidative and lipolytic deterioration resulting from storage is also insignificant. Ghani cake is known to be exceptionally hard and dry and is not prone to mould infestation unless inadvertently wetted.

However, the ghani has disadvantages which are mainly economic in nature. Traditional ghanis have a maximum capacity of about 50 kg per day, and modern powered units only about twice that much. As a result, running costs are disproportionately high. If animals are used, they need to be trained, and they are expensive to feed. Artisan training is also essential. Ghani oilcake as prised out of the unit after crushing is extremely hard and is not accepted by the trade for further solvent extraction, as are expeller oilcakes.

In ancient times, ghani crushers in India were recorded as being a separate caste, and this distinction still persists. Since the start of the twentieth century, as the demand for ghani oil has dropped sharply, younger people have shifted to more remunerative occupations, turning away from ghani crushing as they have from many other artisanal activities in a rapidly changing social, technological and economic environment.

Use of ghani crushing in India has probably stabilised at the current level of subsidised operations. In the future, power-driven devices are certain to displace traditional ghanis worked by animal traction. There may still be room for powered ghanis in India and perhaps even in other developing countries with limited local supplies of raw materials for oilseed extraction, and there may be a place for batteries of power ghanis to multiply oil output from a common shaft in factory operations.

REFERENCES

Achaya, K.T. 1990. *Oilseeds and oilmilling in India: a cultural and historical survey.* New Delhi, India, Oxford and IBH.

Alderks, O.K. 1948. Cooking of meats and recovery of the oil. *In* A.E. Bailey, ed. *Cottonseed and cottonseed products: their chemistry and chemical technology,* p. 615-650. New York, Interscience Publishers.

Bose, N.K. 1975. *The structure of Hindu society,* trans. André Beteille. Hyderabad, India, Orient Longman.

Chaudhuri, J.L. & Selvaraj, K. 1985. Technological developments in ghani oil industry. *Khadigramodyog,* Oct./Nov.: 75-84.

Gujarathi, P.V. 1982. Design and efficiency of power ghani system. *In* V.H. Potty, S. Hariprasad & M.R. Mukunda, eds. *Village oil industry,* p. 91-100. Mysore, India. Central Food Technological Research Institute.

Monier-Williams, M. 1899. *A Sanskrit-English dictionary,* Delhi, India, Motilal Banarsidass. Reprinted 1963.

Nag, T.K. 1982. *Village oil industry. Power-ghani installation: a handbook.* Bombay, India, Khadi and Village Industries Commission.

Patel, J.P. 1943. *Oil extraction.* Maganwadi, Wardha, India, All-India Village Industries Association. 3rd ed.

3

Sugarcane

Sugar in India is mainly produced only from sugarcane, as the hot tropical climate of the country is suitable for cane cultivation. Over the past sixty-five years, the yield of sugarcane in tons per hectare has more than doubled. The quality of sugarcane improves with age, reaches a peak, and then gradually starts declining. A rapid deterioration begins from the moment of harvest. To make the most of your sugarcane crop, the cane has to be crushed within 8 to 10 hours of cutting. Every ton of cane cultivation requires approximately 60-70 tonnes of water under controlled conditions. However, the amount of water required varies according to different soils and climatic conditions. Shortfall in rains can be met through irrigation and sprinkling from canals.

In Kerala, sugarcane is cultivated in the regions along the rivers in Pathanamthitta and Alappuzha districts and in the Chittoor area of Palakkad district. Low-lying areas that include parts of rice fields are also used to grow sugarcane. Sugarcane is also grown in Punjab though the weather here is quite cold. Sugarcane provides an alternative to widely adopted wheat-paddy cropping system. Development of improved varieties and new crop production, plant protection and processing technologies has resulted in the expansion of the sugar industry in the area. Here, seed multiplication of sugarcane is done through tissue culture which increases the speed of maturity and eliminates seed transmitted diseases and pests.

The Sugarcane Breeding Institute has come up with a list of different varieties of sugarcane suitable for commercial cultivation in different States. In Uttar Pradesh, the recommended varieties for early ripening are CoS 687, CoS 8436, CoS 88230, CoS 95435 and CoSe 91232 while sugarcane varieties for mid season and late ripening are Co 1148, CoS 767 and BO 91. The plantation of sugarcane usually starts in the beginning of June. Fertiliser is applied and the field is hoed after applying nitrogen or N. Weeding and application of herbicides is done to keep the field weed free. If irrigation water is scarce, alternate furrow irrigation method may be used. In the second fortnight of June, earthing up is done after the on-set of monsoon. The last dose of nitrogen may be top dressed with the on set of monsoon. Hoeing and weeding is repeated. If there are no rains then mulching by dry leaves and grasses may be conducted to conserve soil moisture. Proper drainage should be provided to avoid water logging in the event of excess rains. Spray 2.5 per cent of urea and KCl in case there is drought for a long period. Sugarcane can be protected against termites through the application of Chlorpyriphos @ 5 litres/ha with 1500-1800 litres water in the furrow during planting. Major Rabi crops such as potato, garlic, wheat, coriander, linseed, lentil and pea may be intercropped with sugarcane. More information about the package of practice for sugarcane.

Sugar Industry in India is well developed with a consumer base of more than billions of people. It is also the second largest producer of sugar in the world. There are around 45 millions of sugar cane growers in India and a larger portion of rural labourers in the country largely rely upon this industry. Sugar Industry is one of the agricultural based industries. In India it is the second largest agricultural industry after textile industry.

As to the statistics there were a total number of 571 sugar factories in India as on March 31,2005 compared to 138 during 1950-51. These 571 sugar mills produce a total quantity of 19.2 million tones (MT). Sugar production in India increased from 15.5 MT in 1998-99 to 20.1 MT in 2002-03. Department of Agriculture and Co-operation, sugarcane production in 2004-05 is estimated at 232.3 MT from 237.3 MT in 2003-04. Sugarcane production is expected to reach 257.7 MT in 2005-06.

The following table shows level of sugar production (In Lakh Tonnes) in Indian States:

State	*2002-03*	*2003-04*	*2004-05 Estimated*
Uttar Pradesh	58.74	46.08	50.32
Maharashtra	61.64	31.99	22.29
Karnataka	17.98	11.57	13
Tamil Nadu	17.04	11.9	9.84
Andhra Pradesh	11.88	8.81	9.75
Gujarat	12.38	10.77	8.32
Haryana	5.99	5.86	4.03
Uttaranchal	4.59	3.93	3.82
Punjab	5.11	3.88	3.37
Bihar	4.21	2.77	2.77
Madhya Pradesh	0.85	0.94	0.85
Other	0.91	1.09	1.58

The sugar production in the states largely depend upon monsoon. From 1998-03 good monsoon resulted a larger production of sugar in the country.

Government of India fixes Statutory Minimum Price (SMP) for sugarcane according to Clause 3 of the Sugarcane Order. This statutory Minimum Price is designed through the consent of Commission for Agricultural Coast and Prices (CACP) and respective state Governments. For the year 2004-05, the rate was fixed at Rs. 74.50 Per quintal with a basic recovery of 8.5%.

The following policy initiatives are taken to boost the Sugar industry:

- Government declared the new policy on August 20,1998 with regards to licenses for new factories, which shows that there will be no sugar factory in a radius of 15 km.
- Setting up of Indian Institute of Sugar Technology at Kanpur is meant for improving efficiency in the industry.
- In the year 1982, the sugar development fund was set up with a view to avail loans for modernisation of the industry.

SUGARCANE CULTIVATION

Sugarcane and sugar-beet are two main crops that contribute approximately 56 and 44 per cent respectively of the total sugar production in the world. Sugarcane (*Saccharum officinarum* L.) is the main source of sugar cane in India and holds a prominent position as a cash crop. It occupies about 1.8 per cent of the total cropped area in the country. The area under sugarcane has been fluctuating from year to year .In the past decade, it has been swinging between 2 and 2.7 million hectares. The maximum crystal sugar produced so far is 4.26 million tonnes in any one year. Uttar Pradesh has the largest acreage under sugarcane, and it is about 57 per cent of the area under this crop in India and accounts for 47 per cent of the annual production in terms of raw sugar. But the production per hectare is highest in Karnataka (Mysore State), followed by Maharashtra and Andhra Pradesh.

Soil

Sugarcane grows best on the medium heavy soils, but can also be raised on the lighter soils and heavy clay, provided there is adequate irrigation available in the former type of soils and drainage is good in the latter type of soils. In northern India, it is cultivated largely on the loams and clay loams of the gangetic and other alluviums, whereas in Peninsular India, it is grown on brown or reddish loams, laterites and black cotton soils. In many places, dark rich clay loams, 120 to 150 cm deep, and lying on a previous substratum of murum (distintegrated traprock) are used for this crop. The provision of proper drainage enables the resumption of cane cultivation on many heavy soils that may have become unfit for this crop owing to water-logging or alkalinity. In some areas in Peninsular India, where the soils are heavy, 70 to 100 cartloads of sand or red earth are added to the soil per hectare to improve its texture. Similarly, in sandy soils, tank silt is added to improve their water-holding capacity.

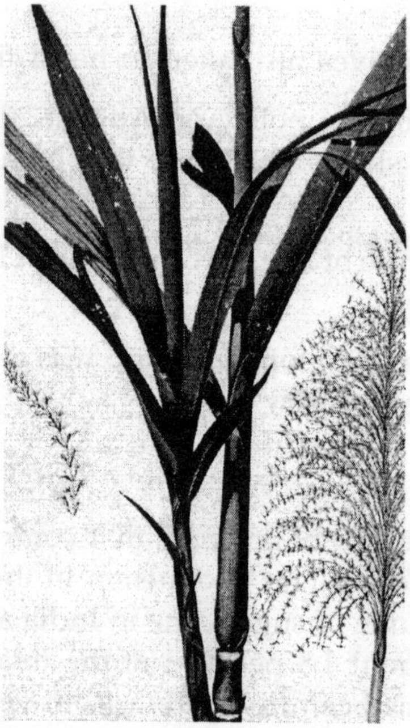

Sugarcane

Climate

Sugarcane is able to grow over a prolonged season. Under warm humid conditions, it can continue its growth, unless terminated by flowering. However, its height is strongly influenced by the age of the crop and seasons. Temperatures above 50º C arrest its growth; those below 20ºC slow it down markedly and severe frost proves fatal. The crop does best in the tropical regions receiving a rainfall of 750 to 1,200 mm. It can also be grown in sub-tropical areas, but where the climate is subject to extremes, and dry season is long, the growth period of the crop is restricted to a bare four months in the year and yields, tend to be distincly lower than those in the tropics. In tracts receiving a relatively high rainfall (1,200 to 1,500 mm), e.g. in parts of Uttar Pradesh, Bihar, Orissa, West Bengal and Assam, The crop is grown without irrigation and thrives on the moisture conserved in the soil. For ripening, it needs a cool, dry season ; but where rainfall is too heavy and prolonged, the quality of the juice tends to be low, and where the weather remains comparatively warm and moist throughout the year, it does not ripen well. In years of drought, the crop becomes a little fibrous and in the dry areas generally, only the harder and more fibrous varieties can be grown. The climatic conditions of Maharashtra and parts of Karnataka, and the Telengana area of Andhra Pradesh are ideal for sugarcane; in some of these areas, its yield per hectare has shown a steady increase owing to improved culture. The acreage under sugarcane has also increased.

Rotation

In northern India, sugarcane is grown in rotation with wheat, cotton gram, lahi (*Brassica napus*), maize, jowar, peas and other crops, and in eastern areas, it is rotated mostly with paddy. Except when ratooned, the crop is not grown on the same land often, than once in three or four years. In southern India, it is usually rotated with paddy, whereas in Maharashtra, the common rotations are sugarcane-ratoon-wheat, or sugarcane-cotton-gram. In many localities, where the crop is grown under irrigation from wells, it is rotated with garden crops, such as potatoes, chillies, sweet-potato, ginger, onion, turmeric, elephant's foot, and even bananas. In some localities, particularly near large towns, onion, radish, coriander, green cucumber and other quick-growing vegetables are grown as companion catch crops on ridges or in interspaces when the cane crop is young. The practice of taking potato, dwarf wheat, sugar-beet, torai, coriander (dhania), berseem and turnip as companion crops in the interspaces of the autumn planting is gaining popularity in the northern region of the country. In the canal irrigated tracts of Karnataka and Maharashtra, a block system of irrigation is practised and under it a three-year rotation is compulsory, the sequence of crop being rice, groundnut, jowar, ragi, sunhemp and sugarcane, as in Karnataka, or sugarcane, fodder, jowar, groundnut, tobacco, cotton and green manure, as in Maharashtra. In the Telengana area of Andhra Pradesh, the rotation sugarcane, ratoon, monsoon paddy, and sunhemp green manure has been found the best.

Preparatory Tillage

Sugarcane requires a very thorough and clean preparation of land. In Peninsular India, where sugarcane soils are mostly clays and clay loams, one or two deep ploughings and one cross-ploughing with a mould board plough are necessary. These operations should be followed by clod-crushing with a discharrow, or a junior cultivator or a beam clod-crusher or even with wooden mallets and stout sticks. In the alluvial loams of northern India, cultivators usually prepare their fields after winter rains, but when these rains fail, the fields are given a preparatory irrigation. After it, when the soil comes into vattar (right stage of moisture) it is cultivated quickly and thoroughly with a desi plough till it is reduced to a fine tilth. In Bihar, the fields meant for cane are left fallow during the preceding monsoon and are cultivated at intervals up to the time of planting cane. All this is done in order to conserve moisture and keep down weeds.

Cattle manure, compost and other slow-acting bulky organic materials are usually applied to the soil and incorporated into it well in advance of planting. This practice helps to decompose the manure and makes the nutrients available. Where cane is green-manured, the leguminous crop should be ploughed under about a month before planting the cane. In northern India, however, where a clod season intervenes, the green-manure crop is ploughed under in September and cane is planted several months later.

Sugarcane is planted either in furrows or trenches. Depending on the method of planting and the fertility of the soil, furrows or trenches are made half to one-and-a-half metres apart. Comparatively narrow spacings are adopted in Nothern India, where yields are usually low, whereas wider spacings are common in Peninsular India where the aim is to obtain high yields. Furrows are made with a padded desi plough in northern India and with a heavy double mould-board ridger in Peninsular India. Furrows made with the desi plough are quite shallow, being about 10 cm deep, whereas those made with a ridger are about 20 cm deep. Where trench-planting is adopted, trenches are dug by hand-labour, using pick-axes and spades, and may be up to 25 cm deep. On large farms, all preparatory cultivation, including ploughing, clod-crushing, harrowing, and ridge-making are mechanised.

The adequacy of moisture is very necessary for the setts. In Peninsular India, arrangements for irrigation are and well before planting. In northern India and in Bihar, care is taken to conserve as much moisture in the soil as possible, so that the planted cane is able to germinate satisfactorily. All the same, as the planting is necessarily done during the comparatively dry season, considerable loss of moisture from the soil and sets takes place, and usually only 30 to 40 per cent of the buds germinate.

Planting Season

In northern India, planting is usually done with the onset of the warm weather and is completed well before the onset of summer. On this basis, the more eastern is the tract, the earlier is the time suitable for planting. Thus broadly speaking, the first fortnight of March is the best time for planting sugarcane in the Punjab and Haryana, February in Uttar Pradesh and January-February in Bihar. In actual practice, however, most of the planting tends to be late, rather than early. On large farms in northern India, where planting work is likely to be prolonged, a part of the planting is also undertaken in October-November. In Tamil Nadu and Andhra Pradesh, cane-planting is done between December and February. In Maharashtra and parts of Karnataka, it is done in December-February for the 12-month crop which is called eksali, in October-November for the 15 to 16 month crop which is called 'preseasonal', and in July-August for the 18-month crop, which is called adsali. Thus in these areas, for this planted almost throughout a period of about 8 months from July to February. This is possible because of the more equable climate, low rainfall, the high elevation and the consequent absence of water stagnation, and the easy availability of adequate irrigation. Under such conditions, cane also makes growth and accumulates sugar, more or less, continuously, and hence earlier plantings give higher yields from giving lower yields, are usually badly attacked by the shoot-borer, which causes gaps in the stand of cane and setback to growth. Since crop planted in different months ripens in succession, by suitable adjustment of planting, the necessary areas in different seasons, and a planned supply of cane to the factory, a satisfactory recovery of sugar in factories can be ensured over a reasonably long crushing season.

Seed

Healthy seed material, free from pests and diseases, and having high viability is essential for establishing the crop in the first instance. The top one-third to half portion of a cane, being comparatively immature, has buds of good viability and is best to be used as seed. The bottom portion is usually richer in sugar than the top portion, and it is best to utilise it for sugar or gur-making. In tropical India, setts for sowing are taken from well-manured, erect and healthy canes. In some places in Maharashtra, Andhra Pradesh, Karnataka and Tamil Nadu, crop nurseries are adequately fertilised, and well protected against pests and disease.

Before planting, the leaves of the cane stalks are stripped off by hand in order that buds on them may not be damaged. These stalks are then cut into 3-bud setts each usually 30 to 50 cm long. If seed is required at a distant place, whole canes should be transported without stripping off the leaves or cutting them into setts. If planting is for any reason to be delayed, but the seed canes have been harvested, they should be heaped in shade under a thick cover of straw or cane trash, and periodically sprinkled with water. Under the dry condition of northern India, the soaking of seed in water for four to six hours before sowing improves and hastens germination.

Under irrigated conditions in Peninsular India, the planted setts are liable to rot, and in some areas they are attacked by the pineapple disease. To prevent rotting and infection with the disease, the dipping of the setts into a suitable fungicide, such as 0.5 per cent mercuric chloride or 0.25 per cent Aretan or Agallol solution or 1 kg of Perenox in 400 litres of water is recommended. Aretan, along with gamma BHC, is also recommended in northern India; the former improves the germination and keeps off fungal attack and the latter keeps off termites is serious, gamma BHC at 1 kg actual ingredient per hectare in emulsion form and diluted 300 times is sprinkled with a watering-can on the setts placed in the furrows.

Depending on the distance between the rows and the vigour and soundness of the buds, 25 to 35 thousand 3-bud setts are usually enough to plant a hectare. According as the cane is thick or thin, 1 1/2 to 3 tonnes of cane by weight is needed to provide about 10,000 3-bud setts. In the Punjab, Haryana and western Uttar Pradesh, 2-bud setts planted in rows, about 60 cm (about two feet) apart, have been found to give the best results. In the case of late planting, the seed-rate is often increased slightly, and the setts are planted a little closer in the row in order to guard against guppy germination.

Methods of Planting

There are three methods of planting, viz., flat-planting, furrow-planting and trench-planting.

1) *Flat-planting.* In the case of this method, shallow furrows are opened with a desi plough, 80 to 100 cm apart; the setts are placed in them end to end and covered with 5 to 7 cm of soil, and the field is levelled with a heavy plank (sohaga).

2) *Furrow-planting.* Furrows are made with a ridger, about 10 to 15 cm deep, in northern India and about 20 cm deep in Peninsular India. Setts are laid end to end or otherwise in the furrows and covered with 5 to 7 cm of soil, leaving the upper portion of the furrow unfilled. Immediately after planting and covering the setts, water is let into the furrow to soak its bed thoroughly and the cane setts are then dropped into the furrow and pressed into the mud by foot. In case of monsoon planting, as a precaution against the stagnant water in the furrow damaging the buds, the setts are placed inclined on the side of the furrow, instead of flat at the bottom of the furrow.

3) *Trench planting.* Trenches, 20 to 25 cm deep and rectangular or trapezoidal in section, are made either by manual labour or with a tractor-drawn rigder; the bottom of the furrow is loosened by digging. Shallow furrows are made in the bed of the trench and the setts are placed end to end or otherwise in the furrows and covered, as in the case of flat-planting. Water is then let into the trenches.

The tractor-drawn sugarcane-planter designed and fabricated at the Indian Institute of Sugarcane Research, Lucknow, is a very suitable device for planting cane in trenches. Trench-making, setts planting applyi ^g fertilisers are done simultaneously with this tractor-drawn planter.

In most parts of northern India and Bihar, and in the Malnad tract of Karnataka, cane is planted on flatland. Furrow-planting is practised in parts of eastern Uttar Pradesh and in Peninsular India, and trench-planting is done in some coastal areas where the crop grows very tall and the strong winds are liable to throw it down and damage it badly.

Planting is done dry or wet. In the case of the dry method, irrigation is not applied at planting or immediately after that, and the planted setts germinate in the comparatively dry soil. In wet planting, on the other hand, the field is irrigated just before of after planting or at both times so that setts germinate in wet soil. Whether the furrows are irrigated just before or just after planting makes little or no difference; indeed, in most cases the planted field gets additional irrigations to keep it wet throughout the period of germination.

Irrigation

In the case of sugarcane, the maintenance of optimum soil moisture during all stages of growth is one of the essential requisites for obtaining high yields. The crop should, therefore, be grown in areas of well-distributed rainfall or under an assured and adequate irrigation. In tropical India, depending on the type of the soil, the seasonal conditions, the variety grown, the method of planting and the rate of manuring, the total water requirement of the crop for optimum growth varies from 200 to 300 cm, inclusive of rainfall. The requirement of an adsali crop is proportionately higher. Where the soil is not retentive of moisture, and where there are no reserves of subsoil moisture, cane requires to be irrigated frequently.

In tropical India, usually one or two waterings are given at intervals of three or four days after planting to help the setts to germinate and the seedlings to establish themselves. Thereafter, in the absence of rains, cane is irrigated every 10 to 12 days during its growing period. In dry areas and in sandy-loams soils, irrigations may be needed at intervals as short as eight days. In deep clay loams, irrigation can be withheld for longer periods, say, up to two or three weeks. Frequent light irrigations, each 40 to 50 mm, adjusted to suit the growing period of the crop and the prevalent weather conditions, are very useful. Towards the time of harvesting, irrigation frequency is reduced, and just before harvest, irrigation is withheld for about a month.

In northern India the summer being drier and hotter the crop needs water more frequently than in southern India, but actually, adequate water for frequent irrigation is not available and irrigations are, therefore, usually given at comparatively long intervals. In canal irrigated areas, the frequency of irrigation depends entirely on the running of the canals. The severity of these conditions is slightly mitigated by the high water holding capacity of the alluvial loam soil. In the Punjab State because of the lower rainfall, drier climate, and slightly coarser soils, the crop gets 8 to 10 irrigations during summer. In central and western Uttar Pradesh three to five irrigations are usually given, and they help the crop to tide over the summer. In the eastern Gangetic areas, cane subsists almost entirely on subsoil moisture and rainfall, and receives no irrigation at all. In the post-monsoon season the crop receives only one or two irrigations or none at all; however these post-monsoon irrigations only help to keep the crop in good condition, for cane does not make growth in winter. In areas where frost occurs, irrigations are applied to save the crop from them. Sugarcane responds to irrigation in northern India as it does in the south and it is profitable to apply frequent light irrigations to the crop during the hot weather. Where irrigation facilities are scarce in summer, trash-mulching in the interspaces of the cane rows is done for conserving the soil moisture.

Interculture and after-care

The first hoeing and weeding should be given to the crop three to four weeks after planting. But in northern India, usually the shoots take a long time to emerge and meanwhile the weeds, particularly the nutgrass come up rapidly in the planted field. This rapid infestation of the sown field calls for at least one blind hoeing which may have to be given within the first one or two weeks. After germination, depending on the field conditions and the frequency of irrigation, two or three more hoeings and weedings may be required during the first three months after planting with the coming up of the crop. Where cane is planted in furrows or trenches these are gradually filled up in the course of interculture. The use of bullock-drawn implements, like blade harrows for mulching, junior hoes for interculture and weeding and ridgers for earthing up and covering the fertiliser, lessen the costs of these operations. The final earthing-up should be completed before the monsoon rains, and should generally synchronise with the application of the final dose of fertiliser. Good care in weeding and interculture helps the clumps to tiller early and form canes. The crop should be so managed that it grows rapidly to form a canopy over the interspaces. This shading helps to keep down the weeds, the moisture loss from the soil surface.

Where the aim is to get crops of 75 tonnes and over per hectare, it is usual to earth up the standing crop in the rows. But whether the crop is earthed up or not it is very desirable to tie up the canes so that they may not sway during winds, and lodge. The best way to do the tying is to bring together the stalks from adjacent rows and tie them together with their own trash and old leaves. In many places in southern India, cane is tied at two or three levels with twist of cane trash, the twistline going from one end of the row to the other; sometimes the stalks are further propped up with bamboos. Many times they sow castor or arhar (tur) seeds around a planted field of cane. This acts as a wind-break.

Manuring

Sugarcane is a heavy feeder. A 70-tonne crop removes from the soil 85 to 100 kg of nitrogen, 180 to 330 kg of phosphoric acid, 60 to 190 kg of potash and 70 to 80 kg of calcium. Adequate manuring, therefore, is essential for sustained high yields. At present only 10 per cent of the sugarcane crop is manured, and in most cases, inadequately.The nitrogen requirement of sugarcane depends upon the soil and climate conditions. It ranges from about 130 kg per hectare in Uttar Pradesh to about 270 kg in Tamil Nadu and 300 to 500 kg in Maharashtra and Karnataka. The general recommendation in most places is to apply half to two-third of the nitrogen in the form of bulky organic manures, such as farmyard manure, compost, green manure or tree leaves, and the remainder in the form of ammonium sulphate or oil-cakes or a combination of the two. These bulky organics are ploughed into the soil sufficiently in advance of planting. In places, sulphitation press-mud is also used as a manure. Oilcakes are usually given at planting and ammonium sulphate in one or two doses at planting or tillering or at the time of earthing up the crop.

The increase in the yield of cane per kg of nitrogen per hectare also depends largely upon the soil and climate. Provided the weather conditions are not unfavourable, farmyard manure usually gives about one quintal ammonium sulphate and about 2.4 quintals of cane per kg of the nitrogen contained in the manure or fertiliser. Green-manuring increases the yield of cane by about 100 to 150 quintals per hectare.

The response of sugarcane to phosphorus has not been very marked in most areas; however, the general indication is that at high levels of nitrogen application, phosphorus is required at the rate of 40 to 50 kg of P_2O_5 per hectare. The states where phosphorus has given perceptible response are Bihar, parts of Uttar Pradesh, Bengal and Maharashtra. In some states, where cane does not give a good response to superphosphate applied to the preceding summer green-manure crop. This practice improves the benefit from green-manuring. The response of sugarcane to potash has been obtained only in localised pockets of light soils. Not much research has been carried out on the trace-element fertilisation of sugarcane.

PEST MANAGEMENT

Sugarcane is vulnerable to several pests and diseases and this problem is more serious in the subtropical regions than in the tropical areas of the country.

The major sugarcane pests are moths, borers, scales, white grubs, termites, white flies, black bugs and mealy bugs. They damage the crop at different stages of its growth. The mechanical, cultural and biological means, and the use of resistant-varieties are some of the means of controlling the pests. Pesticide formulations of chlorinated hydrocarbons, organophosphorus and organomercurials are effectively used to reduce the damage to the crop caused by variety of pests. But the chemical control of moth bores is still not very effective. Some recent formulations of systematic pesticides are however, showing promise in this direction.

Top Borer

Life History: Top-borer (*Tryporyza nivella* F.) is a major pest of sugarcane in almost all the important cane-growing areas in the country. The young larva after hatching, move about one leaf surface for 3-4 hours and then after selecting a suitable spot on the first open leaf on the top, penetrates into the mid-rib from the underside of the leaf, and bores its way towards the axil. The tunnel made in the mid-rib forms the characteristic white steak which changes to red later.

On reaching the axil of the leaf, the larva comes out of the mid-rib and bites into whorl of newly developed leaves forming the spindle. This causes the series of shot holes which are revealed when the leaves open out. It thus gains entry into the spindle and moves down towards the growing point destroying the tender newly-grown top portion on its way. The growing point is killed and the unfurled leaf on the top dries up resulting in the formation of a small, slender, reddish-brown dead-heart. If the top is severed and split opened at this stage, the creamy white-borer larva could be found at the growing point or in a tunnel about 10-15 cm down the tip of the cane. When fully grown, the larva pupates within the tunnel, and about a week or 10 days later, emerges as a silvery white moth. During periods of rush of emergence, these moths can sometimes be seen sitting on the leaves. After mating, female moths lay eggs in clusters usually on tne underside of the leaf; the egg masses are then covered with reddish-brown hair.

The pest passes through 5 to 7 generations in an year depending upon the climatic conditions in a particular region. Under north Indian conditions, it remains active from February to October and passes winter under hibernation in larva stage. The egg-stage lasts for 6 to 13 days, larva stage 26 to 52 days and pupal stage 7 to 11 days during the period of the pest activity. In the initial stages of the crop growth, i.e., before cane-formation has set in, the attack of top borer ultimately results in complete drying up of the attacked shoot. In medium-sized shoots where cane formation has just started, the top portion gradually dries up and the few internodes already formed mostly go waste. In such cases of attack, one or two buds just below the dried top portion may sprout and developed into small-sized side-canes. In comparatively grown-up crop, the borer attack arrests further growth and includes the top buds to sprout and put forth side-shoots giving rise to what is commonly known as the "bunchy top". There is considerable deterioration in juice quality; bunchy top canes (attacked by forth generation of the pest), however, show better quality juice if harvested during November-December, but subsequently deterioration sets in their case also.

Seasonal history and pest carry-over. Under north Indian conditions, the over-wintering caterpillars in the infested cane-tops of the mature crop regain activity in the beginning of February soon after, pupation starts and the moths begin to emerge in later half of February or in the beginning of March. Moth emergence continues till about the end of March. Eggs of the first generation are laid on ratoon sprouts or autumn planted crop, whereever available. This is the main source of carry-over of the pest population to the succeeding plant crops. The moths of first generation emerge in May and lay eggs of the second generation; the caterpillars hatching out of these eggs, attack the shoots of the springplanted crops also.

But because of hot-and-dry conditions prevailing during May (under north Indian conditions) which are unfavourable for top-borer, the attack of the second generation is usually mild. Moths of this generation begin to emerge by the end of June or early July. By this time the cane-formation starts and weather conditions also become favourable for the pest activity. The attack of the third generation in July is, therefore, severe and causes heavy losses. As a result of a study carried out at the Indian Institute of Sugarcane Research, Lucknow, the average loss in weight of canes attacked by the third generation of the pest worked out to about 70 per cent. A large number of such attacked shoots dry up all together.

The fourth generation attack occurs in about the middle of august. Since the crop has attained sufficient growth by then, the loss in yield by this generation is comparatively low. The fifth generation attack appears in September-end or early-October. The crop then by is almost fully grown; the loss in tonnage as a result of attack of this generation, therefore, negligible. The fifth generation caterpillars enter hibernation within the cane-tops towards the end of October or in early November and continue as such till end of January or beginning of February, when the pest regains activity.

In tropical region and in eastern parts of the country where climatic conditions are mild, various generations of the top-borer overlap each other, and all stages of the pest are seen throughout the year.

Infructuous attack of top-borer. During July-September when growth of the crop is normally vigorous and rapid particularly in well fertilised fields, the young larva after boring through the mid-rib of the first open leaf on the top and entering the top portion of the shoot or the cane, fails to reach the growing point. Such shoots escape damage and continue their normal growth. The young larva ultimately dies within the spindle portion.It has been observed that percentage of infructuous attack is less in the varieties susceptible to top-borer than in those comparatively resistant. It has further been observed that in the autumn planted crop, infructuous attack is generally higher than in spring planted crop, probably because of more vigorous growth in the former.

Control:

(i) *Cultural.* Adequate and timely nitrogen fertilisation helps the crop to withstand top-borer attack better.

(ii) *Mechanical.* Destruction of egg-masses and cutting the infested shoots at ground level for destroying the caterpillars present within, prove very helpful in keeping the pest under check, particularly in the early stages of the crop growth before cane formation starts. The egg clusters can be easily spotted in a young crop because of the reddish-brown hairy mass covering them. Young boys can be engaged for collection and destruction of egg clusters and the attacked shoots. These mechanical control measures, to be effective, must be carried out collectively in a locality in the form of a regular campaign.

(iii) *Chemical.* In the north-western sugarcane-growing areas of the country where top-borer has distinct and clear-cut generations, spraying the crop with Endrin or

Endosulfan 0.1% emulsion at the time of egg-laying of the third generation proves helpful in checking serious damage by the pest. Whereever possible, two rounds of spraying, each with half the dose mentioned above, may be carried out weekly. If the weather conditions permit, this treatment may be repeated in about the middle of August against the fourth generation also.

Sprinkling of Endrin or Endosulfan granules from over the top portion of the crop, proves almost as effective as emulsion spraying. A simple device "The IISR Granule Sprinkler" has been designed at the Indian Institute of Sugarcane Research for applying granules in tall sugarcane crops.

Recent studies conducted with some systematic insecticides have shown that application of Carbofuran or Phorate granules in the soil at 2 kg a.i. per ha also proves effective in controlling the pest. Time of operation is an important factor both in regard to follar spraying or application of granules in the soil, and this can be ascertained by keeping a clese watch and tracking the stage of the pest over about a fortnight before the expected rush of moth emergence and egg-laying.

Gurdaspur Borer

Gurdaspur borer, Bissetia steniellus Hmpsn., is another destructive pest of sugarcane in Panjab, Haryana and in some western districts of Uttar Pradesh.

Seasonal and life-history. The caterpillars remain in hibernation during winter, within the sugarcane stubble. They become active after the first shower of rain, and soon after the pupation starts the moths begin to emerge by about the third week of June, and about a fortnight later the pest-attack is noticed in the crop.

Eggs are laid in clusters in and along the groove of the mid-rib on the upper (ventral) surface of the leaf near the leaf-sheath. Newly opened leaves in the top portion of the shoot, are preferred for egg-laying. One cluster generally contains 20-50 eggs. The egg stage lasts for 9 to 13 days.

The young larvae on hatching feed for some time on the bud situated on the top first or second internode and then bore into it (internode) from near the bud, and thus the attack starts. A larva is full-grown in about 6 to 9 weeks after which it enters pupation. The pupal stage lasts for 6 to 13 days.

The pest remains active till the end of November, and passes through 2 to 3 generations in the course of a year. From July to October all stages of the pest are seen simultaneously because of overlapping of generations.

Nature and extent of damage. The attack of Gurdaspur borer starts towards the end of June or beginning of July. At this stage the crop is fairly grown-up and cane formation has already started. Injury by this borer is quite characteristic. The borer larva after entering the internode, moves upwards in a spiral manner making minute a holes on the rind from within the internode. From outside, its passage appears as a dark spiral streak which on closer examinations found to be made up of series of punctures lying side by side like beads in a rosary. Later on, it bores deeper into central portion of the cane and feeds in a single straight tunnel moving upwards. In the initial stages, larvae from egg-cluster enter the same cane and feed gregariously. But after about 10 to 12 days, after biting a hole in the upper portion of the cane they transfer themselves individually to adjoining canes, each larva entering a separate shoot or cane by making a circular incision and then boring upwards in a spiral manner.

In the initial stages of attack when the caterpillars are feeding with the top internodes just below the rind, the side-leaves turn pale. As the attack proceeds further, the entire whorl of leaves at the top dries up, and the attacked cane creases to grow further. The damaged internodes are rendered weak and the stalk breaks off easily when shaken by wind or a passing animal.

In October, the partially of fully grown-up caterpillars leave the top portion of an infested cane and transfer themselves to the lower portion of the same cane or an adjoining cane and then travel downwards boring a straight tunnel (within the cane) towards the bottom portion where they subsequently hibernate.

The damage by this pest may normally vary from 15 to 25 per cent of the total crop, but in years of severe attack, it may be as high as 40 to 50 percent in certain fields. The loss of particularly heavy in earlier stages of attack; most of the attacked canes dry up and are thus a total loss. Canes attacked later during the season manage to survive till harvest but there is considerable deterioration in juice quality in their case.

Control:

— *Mechanical*: When the attack begins to appears in the crop towards the end of June or beginning of July, all the attacked shoots should be regularly removed and destroyed. Such shoots can be easily spotted by their pale and withered leaves in an Otherwise green and healthy crop. This rouging should be done in the initial stages of attack, when the young caterpillars are feeding gregariously. The operation should be continued weekly till about the end of September or beginning of October. The larvae within the affected shoots will thus be killed, and further spread of the pest will be checked. This method of control of Gurdaspur borer has been employed with good results in past several years in Jagadhir and other infested areas of Haryana and Panjab, and also in Saharanpur and Dehra Dun districts of Uttar Pradesh.

The pest remains in sugarcane stubble in the larval stage from December to about the end of May or beginning of June. It can be destroyed during this period by ploughing up the harvested cane fields, and collecting and destroying the stubble. If the attack is on a small scale, only the infested stubble may be dug up and destroyed. These can be easily located immediately after harvest by the presenc of a round tunnel-hole on the cut surface. The uprooted stubble should then b collected and burnt. Fresh setts should be planted in February-March in place c such uprooted stubble if the crops is to be ratooned.

— *Chemical*. Putting in droplets of Endrin emulsion or kerosene oil through the hole (on the cut surface) in the infested stubble, results in high morality of the Gurdaspu borer caterpillars. This operation must be carried out immediately after harvest s that the borer-infested stubble can be easily spotted. The various mechanical an chemical control measure to be effective must be carried out simultaneously by a the can-growers in a particular area in the area of a campaign.

— While majority of larvae travel down towards the stubble during winter month a small percentage is left behind in the stalks also. Due care should, therefore, b taken to select borer-free crop for the seed purposes. In the infested localities, is desirable to soak the cane sets for 2 hours in 0.2 percent Trichlorphon (diptere solution, and plant them after drying in ahead. The treatment will ensure tl destruction of the borer larvae within the setts.

Plassey Borer

Plassey borer, Chilo tumidicostalis Hmpsn., is an important, though sporadic, pest sugarcane in some areas of West Bengal, Bihar and Assam. The incidence of this bor in a field can be easily spotted by the presence of dry crown of leaves on the infeste canes. The injury due to borer in the cane is of two types : primary infestation, ar secondary infestation.

(a) *Primary infestation.* It is caused by the newly hatched larvae congregating in tl top 3 to 5 internodes of a cane. More than 100 caterpillars may be collected fro a single cane. Fresh frass which is shining red in colour, is usually seen coming o of the holes in the top internodes. The leaves of such canes ultimately dry up. Tl affected internodes break off easily. The nodes adjacent to the infested internod throw out elongated sett-roots which envelop the stalk below. Sprouting of sid shoots also sometimes occurs.

(b) *Secondary infestation.* The grown-up caterpillars migrate to the adjoining canes to the lower healthy portion of the canes showing primary infestation and give ri to the "secondary attack". Cane tops do not dry up in this case though individu caterpillars may sometimes bore up to 5 internodes in a single-attacked cane.

Seasonal history and extent of loss. The pest remained active from later half of Februa to about the middle of November, and passed through 5 to 6 generations in a ye There was overlapping of generations from April to October when all stages of the pe were simultaneously met with.The pest exhibits polymorphismin larval stage, and distinct types of caterpillars can be seen. However, a typical larva can be identified by the presence of 4 pinkish-brown stripes on its back and a complete circle of the crochet on its prolegs.During August-September, one life-cycle of the pest is completed in 39-54 days; egg stage lasts for 7 days, larval stage 26-46 days and pupal stage 6-11 days.In severe cases of infestation, losses may vary from 30 to 50 tonnes of cane per hectare in different varieties. Sugar recovery is also appreciably reduced.

Control:

(a) *Mechanical.* Collection of moths at light traps, removal of cane tops showing primary infestation, and collection and destruction of egg-masses, prove effective in bringing down the pest population and its damage.

(b) *Chemical.* Spraying the crop with Endrin emulsion at about 1 kg a.i./ha gave promising results in Plassey area.

Internode Borer

The internode borer, Chilo (*Sacchariphagus*) indicus Kapur, is a major pest of sugarcane in peninsular India causing considerable reduction in tonnage and sucrose content of cane. It has been reported from certain pockets in north India and in Madhya Paradesh and Orissa also. The pest remains active throughout the crop season under south Indian conditions, and its attack is particularly severe in Tiruchirapalli, Tanjore and South Arcot districts of Tamil Nadu where the incidence may be as high as 70 percent in some fields. The special season crop which is planted in July-August, suffers more seriously than the January-February planted crop.

Young newly hatched caterpillars prefer the top portion of the cane for their early feeding. Soon after the emergence they feed by scraping the leaf spindle. Subsequently they bore into the tender cane-crop.

The attack of the pest in young shoots is more commonly found in the late-planted crops and in the ratoons. In grown-up canes the damage by the larvae can be observed only when the leaf-sheath is removed. Fresh excreta can also be seen sticking outside the larval holes on the internodes. The larva is sufficiently grown in size by then and is highly migratory in habit at that stage; it often transfers itself from one internode or cane to another.

Seasonal history. Under south Indian conditions the internodes borer remains active throughout the year, and all stages of the pest are met with throughout the year. On account of overlapping of different stages, it is difficult to determine the exact number of generations in a year under field conditions. Studies in the laboratory revealed that the pest completed 6 generations in a year. The moths lay eggs in masses usually on the upper surface of the leaves. After the first moult which takes place on the third or the fourth day of hatching, the larva becomes more active and starts feeding voraciously on the tender unopened foliage at the top. The grown-up caterpillars bite holes in the internodes through which they enter and subsequently makes tunnels in the stalk.

Before pupation, a fully-grown larva comes out of the tunnel and takes their shelter underneath a tightly fitting leaf-sheath on the cane. It constructs a silken cocoon there in which it later pupates.

This pest is carried over to the succeeding crops though the leaf-sheaths which along with the pupal cocoons sticking to them, are quite often left in the fields as trash after harvest; moths emerging from these pupae fly to the neighbouring fields to lay eggs on the fresh crops. Leaves bearing unhatched egg-masses (and left over as trash in the field) also in the carry-over of the pest. Young ines hatching out from these eggs easily find their way to late tillers or to ratoon sprouts nearby.

In Tanjore area (Tamil Nadu) where cane crops are available in the fields all the year round, and where trash conservation and mulching is a regular agronomic practice, the carry-over of the internode borer to the succeeding crops is very much facilitated.

It has been observed that sugarcane varieties with soft rind generally have a higher incidence of the pest than those with a hard rind. The young larva prefers to bore into the top tender portion rather than the well-formed hard internodes. Varieties with close fitting leaf-sheaths show higher incidence than the loose-fitting and self-detrashing varieties. It has further been observed that the incidence of the borer is less in detrashed crops as compared to crops where no detrashing has been done. Since the newly hatched borer larvae generally remain inside the leaf-sheaths in the initial stages before feeding their way slowly into the cane, and since the grown-up caterpillars usually pupate in between the stalk and the leaf-sheath, varieties with close fitting leaf-sheaths are preferred by the pest. The crop in water logged fields usually suffers more seriously from the attack of this pest, probably because moisture conditions in such fields are favourable for multiplication of the pest.

Control:

(a) *Mechanical.* Detrashing of canes (and burning of the trash) with a view to destroying the larvae and pupate attached with the leaf-sheaths, proves helpful and is generally recommended for controlling the internode borer.

(b) *Chemical.* In preliminary trials conducted in Tanjore area of Tamil Nadu, spraying the crop with Endrin 0.1% emulsion 3 times at 4-weeks interval during December-March gave encouraging results against the pest.

Shoot Borers

There are three important species of shoot-borers, viz., the so-called early shoot-borer (Chilo infuscatellus Snell.), the pink borer (Sesamia inferens Walk.) and the green-borer

(Raphimetopus ablutellus Zell.). All these borers usually attack the crop in its early stages of growth, and hence are called shoot-borers. But the first two species, viz., C. infuscatellus and S.inferens, have been found in some areas to damage the crop as cane-borers also.

The green-borer (Raphimetophus ablutellus Zell.) is active from March to June passing through three generations in a year. The shoot-borer and the pink-borer become a little sluggish after the onset of heavy rains in July-August, but their attack continues till October-November and they pass through 5 to 6 generations each year. The nature of their attack is more or less similar. The newly hatched larvae feed singly or in groups on the tender leaf-sheaths for the first 4 to 6 days. Later, they disperse and each caterpillar bites a hole on the stem just below or at the soil surface, and bores into the shoot ultimately destroying the growing point. The central whorl of leaves dries up and forms the characteristics dead-heart.

The shoot-borer caterpillar is migratory in habit and, after killing one shoot, it transfer itself to another. A single caterpillar may destroy three or four shoots before it pupates. Because of this migratory habit, sometime two or three, or even more holes may be found on the stem of an attacked shoot.

In areas where comparatively high maximum temperatures (350 to 380C) and low remain relative humidity (48 to 52 per cent) prevail till late in the season, i.e. till October as in most parts of Rajasthan, the shoot-borer and the pink-borer caterpillars may damage the crop as cane-borers also. Attack of shoot-borer as cane-borer has been observed in Plassey area of West Bengal also where comparatively high temperatures prevail during October-November. No dead-heart is formed when the attack occurs in the cane stage, and the injury can be located only when the whole stalk is carefully examined for the presence of borer holes on the internodes.

Pupation of the shoot- and the pink borer usually occurs within the tunnel in the shoot or the cane (the pink-borer may sometimes come out of the tunnel and pupate in-between the stalk and the leaf-sheath). The green borer pupates in the soil in an earthern cell which it prepares just before entering pupation.

The incidence of shoot-borers is more pronounced in light soils and under comparatively dry conditions. In severe cases of infestations, more than 50 percent of the shoots may dry up necessitating replanting of the crop. In fields where wheat is taken as companion crop with sugarcane, pink-borer damage in sugarcane is usually severe. The borer which is also a pest of wheat at first attacks the wheat plants during February, and later migrates to the adjoining young sugarcane shoots during March. In some fields the damage in sugarcane in case of companion cropping is so severe that whole of the crop has to be uprooted.

Mortality of tillers on account of shoot-borer attack includes emergence of fresh tillers in the initial stages of the crop growth and this compensates to a certain extent, the loss in tillers. But, if the attack occurs at germination stage or immediately after gemination, the solitary mother shoots are killed outright resulting in the complete failure of the clumps concerned and guppy stand of the crop.

Control:

— *Mechanical.* Timely cutting of attacked shoots at or just below the ground level, ensures destruction of most of the caterpillars present within. The borer larva in an attacked shoot can also be killed by first pulling out the dead-heart and then thrusting the spoke of a bicycle wheel in the cavity formed in the shoot. As in other mechanical control methods, this operation should be carried out simultaneously by all the cane growers in a locality. To check serious loss by pink-borer, in case of companion cropping of sugarcane with wheat, infested wheat plants (with white ear-heads) should be removed and destroyed weekly during February-March.

— *Cultural.* Early planting (by middle of planting) is helpful in checking serious damage by shoot-borer under north Indian conditions. In such crops, tillering phase already started by the time borer attack appears in April. Light earthing at the early stages of the crop growth during May and June also reduces the borer damage considerably.

— *Chemical.* Sprinkling of gamma BHC, Heptachlor or Telodrin emulsion at 1 kg a.i. per ha mixed about 1,500 litres of water over the setts in the furrows at the time of planting of cane, proves effective not only against shoot- and root-borer, but against termites also. As a result of application to these insecticides, there is an appreciable increase in crop yields, particularly in light soils. Telodrin treatment gave the highest increase in yield.

For control of shoot-borers in young ratoon crops, the insecticidal emulsion should be sprinkled (during March-April) on the soil along the sugarcane rows after the soil has been loosened. This will hold good for control of pink-borer in autumn planted crop also when wheat is taken as a companion crop.

Stalk Borer

The stalk borer, Chilo auricilius Ddgn., is perhaps the most destructive of all the moths-borers found in the north Indian sugarcane belt. Like shoot borer, the newly hatched larvae usually feed on the leaf-sheaths by mining into them. About half-a-dozen young caterpillars may be found in a single leaf-sheath. The larvae may feed on the unopened leaves of the central whorl as also on softer parts of the rind. After feeding on the leaf tissues for about a week , the larva penetrates into an internodes by making a fine circular hole of the rind the stalk borer, as the name implies, is primarily a cane-borer damaging internodes. As many as 6 to 8 caterpillars may be seen a single cane; in such cases the entire length of the cane may be found bored through. The caterpillars often migrate from one cane or internode to another. In winter months, late shoots (or the water shoots) with tender succulent internodes, are also attacked; in some cases the borer larvae may travel down to the underground portions of the cane (stubble). It is through the late shoots (which are sometimes left over in the field at the time of harvest) and the stubble (and their sprouts), that the pest population is carried over to the succeeding crops.

The attack of the pest in early stages of the crop growth results in the formation of dead-hearts resembling those of shoots ultimately dry up. In mature crop the quality of juice deteriorates appreciably as a result of damage to the internodes. In severe cases of infestation, 75 to 100 per cent of the canes may be attacked and over 30 per cent of the total internodes may be found bored through. Autumn planted crop suffers more seriously from its ravages than the spring planted one. This may be attributed to better growth of the crop and earlier start of cane-formation in the autumn planted crop. In some areas of Bihar and West Bengal, the sugarcane is planted in July or August. In such cases, the crop is fairly grown up by April-May and therefore it can serve as a good medium for the stalk borer to tide over the critical period (April to June) in its seasonal history. Heavy incidence of the pest was actually seen during April, 1962, in the crop planted in July at the sugarcane Research Station, Pusa (Bihar), whereas the spring planted crop in an adjoining field was found particularly free of the pest attack at that stage.

Seasonal history. Pupation among the over-wintering larvae starts in the later half of January or beginning of February when there is slight rise in temperature. The moths begin to emerge by the end of January and there is rush of emergence during February-March these moths lay their eggs on the ratoon sprouts or October-planted crop or on the late shoots left over the fields at the time of harvest. The attack of the pest during March-June, when the spring-planted crop is in shoot-stage, is generally mild.

In July, with the onset of mansoon, the pest becomes more active. Cane-formation also starts at this stage and serious infestation occurs from August onwards. The pest passes through 5 or 6 generations in a year, the main periods of their appearance being

(i) February-end to June-end,

(ii) Middle of June to beginning of August,

(iii) July-end to September-end,

(iv) Middle of September to the beginning of November, and

(v) End of October to the beginning of March.

Total life-cycle is completed in 32 to 58 days during March to October, and 89 to 114 days during November-February. The pest remains active in larval stage during winter months and continues to damage the grown-up crop. Late shoots are also infested from October onwards. There is no hibernation in this borer. In rare cases fresh egg-masses and newly hatched larvae are also seen on late shoots. Study of stalk borer incidence and multiplication in relation to weather conditions has shown that after the cane-formation has started in July, moderate conditions of temperature (30⁰ to 35 ⁰C) and relative humidity (46 to 76 per cent) are favourable for activity of the moths. Prolonged dry spells during moth-emergence (and egg-laying) of various generations followed by heavy showers during larval development within the canes, help in the multiplication of the pest.

It has further been observed that attack of the pest is usually more serious in water-logged and ill-drained fields. On account of the migratory habits of the larvae, lodged canes offer better conditions for their movement and development, and these are therefore more heavily attacked. High doses of nitrogen (above 100kg N/ha) make the canes soft and succulent thereby heavy infestation of the pest.

Control:

(a) *Mechanical*

(i) Stripping of dry leaves from the canes once in September or October, proves helpful in checking serious damage of *C. auricilius.* The borer larvae which are migratory in habit, find it difficult to move about on the exposed and slippery internodes and to transfer themselves from one cane or internode to another.

(ii) Removal and destruction of late shoots or water shoots at the time of harvest is also recommended as a mechanical control measure because late shoots serve as the main source of carry-over of the stalk borer to the succeeding crops.

(iii) Moths emerging from the pupae of over-wintering caterpillars lay their eggs on the fresh ratoon sprouts on which the attack of the first generation usually occurs. Cutting-off these sprouts at the ground level once or twice during February-March will ensure destruction of egg-masses as also of young caterpillars infesting these shoots, and thus the carry-over of the pest to the next crop will be checked. Eradication of Johnson grass is also necessary to prevent the carry-over of the pest population from crop to crop.

(b) *Chemical.* Chemical control of stalk borer is rather borer because of its peculiar feeding habits and overlapping of generations. Laboratory studies conducted at the IISR, Lucknow, revealed that if grown-up caterpillars of stalk borer are exposed to surfaces treated with 0.1 percent Endrin emulsion immediately after spraying, there is 100 percent morality among them. But if such exposure is made even a single day (24 hr) after spraying, the morality is reduced to 50 percent. In case of newly hatched larvae, however, there is complete morality till as long as 16 days of such spraying. On the basis of these findings and subsequent field trials, spraying of Endrin fortnightly was recommended during July-September when the pest activity is at its maximum and all stages are met with in the field.

Army Worms

The army-worm, Pseudaletia (Cirphis) unipuncta Haw., a pest of grasses in general, often damages the cane crop in its early stages of growth. Its attacks has been reported from Bihar, West Bengal, Tamil Nadu, Maharashtra, Karnataka, Madhya Pradesh and Uttar Pradesh. Besides sugarcane, its infestation has also been observed on wheat, jowar, maize, oats, peas, paddy and ragi.

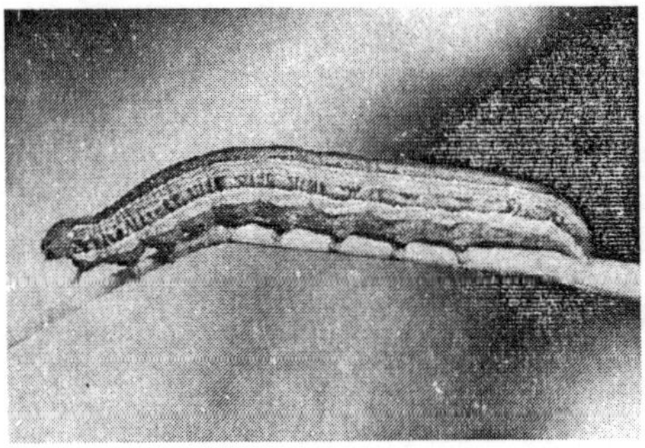

Army-worm

The caterpillars damage the crop by feeding on leaves. The larvae of the first and second instars feed by scraping the sugarcane leaf. Maximum damage is done by the grown-up caterpillars which are voracious feeders. They are nocturnal in habit. In the daytime, they remain hidden under the thrash or in the loose soil upto a depth of 5 to 8 cm, or under the clods in the fields. They come out at night and start feeding from the edge of the leaf eating away all the green portion leaving only the mid-rib behind.

The peak period of attack in tarai area of Uttar Pradesh and in Sehore region of Madhya Pradesh was observed to be from April to June, after which the pest activity subsided with the onset of monsoon. Under Kolhapur (Maharashtra) conditions, the pest attack appears in August immediately when the rivers in the vicinity (Panchganga and Krishna) are flooded.

Seasonal and Life-history The eggs, which are covered with greyish hairs, are laid in clusters inside the leaf sheaths of sugarcane plants. These may also be laid on trash provided at the base of plants. Egg stage lasts 4 to 5 days.

The newly hatched larvae crawl to the leaves on which they feed by scraping the sugarcane tissue. There are 5 to 6 larval instars. Total larval period is 21-35 days. When fully grown the larva prepares an earthen cell (having a loose silken lining from within) in which it pupates. These cells are formed at about 2 cm to 5 cm in-depth in the soil. The pupal period continues for 7-10 days. The moths remain hidden under the trash or the sugarcane leaves during the day time. The total life cycle is completed in 40-55

Control:

(i) *Chemical.* Dusting the crop with 10 percent BHC or 5 percent Aldrin or Malathion at 25-40 kg/ha (depending on the growth of the crop) proves effective against the pest.

(ii) *Cultural.* the trash in an infested field should be burnt to destroy the eggs laid underneath, and field should later be ploughed to expose the pupae to the predatory birds.

Sugarcane Mites

Six species of mites have been recorded infesting sugarcane crop in India. These are:

(i) Oligonychus (Paratetranychus) indicus Hirst.

(ii) Schizotetranychus and ropogoni Hirst.

(iii) Tarsonemus spinipes (rust mite)

(iv) Agistemus flescheneri Summer.

(v) the blister mite, *Eriophyid sp.*; and Scirus sp.

The first two species namely, *O. indicus* and *S. andropogoni* occur commonly though sporadically, and serious infestations by them have been reported both in peninsular India and in the north Indian sugarcane belt.

O. indicus

Usually known as the red leaf mite, it attacks the crop in the pre-monsoon period (May-June) when temperature is high and the relative humidity low under the north Indian conditions.

Nature and extent of damage. The adult female mites settle on the under surface of the leaves and weave out fine webs around them within which they lay eggs thus a number of colonies are formed. As a result of continuous feeding by the adults and nymphs, photosynthetic activity of the leaves is adversely affected and they turn reddish hence the name red-leaf mite. In cases of severe infestation, such leaves dry up altogether and later fall off. Sugarcane varieties with soft leaves are move susceptible to the attack than those with hard and tough leaves. Besides sugarcane, the pest has been reported infesting *Saccharum spontaneum, Sorghum halepense, S. vulgare, Andropogon sorghum, Panicum javanicum* and *P. distachyum.*

Seasonal and Life-history. The pest breeds on sorghum halepense (baru) throughout the year, but is more active during the summer months, viz. April to June when sugarcane and other host plants are also attacked. The egg, larval and nymphal stages last for 1 to 3 days, 3 to 4 days, and 2 to 4 days respectively, and the total life cycle is completed in 6 to 11 days during April-June.

S.andropogoni.

It is commonly known as web-mite because of the whitish webs which are formed all over the ventral side of the leaf.

Nature and Extent of Damage. Six to eight rows of webs may be found more or less parallel to the mid-rib. The web population is higher towards the tips. In cases of severe infestation, the number of webs per leaf may be as high as 500 or even more. The webs when, newly formed are whitish and, more or less, oval. After some time, they turn brown and are ultimately blown off from the leaf surface leaving behind whitish patches caused as a result of feeding by the mites. They feed by scraping caused as a result of feeding by scraping the epidermis and sucking the juice. The population of the mites varies form web to web. In some cases all the stages, namely egg, nymph and adult are seen. Occurrence of only one or two stages in a web is also common.

Seasonal and life history. Under tropical conditions the activity of S.andropogoni starts in November and the mite population reaches its maximum in April weather the warms up considerably. In north India, the pest activity starts with the onset of monsoon rains in July and the highest incidence occurs during September-October. There is a decline in its activity during winter months and the population remains low till June next. A number of overlapping generations are passed in the course of a year.

The nymph passes through 3 instars before attaining maturity. The duration of 1st, 2nd and 3rd instars is 2-3 days, 3-4 days, and 3-4 days respectively, total duration of the nymphal stage being 8-11 days during September-October.

Mating takes place immediately after the last moulting of the female deutonymph. Egg laying starts about 24 hours after mating. Ovi-position period varies from 6-9 days; and 1-2 days after completion of egg-laying the mother-mite is dead. The males die soon after mating. The eggs are laid singly within the webs and these get stuck up on the leaves; 30 to 60 eggs may be laid per female.

One life cycle of the pest is completed in 11-24 days during September-November, and 31-48 days during November-February.

Control:

(i) *Mechanical.* Timely removal (during March-April) and destruction of germinaceous weeds like baru (Sorghum halepense) etc., in the vicinity of sugarcane fields, and clipping of infested sugarcane leaves in the initial stages of pest infestation, prove helpful in checking serious damage to the crop by Oligonycims indicus.

(ii) *Chemical.* spraying the crop with lime-sulphur-wash, fish oil rosin soap or parathion is effective against O.indicus, and with Kelthane, Parathion or Chloribenside+ Malathion Miscible liquid spray against Schizotetranychus andropogoni.

Beetles

Seasonal and life-history. Both the species have one generation in a year. Beetles which remain hidden up to a depth of about 1 metre in the soil during winter months (October-March) become activated by an early shower of rain some time during April-May when they begin to merge. In H. serrata, the emergence is completed by the end of May or beginning of June, while rush of emergence of H. consanguinea beetles takes place with the onset of monsoon in later half of June or beginning of July. These beetles are active at night. H.consanguinea beetles are attracted to light, but no such attraction has been noticed in H.serrata.After mating, the females lay eggs singly in the soil at a depth of 5-10 cm. the eggs hatch out in 7-10 days and the young grubs begin to feed on the tender roots of grasses and sugarcane, and other vegetable and organic matter below the soil surface. As they grow, they slowly travel downwards.

The peak period of activity of the grubs is July to September when maximum damage to sugarcane takes place. The fully grown grubs stars migrating downwards into the soil of pupation. In H.consanguinea, pupation takes place during September-October, while in H.serrata, it occurs during October-January. Adult emergence starts in the former species in the last week of September and is completed by the end of November (under north Indian conditions). In H.serrata, adult emergence takes place during November-February (under tropical conditions). The larval stage lasts from 8 to 10 weeks in H.consanguinea and from 20 to 25 weeks in H.serrata. The pupal stage occupies 2 to 3 weeks in both cases.

The adult beetles remain in a quiescent condition within the soil till April-May when emergence in the open starts. They are active at night and feed on the foliage of nearby bushes and trees like Ficus glomerata, Zizyphus jujuba, Azadirachta indica etc. So far no damage has been recorded on sugarcane crop by the adult beetles.

Nature and Extent of Damage. Nature of damage in both the species is almost identical. The grubs feed on the roots and rootlets of sugarcane below the soil surface, and eat away a major portion of the root system; sometimes they scoop the bottom portion of the cane stalk also. In case of severe infestation, the whole root system gets completely depleted and this in turn deprives the cane-stalk of its nutrition ultimately resulting in its drying up. The outer leaves dry up first and as the attack progresses, the entire shoot dries up and gets dislodged easily.

In the initial stages the attack of the pest occurs in patches, but later on if the infestation is severe, the damage spreads to the whole field. Each clump may harbour upto about 2 dozen grubs in its root system, and if all the clumps are seriously infested, the whole crop dries up by August. In case of light infestations, particularly in crops with good growth, signs of drying may be observed in the initial stages, but by September the crop recovers by putting forth fresh rootlets. The early season crop (October to January) is able to withstand the attack better than late-season crop or the spring planted crop.

Control. The attack of the pest is noticed when the crop begins is to dry up either as a whole or in patches in August. Even then the actual of the damage is known only when the soil is carefully removed around the clumps and the grubs are seen in the root zone. By then they are fairly grown-up and have travelled to a depth of 10-20 cm below the soil surface. It is too late at this stage to take any precautionary or control measures against the pest the control measures have, therefore, to be taken up in the initial stages of the infestations, i.e. during June-July. The following measures are usually recommended.

1) *Cultural:*

   a) Flooding the fields at frequent intervals. This helps to reduce the damage since the grubs temporarily migrate downward when the field is flooded.

   b) Planting of varieties with a vigorous and prolific root system because a luxuriant crop is able to withstand the attack of the pest better by putting forth fresh roots.

   c) Early season planting or autumn planting of sugarcane; in early planted crops the root system becomes fairly well developed by June-July when the young grubs first appear. Hence, this crop is able to withstand the pest attack better.

2) *Mechanical.* Collection of adult beetles (of *H.consanguinea*) by means of light traps or by hand-picking from beneath the electric or Petromax lamps in case special light traps are not available. Adults, (of both species) can also be collected by vigorously shaking the trees and bushes like mango, guava, gular, peepal, ber, neem, Arjun, etc., on which they congregate after sunset during June-July the beetles so collected can be destroyed by immersing them in water having a layer of kerosene oil on top.

3) *Chemical.* Destruction of young grubs by mixing BHC 10 percent dust at 50 kg/ha in the top 10-cm layer of the soil in-between and along the rows of the young sugarcane crop. This treatment should be given twice :

   (i) once just at the time of first rush of adult emergence which occurs after early showers of rain in the end of May or beginning of June, and

   (ii) once in July. The dust should be evenly applied over the soil surface and then worked upto it with a spade or a khurpa for a uniform distribution of the dust, an ordinary cigarette tin which has been perforated at the bottom and on the sides, proves very helpful. Since the newly emerged grubs are present near the soil surface, these are killed by mixing BHC dust in top layer of the soil.

In certain areas of Maharashtra and Andhra Pradesh where adsali crop is grown, attack of the pest to the newly germinating crop (during July-August) can be checked by application of Heptachlor emulsion at 2-3 kg a.i./ha over the setts in the furrows at the time of planting, or at the base of the young shoots (in case of early plantings) and then working it up with the soil.

Besides the above referred two species there are others like Heteronychus robustus Arrow., Alissonotum sp., etc. which attack the crop in adult stage. Heteronychus robustus Arrow., damages the young crop in several areas in Uttar Pradesh and Bihar. The attack is sporadic occurs in small pockets only. The shining black bettle makes a cut on the stem and eats away the plant tissue at the point of attack. This results in immediate drying up of the central whorl of leaves, and sometime later the entire shoot dries up. In serious cases of infestation up to 20 percent of the shoots may be found attacked. Collection and destruction of beetles, by immersing them in kerosenised water, proves helpful in checking serious damage.

Black-bug

Two species of black-bug (i) *Blissus gibbus* F. and (ii) Macropes (cavelerius) excavatus Dist., are reported to the infest sugarcane crop in north India. The latter species, M. excavatus commonly known as lygaeid bug is a serious pest of the ratoons in Punjab, Haryana and western Uttar Pradesh. This species has been studied in greater detail and a brief account is given below.

Nature of Damage. The nymphus and adults of M.excavatus congregate in the whorl of the central leaves and under the sheathing bases of the side leaves where they suck the cell-sap. The attack is severe in young ratoon crops in fields where the trash had been left over after harvest. The affected crop turns pale and the individual leaves show deep reddish-brown spots. In cases of severe attack growth of the crop is checked; the quality of gur is also adversely affected. The pest prefers varieties with loosely attached leaf-sheaths.

Seasonal and Life-history. M.excavatus passes winter (October-March) as adults. They remain in the soil round the sugarcane stumps up to depth of 5-8 cm, feeding and laying eggs. The nymphus appear in March-April; from June to the end of October the pest is present in all stages. The adults may live for 3 to 4 months. Mating takes place after about a week of emergence. In summer the eggs are laid in clusters on the inner side of sheathing bases of the side leaves and in winter singly (occasionally in pairs) close to the roots of the plants in the soil; the number of eggs laid by a female may vary from 50 to about 500. The duration of egg-stage is 9-17 days during May-September there are 5 nymphal instars. Total duration of nymphal stage is 5 to 7 weeks in March-July and 4 to 6 weeks in January-October the pest passes through three generations in a year.

Control:

(i) *Cultural.* Burning of sugarcane stubble during March-April when pest is usually found in huge numbers in fresh ratoon sprouts, proves helpful in destroying the pest in the early stages of its activity. The trash and the dry leaves left in the field after harvesting the crop, can be utilised for this purpose.

(ii) *Chemical.* Spraying the crop with 0.25 percent BHC suspension, or 0.025 percent Endrin or Endosulfan emulsion or application of 2 percent Endrin granules or Sevidol granules at 10 kg per hectare, is reported to have proved effective against this pest. Since the bugs usually congregate in the central whorl of leaves, application of the insecticide should be so arranged as to reach that particular portion of the plant.

Mealy-bugs

Three species of mealy-bugs occur in India; these are :

(i) Saccharicoccus sacchari Ckll.

(ii) Pseudococcus saccharifolii Green and

(iii) Ripersia sacchari Green.

Among them, S, sacchari commonly referred to as the pink mealy-bug, is the most important, and is met with in almost all the major cane-growing areas of the country.

Seasonal and life-history. The bugs are usually found embedded in a white waxy mass near the joints of internodes underneath the leaf-sheaths. Sometimes these are also seen on exposed portions of the internodes the eggs are laid in the waxy mass underneath the leaf-sheath. One female may lay 400 eggs. They hatch in 4-7 days depending on the temperature and humidity. Youngs nymphs on emerging remain embedded in the waxy material for a day or two after which they begin to move about. They soon get settled on fresh internodes near the joints, insert their mouth part into the plant tissue and begin to feed on the plant sap.

Like the adult female, the young nymphs are oval and pink but are flattened and much smaller in size, without mealy covering. The adult male is wingless. One life-cycle of the pest is completed in about a month, and the broods follow in quick succession. There is overlapping of generations, and all stages of the pest can be seen all through the crop season.

Nature and Extent of Damage. The attacked crop looks pale and sickly because of withering of leaves, and can be spotted from a distance. Severe infestation may cause drying of the affected leaves. Because of continuous de-sapping by the bugs, cavity like depressions are formed on the internodes under the leaf-sheaths. The buds also gets badly damaged and looked atrophied. The juice quality is adversely affected in case of seriously infested canes, and loss in sucrose content may be abut 25 percent. Varieties with loose fitting leaf-sheaths are more seriously attacked.

Control. spraying the crop with 0.05 percent Folidol emulsion or 0.1 percent Malathion emulsion proves effective in controlling the pest. Since the Folidol has acute toxic effect, use of malathion is usually recommended. As the nymphs remain almost hidden under the leat-sheaths, the spraying should be specially directed towards the cane stalks (preferably after detrashing) so that it can trickle down the internodes into the underneath the leaf-sheaths.

To prevent the spread of infestation to new areas, it will be necessary to adopt quarantine measures for ensuring the despatch of pest-tree seed material. Stripping of the dry leaves at harvest also proves helpful provided it is done thoroughly. Chemical treatment of the infested seed-cane with Folidol of Malathion emulsion is also effective. The ants (black ant) which usually carry the female bugs from one plant to another and disseminate the pest may be controlled by dusting the crop with 5 percent Chlordane.

Grosshoppers

Of the fifteen species of grosshopper recorded attacking sugarcane crop in India, the following 6 are important.

1) colemania spheneroides B.-Deccan wingless grosshopper.
2) Hieroglyphus Bol.-Commonly known as the rice grasshopper.
3) H. nigrorepletus Bol.-Phadka grasshopper.
4) H. concolor Wlk.-Recorded in tarai area of Uttar Pradesh.
5) Gastrimargus marmoratus Thu.
6) Choroedocus illustris-Reported from Coimbatore eating sway the growing point o the sugarcane shoots.

Among these *H. banian* has a much wider distribution and has been reported to occu as a pest in almost all the states in India. The first and second stage hoppers are gregariou and feed mostly on grasses. In the third stage they migrate to sugarcane and feed ther till they mature. They eat away the green tissues of the leaf leaving only the mid-ri behind.

Control

Dusting the crop with 10 percent BHC at 25 kg/ha, or spraying with Aldrin emulsio at 225 g/ha active ingredient is reported to be effective.

Non-insect Pests

Among the non-insect pests of sugarcane, two species of mites described earlier, namely Oliganychus indicus and Schizotetranychus androprgoni (Arichnida), thought sporadic in occurence, cause considerable damage to the crop in some regions. Rats, jackals and squirrels also come under this category.

Rats

Rats cause enormous damage to agricultural crops as well as stored grains and other household products. The following species are common and damage crop in various parts of India:

1. Bandicota bengalensis
2. Rattus morvegicus Berken.
3. Rattus rattus
4. Bandicota indica Bech.
5. Gerbillus gleadowi
6. G.indicus Desm.
7. Golunda ellioti Gray
8. Legada (Mus) Booduga Gray
9. Tatera indica indica Hard.

Control. Since rats damage all types of crops, it will be desirable to make a collective and co-ordinated effort for their destruction in a particular area or locality the following measures are suggested for their control :

(i) *Survey.* First step is to have a thorough survey carried out for determining the 'live' and 'dead' burrows in the locality.

(ii) *Pre-baiting*: Normal food articles (gram, etc.) wrapped in small paper-packets should be kept at the entrance point of 'live' burrows for 2 or 3 days to familiarise the rates with unpoisoned food.

(iii) Poison-baiting when the 'pre-bait' is freely accepted, the poison baiting should be restored to with zinc phosphide in the ratio of 1:49 with grain and a little vegetable oil (for 'live' burrows only).

(iv) *Fumigation.* Fumigation of 'live' burrows with aluminium phosphide tablets also proves effective

(v) *Use of rodenticides.* In thickly growncrops like sugarcane where rat burrows are difficult to locate, use of blood coagulant rodenticides like Warfarin or Ratafin (one part mixed with 19 parts of wheat, rice, maize or any flour and a tablespoon of vegetable oil) is recommended. In this case, no 'pre-baiting' is necessary.

Aphids

Aphids cause damage to sugarcane crop, generally, as vectors of a number of virus diseases. Longiunguis sacchari Zehnt., L. indosacchari, Aphis idiosacchari and A. maidis are reported to act as a vector of grassy shoot diseases; A.maidis acts as vector of mosaic disease also. Sometimes, these aphids damage the crop directly also. Heavy infestation of L. sacchari and L. indosacchari was reported from a sugarcane field in Koothanallur area of Tanjore district during November-December, 1960. Another species of aphid, Tetraneura javensis v.d. Goot was reported infesting the roots of sugarcane in the field near Coimbatore in 1957. T. hirsuta Baker and Geoica spatula that have also been recorded on roots of sugarcane.

Control. Spraying the crop with 0.1 percent Malathion emulsion or 0.03 percent Phosphamidon emulsion, proves effective against the aphids.

Scale Insects

The following species of scale insects have been recorded in India: Acanthyomytilus sacchari Hall., Duplachionaspis divergence Green, Melanaspis glomerata Green, Odonaspis saccharicaulis Zhnt., and Temnaspidiotus kellyi Brein. Among them, Melanaspis (Targionia) glomerata Green is by far the most common species.

Nature and extent of damage. Infestation of the scale insect is mainly concentrated on the internodes which usually remain covered with leaf-sheaths. In severe cases of infestation, almost the entire surface of an internode may be covered with hundreds of any little greyish-black scales. In such cases, infestation may be seen on leaves also. The pest causes considerable damage whenever it occurs in a serious form. Both tonnage and quality of the crop are adversely affected. About 13 percent loss in cane weight, 28 percent in brix., 47 percent in sucrose content, and 26 percent in purity were reported as a result of a study conducted at S. B. I., Coimbatore.

The pest may be seen in association with wilt (Cephalosporium sacchari Buttler), and in cases of heavy infestations the crop may dry up together. Varieties like 'Co. 740' with enveloping and adhering leaf-sheaths are more seriously infested. Ratoons and adsali crops usually suffer severe damage. The pest has been reported to infest Saccharum also at Pusa (Bihar) and at Coimbatore (Tamil Nadu).

Seasonal and Life-history. The young tiny nymph which looks like a yellowish speck, comes out from beneath the mother-scale, crawls about for some time and then settles down after selecting suitable spot preferably on one of the internodes. It inserts its needle like-mouth parts into the plant tissue and goes on sucking cell-sap continuously. Once nymph gets settled like this, it remains stationary throughout its life if it is a female, or till it matures into a winged adult in case of a male. The formation of the scale (protective covering) starts soon after the nymph gets settled; this covering scale becomes thicker and increases in size as the individual underneath grows. The female scales are round and circular, whereas the male scales are usually oval and oblong.The male after coming out from underneath the scales, begin to move about on the infested internodes, and it is at this stage that the mating with females takes place. The later begin to lay eggs after some time which hatch out as nymphs (underneath the mother scale), and thus the cycle continues. One female may lay about four hundred eggs in its life time. The pest remains active practically throughout the year under south Indian conditions and there are 4 to 9 overlapping generations in the year. One life-cycle is completed in 18 to 25 days in males and 39 to 46 days in females.

Preventive and control measures. The pest usually gets introduced to a new locality through the infested seed-cane. Once established, its further spread in the area can be brought about by wind, birds, animals, etc. which might carry the tiny crawler nymphs from one shoot or field to another. Due care should be exercised in selection of seed material, particularly for such areas as are so far free from infestation. In infested localities, the following measures are recommended.

1) *Segregation of heavily infested pockets:* A thorough survey should be carried in the infested areas, and pockets of severe infestation clearly marked out. Such pockets should be isolated by removing the cane crop from about 4-5 m wide belt around the infested crop to prevent further spread of the pest to the adjoining fields. The crop in the infested pockets should be harvested and crushed as soon as the crushing season starts.

2) *Stripping of dry leaves:* The dry leaves as well as the green leaves bearing scale infestation, should be removed from the infested crops at fortnightly intervals and destroyed by burning.

3) *Treatment of infested crop:* Malathion emulsion (0.1 percent) or fish oil rosin soap emulsion (0.1 percent) should be sprayed twice or thrice at 2 to 3 weeks interval. Border crop in the surrounding fields should also be given this treatment as a precautionary measures.

   In varieties like 'Co.740' (in Maharashtra) where detrashing results in bud-sprouting, application of systematic insecticides like Dimethoate (rogor) as foliar spray or with irrigation water, is reported to be effective. Application of Disyston granules in the soil after internode formation has started, has also given encouraging results.

4) *Treatment of seed-cane:* The cane setts before planting should be dipped in 1.0 percent fish oil rosin soap emulsion or rubbed with a cloth piece soaked in 0.1 percent Malathion emulsion, to destroy the sticking scales, if any. Care should be

taken not to injure or damage the buds. Discard the badly infested crop so far as its use as seed-cane is concerned. Ratooning should also be avoided in case of seriously infested crops.

Biological control. An indigenous parasite, Anabrolepis mayurai and a predator, Pharoscymnus horni, are reported to have exercised appreciable check or multiplication of M. glomerata in certain fields in East Godavvari district. Chilocorus nigritus was another promising indigenous predator seen in the area. Two species of exotic predators, Sticholotis madagassa and Chilocorus cacti, were introduced through CIBC, Bangalore, and were released in some fields after laboratory multiplication. Recovery surveys of release-sites later indicated that *S. madagassa* had successfully survived (the summer months) and was predating well on M.glomerata in the field.

Termites

The common species of termites attacking sugarcane crop in India are Odontotermes obesus 'Ramb', *O. assumithi* Hlmgr, *O. taprobanes* Wlk., *O. bengalorensis* Hlmgr., *Eremotermes paratoxalis* Hlmgr., *Trinervitermes biformis* Wasm., *Microtermes obesi* Hlmgr., and *Coptotermes heimi Wasm.* Their attack usually occurs soon after planting of setts in the field the cut-ends and the buds are the first to be attacked resulting in poor germination and guppy stand of the crop. In severe cases of infestation, germination is so badly affected that it becomes necessary to replant the crop. Post-germination attack is also common, particularly in loamy soils and dry areas where proper facilities for irrigation are not usually available. Outer leaves dry up first, and as the attack spread further, the whole plant withers. In most areas, the severity of termite attack subsides after the onset of monsoon rains, but in criteria regions, e.g. in eastern Uttar Pradesh and Bihar, post-monsoon damage in crop is also commonly met with.

Control. Application of 1.0 kg (a.i) per hectare of BHC, chlordane, Heptachlor, Aldrin or 0.5 kg Dieldrin over the cane setts in furrows at the time of planting provides protection of the setts, and wards of terminate attack in the initial stages of the crop growth. While residual toxicity of BHC lasts for a period of about 2 to 3 months only, application of Chlordane, Heptachlor, Aldrin or Dieldrin is reported to have proved effective against the post-monsoon attack of the pest also.

REFERENCES

Bailey, L. H. and Bailey, E. Z. 1976. *Hortus Third: A Concise Dictionary of Plants Cultivated in the United States and Canada.* MacMillan Publishing Company, New York.

Boddey, et.al., "Biological nitrogen fixation associated with sugar cane", *Plant and Soil,* Volume 137, Number 1 / November, 1991.

Dong et al., "A Nitrogen-Fixing Endophyte of Sugarcane Stems" (A New Role for the Apoplast), *Plant Physiology,* 1994, Vol. 105, Issue 4 1139-1147.

Watson, Andrew. *Agricultural innovation in the early Islamic world. Cambridge University Press.* p. 26-27.

4

Fiber Crops

Fiber crops are field crops grown for their fibers, which are used to make paper, cloth, or rope. These crops are generally harvestable after a single growing season, as distinct from trees, which are typically grown for many years before being harvested for wood pulp fiber. In specific circumstances, fiber crops can be superior to wood pulp fiber in terms of technical performance, environmental impact or cost. There are a number of issues regarding the use of fiber crops to make pulp. One of these is seasonal availability. While trees can be harvested continuously, many field crops are harvested once during the year and must be stored such that the crop doesn't rot over a period of many months. Considering that many pulp mills require several thousand tonnes of fiber source per day, storage of the fiber source can be a major issue.

Botanically, the fibers havested from many of these plants are best fibers; the fibers come from the phloem tissue of the plant. The other fiber crop fibers are seed padding, leaf fiber, or other parts of the plant. Fibre is a class of materials that are continuous filaments or are in discrete elongated pieces, similar to lengths of thread. They are very important in the biology of both plants and animals, for holding tissues together. Human uses for fibers are diverse. They can be spun into filaments, string or rope, used as a component of composite materials, or matted into sheets to make products such as paper or felt. Fibers are often used in the manufacture of other materials. Synthetic fibers can be produced very cheaply and in large amounts compared to natural fibers, but natural fibers enjoy some benefits, such as comfort, over their man-made counterparts.

Fiber crops are field crops grown for their fibers, which are used to make paper, cloth, or rope. These crops are generally harvestable after a single growing season, as distinct from trees, which are typically grown for many years before being harvested for wood pulp fiber. In specific circumstances, fiber crops can be superior to wood pulp fiber in terms of technical performance, environmental impact or cost.

There are a number of issues regarding the use of fiber crops to make pulp. One of these is seasonal availability. While trees can be harvested continuously, many field crops are harvested once during the year and must be stored such that the crop doesn't rot over a period of many months. Considering that many pulp mills require several thousand tonnes of fiber source per day, storage of the fiber source can be a major issue.

Botanically, the fibers havested from many of these plants are bast fibers; the fibers come from the phloem tissue of the plant. The other fiber crop fibers are seed padding, leaf fiber, or other parts of the plant.Fibre is a class of materials that are continuous filaments or are in discrete elongated pieces, similar to lengths of thread. They are very important in the biology of both plants and animals, for holding tissues together. Human uses for fibers are diverse. They can be spun into filaments, string or rope, used as a component of composite materials, or matted into sheets to make products such as paper or felt. Fibers are often used in the manufacture of other materials. Synthetic fibers can be produced very cheaply and in large amounts compared to natural fibers, but natural fibers enjoy some benefits, such as comfort, over their man-made counterparts.

JUTE

Jute, the 'Golden Fiber' as it is called, is a plant that yields a fiber used for sacking and cordage. Known as the raw material for sacks the world over, jute is truly one of the most versatile fibers gifted to man by nature that finds various uses in the form of Handicrafts. Next to cotton, jute is the cheapest and most important of all textile fibers.

Jute is the cheapest lignocellulosic, long vegetable bast fibre available annually in the world. Jute cultivation provides work for millions of farmers, landless laborers, industrial workers and provides jobs for many others, indirectly.

Rapid expansion of jute into a wide range of life style consumer products has been made possible due to the versatility of Jute. Spinning of high quality yarns and weaving of light-weight fine-textured fabrics of uniform structure in exotic colours and designs are made in both jute factories and hand loom sectors. With vastly improved bleaching, dyeing and finishing processes and by blending jute with other natural or synthetic fibers, the finished jute products now ensure feel, luster, abrasion resistance and aesthetic appeal.

Botanical Features

Jute (*Corchorus capsularis* L. and *C. olitorius* L.; n=7), a bast fibre, is obtained from one of the most important cash crops of eastern India Exported as manufactured goods and as raw fibre, it earns foreign exchange around 3,000 million rupees. The crop is grown in West Bengal, Assam, northern Bihar, south-eastern Orissa, Tripura and eastern Uttar Pradesh on 0.80 million hectares annually and the average production of dry fibre is 6.5 million bales.

Corchorus capsularis

Currently, about 82 per cent of the total annual production is used by jute-mills in India, 8 per cent is retained by cultivators and the remainder is exported to Europe and the USA. The fibre is chiefly used for manufacturing hessian, sacking and carpet backing. It is used for storing and transporting grains, pulses, spices sugar, cement, fertiliser, minerals, cotton and wool all the world over. Jute is also used for making mats, tarpaulins, ropes and twines. Woollenised jute is used for manufacturing cheaper rugs, and mixed with cotton it is used for producing decorative cloth, curtains and upholstery. The jute sticks are largely used as fuel and also for making gunpowder charcoal. Since recently, the paper industry has been using it as a raw material for coarser paper. Resin-bonded pressed jute sticks make durable hard board.

Climate

Jute requires a warm and humid climate, with temperature fluctuating between 24°C and 37°C, the optimum being around 34°C. The permissible diurnal variation in relative humidity favourable to growth is between 57 and 97 per cent. Incessant rain water-logged conditions are definitely harmful. Whereas the capsularis varieties can stand water logging to some extent. In the seedling stage, water-logging is not tolerated by both the species.

Areas receiving early showers in late February or early march, followed by a relatively short spell, are suitable fore sowing the capsularis varieties. Olitorius varieties, until the introduction of 'JRO 878' and 'JRO 7835' have been sown with showers in April or even in May. The monsoon sustains the subsequent growth. Rain below 1,000 mm is not helpful. Alternate sunshine and rainy days are most conducive to growth. The dry spell proceeding the break of the monsoon enables the essential weeding and thinning operations to be carried out.

Soil

The new grey alluvial soil of good depth, receiving silt from the annual floods, is nature's best gift, but jute is widely grown in sandy loams and clay loams, with varying soil-management practices. Sandy soils and heavy clays are unsuitable. Soils with a low pH give a poor crop, the optimum pH being around 6.4.

Cultivation

Jute requires a clean, clod-free field with fine tilth. The land is, therefore, ploughed, cross-ploughed, and planked several times. All weeds are thoroughly removed.

Sowing

The sowing of capsularis varieties starts in late February in low-lying areas that retain moisture of the previous flood or monsoon.

Sowing in midlands and highlands starts with showers in March or April and continues till early June in the western part of the jute belt. For broadcast sowing, the seed-rates are 10 and 7 kg per hectare for capsularis and olitorius varieties, the germinability of seed being 80 per cent or above. The high seed-rates ensure an even emergence of seedlings, which are thinned out in broadcast plots to a spacing of 10 cm. This operation is preferably carried out in two instalments, once when the seedlings are about 10 cm and subsequently when they are about 15 cm tall. While thinning, care is taken to remove all weeds with a hand-hoe. These operations account for 30-40 per cent of the cost of cultivation.

Row-cropping has proved advantageous and a single row seed-drill is used to sow the capsularis varieties 30 cm apart and the olitorius varieties 20 cm apart. The plant-to-plant spacing is manually adjusted between 5 and 7.5 cm. Sowing is always done shallow, i.e. not exceeding 3-4 cm.

Cropping pattern under Irrigation

In jute-paddy potato, jute-paddy-wheat, (jute-moong)-paddy-potato, cowpea-jute-potato, and jute-paddy-berseem rotations, jute may need one pre-sowing irrigation and one sustaining irrigation cropping pattern under rainfed conditions. Jute-paddy-mustard, jute-paddy-pulses or jute-paddy rotations are possible under this cropping pattern.

Nutrition and Manuring

The gross uptake of N, P, K, Ca and Mg by the capsularis variety 'JRC 212' is 84, 16, 147, 84 and 29 kg respectively and by the olitorius variety 'JR0 632' is Ill, 28, 164, 124 and 25 kg per ha, respectively. The resulting fresh crop weighs 40 to 50 tonnes, yielding 20 to 25 quintals of dry fibre, which is around 5 per cent of the fresh weight of the crop.

Acid soil require amendment with 3-7 tonnes of lime in 3 years. Compost or farmyard manure at the rate of 4-7 tonnes per hectare promotes proper growth, specially when fertilisers are not available. P and K are applied as basal nutrients, whereas nitrogen is better top-dressed in two instalments-N (40-80 kg per ha for capsularis varieties, 20-60 kg per ha for olitorius varieties)-P (half of the quantity of N) and K (quantity equal to N). If the jute crop is preceded by potato, no fertilisation is recommended; if it is preceded by paddy or wheat, P and K are not so much required. Nitrogen is the most important nutrient required. The ash of water-hyacinth is a rich source of potash. The application of magnesium on the North Bank of Brahmaputra and in northern Bengal produced good results.

In case of short supply of nitrogen, foliar feeding with nitrogen at the rate of 15 kg of urea per ha at 10 per cent concentration, with a low-volume power sprayer in 2 instalments between 35 and 60 days' age of the crop proved useful. Capsularis varieties respond better up to 30 kg of urea.

Interculture

Weeding, specially in the early stage, is a must. For a broadcast crop, the pre-sowing application of 2,2,3,3 tetrafluorpropionate of sodium proved to be beneficial. A post-emergence application of mono-sodium methanarsonate with or without 2,2 dichloropropionate, as a directed spray, proved to be beneficial in row-cropped fields. Yields, with two hand-weedings, however, are significantly superior and constitute the prevailing standard practice. Smothering the weeds between rows with a wheel-hoe saves labour and helps to mulch the soil.

Major Pests

The female of jute apion (Apion corchori Marshall) lays an egg at a node where a knot is formed; it makes one puncture or several on the same plant. The jute semilooper (Anomis sabulifera Guen.) eats tender leaves and is widespread ; it usually comes in three waves between late June and August. The vehow mite (Hemitarsonemus latus Banks) sucks the sap from young apical leaves which curl, and the growth of the plant is arrested. The hairy caterpillar (Diacrisia obliqua, Walker), when young, is gregarious on a single leaf and later scatters out and feeds on the foliage.

The root-knot nematode (Meloidogyne incognita Kofoid and White) attacks the capsularis varieties more than the olitorivg varieties and induces chlorotic symptoms and arrests growth. What provides foci for root infection bataticola (Taub.) The pest is mostly confined to Assam.

Minor Pests

The indigo caterpillar (Lapltygma exigua Habn.), the leaf-miner (Trachys dasi Thery.), the red mite (Olygonychus coyfeae Neither), the jute stem- girdler (Nupserha bicolour postbrunnea , and the leaf-eating weevil (Myllourus discolor Boheman) are not of commercial importance.

Control Measures

DDT and Gammexane, with a limited efficacy in the field, are unprofitable. Ethyl parathion and Endrin are highly efficacious against all major pests. But because of their hazardous nature, they have been withdrawn. Endosulfan, Leptophos, Phosalone, Fenitrothion, Chlorfenvinphos are effective to different extents against the major pests; Monocrotophos and Tamaron on semilooper; lime sulphur on yellow mite and Heptachlor and Aldrin dust on crickets and Dicofol on red mite, are effective. Timely sowing and clean cultivation guard against afion; hand-picking of the leaves with gregarious clusters young hairy caterpillars controls the pests.

Diseases

The important diseases are:

1) seedling blight, stem-rot, collar-rot, and root-rot caused by Macrophomina phaseoli (Maubl.) Ashby, both seed-and soil-borne, secondary spread through air-borne pycnospores. The predisposing factors are low pH of the soil, the lack of adequate potash water-logging, high temperature and excess nitrogen. Seed treatment prevents seedling blight, whereas soil amendment with lime and potash retards its incidence. Jute-paddy and jute-paddy-wheat rotations are useful in reducing the incidence of the disease.

2) Hooghly-wilt, caused by a microbial complex, including Macrophomina phaseoli, Fusarium solani. Pseudomonas solanacearum, is confined to the jute potato fields only; control through rotation with paddy and legumes is possible.

3) Colletotrichum gloeosporioides causes anthracnose in Assam.

4) Leaf mosaic is caused by a pollen-transmitted seed-borne virus, also by white fly Bemisia sp. in Assam, specially in capsularis varieties.

5) Physoderma stem gall on the seedlings of olitorious varieties occurs in Uttar Pradesh

and Tripura.

6) Soft rot of stem Pellicularia rolfsii (Sace.); the disease can be controlled with deep ploughing, and clean cultivation.

7) Nematodes provide foci for infection by Macro phomina phaseoli in sick soils.

Harvesting

Jute may be harvested any time between 120 and 150 days. Early harvesting gives finer fibre of good quality, whereas late harvesting gives a larger yield but a coarser fibre. The compromise between quality and quantity is found in harvesting at the early pod stage or around 135 days of cropping. The other consideration for early harvesting is to accommodate paddy-transplanting in accordance with the cropping pattern chosen.

Harvesting is done by cutting the plants at or close to the ground level. In flooded land, plants are uprooted. The harvested plants are left in field for 2 to 3 days for the leaves to shed. Next, the plants are tied into bundles, 20-25 em in diameter, and the branching tops are lopped off to rot in the field.

Retting

The bundles are kept standing in water, 30 cm deep, and later placed side by side in retting water, usually in 2 to 3 layers and are tied together. They are covered with water-hyacinth or any weed that does not release tannin and iron. The float is then weighed down with seasoned logs or with concrete blocks or are kept submerged (at least 10 cm below the surface of the water) with bamboo-crating. Clods of earth used as a covering material or as weighing agent produces dark (shyamla) fibre of low value.

Gently flowing, fairly deep, clear and soft water is ideal for retting. The optimum temperature is around 34°C ; ditches, tanks and pools are also used for retting. Incomplete submergence produces 'croppy' fibre of extremely low value. Most of the defects in fibre are due to faulty retting. Over-retting results in 'dazed' weak fibre.

Retting is a microbiological process and, therefore, the end-point is determined by inspecting a few plants each day from the tenth day onwards. If fibre slips out easily from the wood on pressure from the thumb and fingers, retting is considered complete.

Extraction

The fibre, when extracted separately from each reed (stem) with fingers ; is sleek, clean and free from entanglement. By the beat-break-jerk method, ten or twelve reeds are taken at a time; their stiffer root-ends are beaten with a mallet to loosen the fibre. The bundle is then broken in the middle and the fibre is loosened. By gripping this loosened fibre in the middle, the broken bundle is jerked in water so that the sticks slip off. The fibre is then washed in clean water, wrung and eventually spread to dry, preferably in shade or mild sun. This second method often leaves the broken sticks and makes fibre somewhat entangled, resulting in 'sticky', fibre.

The extraction of fibre from the green stem with a machine, followed by a short-period retting has proved to be successful. Adaptive research continues at the Jute Agricultural Research Institute.

Yield

The national average is 13 quintals, per hectare. In general, an olitorius variety gives better yield than a capsularis variety, and the former is a more efficient user of nitrogen. Under the National Demonstration Programme, the average yield has been improved to 27 quintals in the case of the olitorius and to 20 quintals in the case of the capsularis varieties, the maximum potential being 40 quintals and 37 quintals, respectively.

A strong positive correlation exists between the plant height (PH) and the yield, between the base diameter of the stem (BD) and the yield, also between PH and BD. Hence, taller and stouter plants give higher yield. However, if the yield from a smaller number of stout plants is equal to that from a greater number of thinner plants, the quality in the case of the latter is always better and, therefore, preferred. The leaves of both the species are used for culinary purposes, olitorius being sweet and capsularis being bitter, but the latter is considered to be good for the stomach.

Varieties

The capsularis varieties cover 60 per cent of the jute area, whereas the olitorius varieties cover 40 per cent of the area, mostly in the southern West Bengal.

Released capsularies varieties

'JRC 321' (Sonali) has a light coppery-red stem. It is suitable for low-lying areas for early sowing from late February to mid-March. It comes to 50 per cent flowering in about 125 days (variable); it is fast-growing and has good-quality fibre. It yields 20-25 g per hectare in northern West Bengal and Bihar.' JRC 212' (Sabuj sona). This variety has a full-green stem. The sowing time is from late March to mid-April. It is suitable for medium and highlands where inundation does not occur. It is particularly good for Assam, Bihar, northern West Bengal, Uttar Pradesh and Orissa. With 60 kg N per ha, it yields around 30 q per ha.

'J.RC 7447' (Shyamali). This variety has a full-green stem, and is an X-ray derivative of 'JRC 212'. It responds better to higher doses of N above 60 kg per ha; it is suitable for Orissa, Nadia, Murshidabad and 24-Parganas.

'D 154'. It is an old standard variety, originally selected in 1918 from 'Kakya Bombai'. It is a hardy variety and can adapt itself to all conditions. It yields 20-25 q per ha.

'JBC 1108'. A pre-release variety. It yields better than 'D 154', equalling 'JRC 212' at places, and is fairly resistant to stem-rot and root-rot.

'JRO 632' (Baisakhi tossa). This variety has a full-green stem; its sowing time is from 15 April onwards on midlands and highlands. It takes 130-140 days to flower; its seed colour is steel grey; its yield potential is 40 q per ha ; its yield averages 32 q per ha with 40 :20 :40 "NPK plus farmyard manure. Its fibre is of good quality.

'JRO 878' (Chaitali tossa). The stem of this variety is red-pigmented. Its sowing time is the second week of March to mid-April on highlands. It is non-lodging and has non-shattering pods. It does not flower prematurely. It is suitable for West Bengal and Orissa. Its Yield Potential is 38 q and the average yield is 32 q per ha with 40 N : 20 P : 40 K plus farmyard manure.

'JRO 7835' (Basudev). The stem of this variety is green. It is sown in early March to early May. It is non-lodging and bears non-shattering pods (It does not flower prematurely. It is suitable for all areas with highlands). It is high-yielding and produces around 34 q per ha with 40 : 20 . 40 NPK plus farmyard manure.

'JBO 620'. The stem of this variety is red-pigmented. It yields fibre of superior quality. It yields 3-4 quintals less than 'JRO 632' and, therefore, has gone out of cultivation.

'C.G.' (Chinsurah green) is a Poor Yielder and has been withdrawn. The pre-release variety 'JRO 524' (Navin) ,yields as high as, or more than 'JRO 7835', with added qualities of fine fibre and a stem that rets quickly.

Seed Selection

Since jute is harvested much before seed formation, special arrangements to raise the seed-jute Crop. The seed crop is sown later than the fibre crop for seed production. The seed crop is sown later than the fibre crop. Some interculture is necessary to keep down weeds. Unlike the fibre crop, the seed crop is spaced wider, so that branching is promoted. The induction of branching by clipping the tips at the right time and with right stage of growth, with assured nutrition for the development of branches, has been successfully practised. This practice, however, would not be resorted in the areas with low rainfall and where land is not very fertile. Natural cross-pollination being low genetic mixture is avoided by alternating capsularis and olitorius seed plots. The crop requires much less nitrogen for proper growth. The average yield of seed is 3.8 to 5.0 q per ha in the case of capsularis varieties and 1.7 to 2.5 q per ha in the case of 'JRO 878' and 'JRO 7835' and around 3 q per ha in the case of 'JRO 632'. After harvesting the crop the seed is dried in the sun for about 4 days and stored in gunny bags lined with plastic cloth and kept in dry godowns, avoiding contact with the floor.

Corynespora casseicola and powdery mildew attack pods and capsules. These pathogens are controlled with copper fungicides. Die-back (Diplodia corchori) on the seed crop is rare, but serious.

Defects in the Fibre

The defects in the fibre are its being barky and croppy (incomplete submerging) or incomplete retting ; specky (apion damage) ; leafy (incomplete retting or washing) ; and sticky (defective extraction). The fibre becomes dirty by using clods of earth and muddy retting water and careless washing. The plants become mossy with adventitious roots emerging from the stems. The fibre becomes shyamla as a result of iron-tanning reaction.

'Roots are, caused by more than one factor but the main one is 'incomplete separation' of fibre at the base of the stem during retting'.

Quality

The quality of the jute fibre among varieties differs on the basis of anatomical conditions and is genetically controlled. Coarser and light-body fibre is obtained from sandy soils whereas clay-loam soils with silt give fibre of superior quality. Climate and the nutrition pattern also affects the fibre. But the most important single factor is 'retting' which, if faulty, mars the positive contributions of the variety, soil, climate etc. Under-retting gives coarse and over retting dazed and weak fibre.

Marketing

Jute Corporation of India, Calcutta, has the objective of stabilising the price line, both in the interest of the farmers and the consuming mills.

Grading

Grading system for white jute (capsularis) has 8 classes, viz., WI to W8, on the basis of length, strength, fineness and lustre, and freedom from enlargement and 'roots' (rejectable basal portion). For Tossa jute (olitorius) there are 8 classes, viz. TDI to TDS.

Jute handicrafts

The traditional excellence of Indian craftsmen & artisans is reflected in a wide range of jute handicrafts of utility, decoration and novelty. A beautiful mosaic of Jute handicrafts that comprises of carpets & floor coverings, wall coverings & window dressings, table mats & table-wares, swings, hammocks, office bags, table covers, table mats, pen and pencil stand, wall hangings & room decor, bedspreads & bedroom slippers, blankets & shawls, skirts & jackets, shopping & travel bags and a host of other home textile made ups.

The jute handicrafts of Madhya Pradesh are a unique blend of aesthetic sense and utility. Craftsmen at Hasta Shilp Vikas Nigam have experimented with the fiber. Every artifact emanates a unique charm exclusive to it and the goods so produced are marketed through the MP state emporium, "Mriganayani". Bhopal, Raipur, Indore and Gwalior are major centers for producing jute handicrafts.

Indian Jute Industry

Today, jute can be defined as an eco-friendly natural fiber with versatile application prospects ranging from low value geo-textiles to high value carpet, apparel, composites, decoratives, upholstery furnishings, fancy non-wovens for new products, decorative color boards etc. Jute with its unique versatility, rightfully deserves to be branded as the "fiber for the future".

In coming decades, a number of jute mills and mini-jute plants will be seen engaged in a big way in production of jute and jute blended yarns, specially of finer accounts, through various routes of yarn manufacturing. The uses of jute are manifold although the traditional use remains in packing as sacking, hessian and carpet backing. These light weight yarns are to be used in value added textile applications like upholstery, furnishing, garments and bags etc are just a few examples.

Jute is being cultivated in India for centuries. the landmark in the history of jute industry in India dates back to 1854 when the first jute mill as set up by George Auckland at Rishra in Hoogly district of West Bengal. Jute is predominantly a crop of eastern India. The major jute growing states are West Bengal, Assam, Bihar, Orissa, Tripura, and Uttar Pradesh.

Currently the production of the fibre in India is around 100 lakh bales and about 73 jute mills are operating in the country at present. Besides, there are several small scale industries in the decentralised sector producing handicrafts, decoratives, twines, pulp & paper from jute and allied fibers and particle board from jute stick.

Industry Milestones

01. The largest producer of raw jute in the world.
02. The biggest manufacturer of jute goods in the world.
03. The capacity to manufacture the widest range of products, tailored to customers' requirements.
04. Comprehensive and world class research and development facilities in the area of jute agriculture, product development and machinery design.
05. Highly quality conscious employing most modern facilities for TQC.

06. Export history of raw jute and jute manufactures dates back to over 100 years.
07. Highly developed infrastructure like ports, shipping facilities etc.
08. A vibrant jute industry oriented to provide maximum customer satisfaction.
09. Indian Jute Industry has the capacity to produce and meet the International demand for food grade Jute bags and cloth.
10. Food grade Jute Bags and cloth manufactured in India - are a boon in the context of global environmental and ecological concern. Indian food grade jute bags have a twin edged advantage: preservation of food in the most natural way.

Jute Growing Areas

Jute is a rainy season crop and hence sown from March to May according to rainfall and type of land, and harvested from June to September depending on whether the sowings are early or late.

Major Growing Areas

Jute thrives best in damp heat, and the climatic conditions prevalent in West Bengal in India are well suited for its cultivation. Indian states ideally suitable for the cultivation of jute are:

- West Bengal
- Bihar
- Assam
- Orissa
- Uttar Pradesh

Mesta, or Kenif, botanically known as Hibiscus Cannabinus, is also grown in these areas as a textile fiber. Mesta is a coarser, more brittle fiber, and is used by the jute mills in admixture with jute to obtain certain desired properties.

Research and Development

Research and development by reputed organisations in India, have led to the strengthening of the natural attributes of jute. With technological support coupled with government's developmental efforts, the golden fiber, today, has entered into a varied range of applications generating employment, contributing to sustainable human development and leading to a cleaner and healthier environment.

Intensive R & D efforts through various research institutes mainly funded by Government of India have resulted in remarkable breakthrough in product diversification. Prominent Products are:

a) Hydro carbon-free Jute Products for packing food grade products like cocoa, coffee, shelled nuts etc.

b) Major R & D works in DREF Spinning System.

c) Production of jute blended yarns with jute as the major component-using Jute/Viscose, Jute/Acrylic, and Jute/Cotton—jute content varying from 60 to 70% by weight. Such yarns suitable for blankets, shawls, pullovers and as also thick dress materials.

d) Production of fine blended yarns using Jute/Cotton, Jute/Silk, Jute/Wool etc. for apparel use Jute Content is 30-40%.

Jute based fabrics used particularly in areas of furnishing, upholstery, carpets, blankets and other home textiles. R & D Institutions have developed Jute Reinforced Composites in the form of flexible, semi rigid sheet and rigid board to substitute wood, timber, and plywood.

UNDP jute program has also facilitated diversification of jute sector by developing new technology, promotion of employment opportunities by encouraging new entrepreneurs to set up production units and development of indigenous machine manufacturing sector. Some of the on-going projects are use of jute for paper production; jute based needle-punched carpets, development of multi-component yarn from wool, jute and other fibers for floor coverings blankets and knitwear.

Diversification

In order to enable jute-based products to make successful inroad into the textiles sector, jute has been largely experimented with. In view of challenge faced from cheaper prices of synthetic substitutes, more thrust has been given on diversification of jute products.

Major thrust areas of value-added diversified jute products include Jute Handlooms and Handicrafts, Non-woven and Industrial Application, Jute Rigid Packaging, Decorative products and Geo-Jute etc.

MESTA (HIBISCUS CANNABINUS & HIBISCUS SABDARIFFA)

Mesta is common word used for both *Hibiscus cannabinus* and *H. sabdariffa* which produces good fibre of commerce. These two species belong to the family Malvaceae with chromosome number 2n=36 and 2n=72 respectively. *H. cannabinus* is f) popular in the Western world as kenaf. *Hibiscus cannabinus* is known by various names in India such as Bimli, Deccan hemp, Gogu, Channa, Ambadi, Gongkura, Sunkura, and Sunbeeja etc while *H. sabdariffa* is known as roselle, java jute, Thai jute, Pusa hemp, Tengrapat Lalambadi, Chukair, Yerragogu, Palechi and Pundibeeja etc. Besides India the mesta is grown mainly in Argentina, China, Cuba, Egypt, Hewti, Guatamala, Italy, Iran Indonesia, Mozambique, North Africa, New Guina, Peru, Spain, South Africa, Southern Part of Zimbabwe, Thailand, U.S.A and Russia.

The two major fibre yielding species of *Hibiscus* i.e. *H. cannabinus* and *H. sabdariff* have the following characteristics.

***H. cannabinus* (Kenaf)**

The species *cannabinus* has been found to grow so far north and south of equator in Afric and undoubtedly varies in the same way whenever the plant is found growing wild, is probable that the mixtures merge. Generally, the species grows best in the warme regions between the latitude of 300N and 300S. It does, however, grow wild or i successfully cultivated at latitudes much farther from the equator for instance on th Southern shores of Caspian sea in Southern Russia, in Manchuria and Korea.

H. cannabinus

Since the *cannabin us* is cultivated at some distance north and south of equator, it is lear that the plant is capable of withstanding a considerable degree of cold. In the quatorial range of Africa the wild plants are found growing best in places where the emperature ranges from 15°C to 30°C. Variations are existing in different species of nesta and even in different varieties of same species. For example in case of *I.kitaibelifolius*, the plants possess a short flowering period which covers all the activities ncluding reproduction. From the appearance of first flower on the plant, the plant ractically ceases to grow. Another variety of *cannabinus* collected from Java was found have a longer flowering period and the plants continued to grow even after the nitiation of flower. Many exotic collections have been found to very much in their lowering and growth behaviour.

The plant is hermaphrodite, annual, producing large cream coloured flowers haracterised by a reddish purple or scarlet throat. The flowers are short lived opening n the early hours of morning before sunrise and closing by noon of the same day. While *nnabinus* is generally considered as seif pollinated, but a small amount of cross ollination may be effected by bees and other insects. The colour of the stem is generally reen. However, some types with reddish stem are also found. The reddish colour enerally appears after the plants attain a certain age in some of the types. This has been ttributed due to sunlight rather than a varietals character. The plants when mature may ttain a height or four meters with a basal diameter of nearly 1.5 to 2.5 cm.

Most variable character in mesta is the leaf. Both compound and simple leaves may e found on the same plant or same type may have all of one leaf shapes. Leaf shapes so vary with the type of variety. Generally variety 'viridis' possesses entire leaves while e variety "vulgaris" has divided leaves. The variety 'simplex' has red stem with ivided leaves while the variety 'purpureus' possesses purple stem and divided leaves.

The shape of the leaves also arise on the same stem. They are of entire shape at the ottom, may be trilobular at slightly higher point and five to seven lobed at the top of e stem. The arrangements of leaves also differ in the varieties. In some varieties the etiole is at right angle to the stem while in other it is having an angle of 60° or 30°.

The varieties of *cannabinus* also differ in their flowering behaviour. This difference epends on the type of variety used for different purposes. The fibre yielding varieties ower with the onset of short days. For normal growth a day length of 13 ½ hours is ood for cultivating variety for fibre purpose. The harvesting of crop depends generally n the formation of buds or opening of flowers on the plant.

Photoperiodism plays an important role in the cultivation of *cannabinus* in the world. e *cannabinus* types do not flower in Cuba until the day length is shortened to about to 121/2 hours. The short days do not begin at the latitude of 230N until the beginning September, no matter when the seeds are sown during the period from April to ugust. The plants will not flower till September-October. For the purpose of fibre production it is considered desirable to obtain plants with longer stems without branche of fruiting stalk, since these interrupt the continuity of the fibres and sowing must begi when the days are longer to coincide with the proper amount of rainfall.

H. cannabinus is capable of adapting itself in variety of soils and climatic conditions The crop can not, however, tolerate frost. The crop can be grown up to the height of 300 ft above sea level at a latitude of 450N to 48 N in Russia and to a latitude of 300S in Sout Africa. For its good growth, a well drained sandy loam soil most suited. Mesta does no grow in waterlogged conditions. It, however, requires a rainfall of 50 to 65 cm durin its growth period.

The seed capsules are cylindrical, pubescent bearing from 18 to 20 seeds per capsul The seeds are grey in colour. The seed is badly affected by the temperature of soil. soil temperature of 15 to 20°C is most suitable for the germination of seeds.

H. sabdariffa (Roselle)

The flowers of *sabdariffa* are comparatively smaller than those of *cannabinus*. Th predominant types have cream to light yellow flowers, having a scarlet to magenta thro and a green or slightly reddish stem depending on the variety. In India the *sabdariffa* called 'roselle' and all the *sabdariffa* types are classified into four main groups accordin to the extent of pigmentation present on the stem. They are full green, green pigmente green light red and red. The leaves in 'roselle' are generally palmate, deeply lobed an alternately borne on the stem. The plants are normally non branching and attain a heig of nearly 3 to 3.5 mts with a basal diameter of 1.0 to 2.0 cm.

Many wild forms of *sabdariffa* &re found in Uganda which closely resemble *H* *machowii* and is considered as the immediate wild progenitor of *sabdariffa*. The actual fib forms have developed from edible types through a long process and may b interogressed by *H. asper* and *H. machowii*. The tall fibre yielding type was isolated an described by P. J. Webster (1914) in Thailand as *H. sabdariffa* var. *altissima* from some see received from Gold Coast (W. Africa). This variety was accidentally introduced into Ind as a single seed in admixture with some of other consignment of seeds from Jav (Indonesia). It was first described by Khan as a new type of roselle hemp. It is clear th the species of *sabdariffa* consists of two distinct types, wild and cultivated. The calyc of the wild types are fleshy and a are used for making jellys and jams and produce fib of inferior quality. The tall types with less branches are cultivated for fibre purpose ar they belong to *H.*sabdariffa var. altissima.

Sabdariffa types are best cultivated in between 100N and 300S, where the temperatu is not less than 100C together a rainfall sufficient to give at least 10 inches during ea month of the growing period. The air too must be still and humid to bring about a rap and healthy growth. Strong winds, prolonged period of cool foggy weather interrupt ar check its growth. The *sabdariffa* types in general are not responsive to photoperiodis although the plants of *sabdariffa* types tend to be short day. These types can be grown in soils having a pH of 4.4 to 7.8. Both *H. cannabinus* and *H. sabdariffa* types require following factors for their proper growth.

1. Enough moisture in soil during the growing period.
2. Rainfall should be at least 100 mm or more per month during the crop cycle with a fairly uniform temperature.

The two species of mesta resemble more or less in most of the characters as described above. However, they differ in some characters which help in the identification of these two species. The following characters may help in this direction:

I. The apex of the epicalyx (bractiole) is entire in cannabinus and inconspicuously channeled in sabdariffa.

II. The sepal nectary (gland) in cannabin us is prominent (swollen) while it is much less conspicuous (shrunken) in sabdariffa.

III. The stem of sabdariffa is flexible while it is more or less rigid in cannabinus.

Mesta in Relation to India

India had to loose about 80% of total jute production area at the time of partition of the country during 1947. The jute crop needs a specific set of climatic conditions, therefore, the cultivation of jute could not be extended beyond the states of West Bengal, Assam, Bihar, Orissa, and parts of U.P. and Tripura. As a result the production of jute fell below the requirement of mills. Mesta can, however, be grown even in those areas where jute is not grown under wider climatic and soil conditions with much less care. This helped the country to expand more area under mesta. At present Mesta is grown in an area of more than 26 lakh hectares with a production of more than 12 lakh bales.

Cultivation of Mesta for Fibre Production

A warm humid climate is considered most suitable for growing both the species of mesta. Both grow well in the drier rain fed areas. The *sabdariffa* being better drought resistant type. In areas where the rainfall is 50 to 90 mm, the *cannabinus* mesta suits better by virtue of shorter duration and comparatively faster growth than the other. None of them can stand water logging and both are grown in kharif season in India. As mentioned earlier they require rich loamy soils although they may grow in a variety of soils including new and old alluvium & lateritic loam. The acid soils are not suitable without proper amendment. The *sabdariffa* types develop chlorosis with high pH of the soil.

In India *sabdariffa* mesta is generally grown in larger parts covering areas from Karnataka to Tripura including Maharashtra, Andhra Pradesh, West Bengal, Bihar, Orissa and Meghalaya. In Tripura and Meghalaya the mesta is grown in highlands either as a pure crop or in mixture with rice. In West Bengal and Bihar it is grown in sandy to sandy loam marginal lands. In Orissa it is grown in the hilly districts of Koraput and Kalahandi. Andhra Pradesh has maximum area under *sabdariffa* mesta in the country. Inspite of the fact that mesta can grow in a wide range of climate and soil conditions, its cultivation has been restricted due to certain inherent constraints. These constraints affect the fibre yields in this crop. Some of the major constraints are:

I. *Rainfed crop:* The crop in general is rain fed. This has adversely affected the cultivation to a great extent. The uncertain rainfall affects its time of sowing and other intercultural operations and also the growth of plants.

II. *Marginal crop:* Mesta is generally grown by marginal and small farmers. They are poor and cannot afford to invest more money in growing this crop. Moreover, it is market oriented crop and as such 95% of the crop is necessarily sold and only 5% is retained by the farmers for their own use.

III. *Non-availability of good seed:* There has been difficulty in producing good quality seeds of improved varieties in the country. The standard and high yielding varieties have not been notified for release. This has adversely affected the breeder & foundation seed production. The present level of the production of quality seeds of mesta is hardly 5% of the total requirement.

IV. *Broadcast crop:* Mesta is generally sown by broadcast method. Line sowing is hardly followed. The farmers also do not give much importance to this crop.

V. *Inadequate retting facility:* Since the mesta crop takes more time to harvest, the water in the retting tanks is dried by that time. The plants have to be kept after harvesting till the next season of rains. This adversely affects the fibre yield as well as fibre quality.

VI. *Marketing:* It is most unfortunate that the country has not been able to develop good market for mesta fibre. The farmers generally do not get remunerative price of their produce. This discourages them to grow mesta next time.

Land Preparation

The land is ploughed and cross ploughed thrice to four times depending upon the type of soil. The ploughing is followed by laddering. This helps in pulverising the soil properly. The ploughed field is made free of weeds and stubbles and leveled properly. The pulverised soil helps in the proper operation of seed drills etc. It is advised to apply organic manure before the land is prepared. Recommended dose of inorganic fertiliser is also applied as a basal dose after the preparation of land.

Sowing

In India, mesta is normally sown by broadcasting method. However, under the extension and transfer of technology programme line sowing has been advocated. Line sowing has been found to have certain advantages such as:

i) Plant growth is uniform since uniform spacing is maintained in the field.

ii) Intercultural operations like weeding, hoeing etc. are easier to attempt.

iii) Application of fertilisers and pesticides etc. also become easier.

iv) Higher yields are obtained in line sowing. Less quantity of seed is required in line sowing.

While sowing it is essential to adjust the distance from seed to seed in the field. The can be managed by adjusting seed drills. For broadcasting sowing, the seed is scattered in the field in crosswise manner and then the laddering is done to cover the seeds. Generally a sowing depth of 2.5 to 3 cm is maintained for good germination. Proper moisture in the soil at the time of sowing is also very much essential. Soil moisture of 20-32 percent is good for the germination of seeds. A spacing of 25 to 30 cm between rows and 7 to 10 cm between plants is recommended for good yields. In case of broadcasting sowing, plant to plant spacing is maintained at 12-15 cm by thinning. It is recommended to treat seeds with any fungicide to prevent the attack of diseases.

Seed quantity and seed rate varies in between two species of mesta in India. A seed rate of 15 to 17kg/ha under broadcast and 13-15 Kg./ha in line sowing in *cannabinus* has been recommended. Similarly seed rate of 13-15kg/ha under broadcast and 11-13kg/ha under line sowing is recommended in case of *sabdarffa.*

Manuring

Balanced application of fertiliser at proper time is very important for a crop. A balanced fertiliser dose also helps in the improvement of physical properties of soils as well. A proper balance between organic and inorganic doses is important for proper plant growth.

Organic fertiliser in the form of compost at the rate of 10 to 12 tones per ha is recommended. Nowadays it is difficult to get so much of compost. It is, therefore, advised to atleast, apply 4 to 5 tones of compost to the crop for better growth. It is advisable to apply organic manure before the land is prepared for sowing. Application of inorganic fertiliser to the crop of mesta is also important.

Nitrogen is essential for cell enlargement and for division of cells at faster rate. Phosphorus helps in proper root development. Potassium is most important in inducing drought tolerance in this crop. This element also induces resistance to pests and diseases. Besides, it helps in the development of fibre cells properly. A fertiliser dose of N, P and K at the rate of 40:20:20 is recommended for higher fibre yields.

The inorganic fertiliser is applied in two split doses of N, P and K. The nitrogen is applied as a top dressing into two split doses one at the time of first weeding and another half after six weeks of sowing. The P and K are applied as basal dressing before sowing. The nutrient uptake by a crop grown in one hactere of land has been studied. It has been found that the crop of *cannabinus* can lift easily about 172 kg of nitrogen, 13 kg of phosphorus, 72 kg of potassium and 150 kg of calcium. This clearly indicates that a crop of *cannabinus* mesta responds quite well towards the application of nitrogen and calcium. Similarly *sabdariffa* lifts 100 kg of nitrogen, 31 kg of phosphorus, 148 kg of potassium and 153 kg of calcium. The *sabdariffa* needs more potassium and phosphorus as compared to *cannabinus*. Requirement of other nutrients is more or less same.

Intercultural Operations

Weeding, thinning and hoeing are three major intercultural operations attempted in mesta crop. Among these weeding takes nearly 25-30% of the total cost of production. If the weeding is not done at proper time, whole crop is adversely affected.

The growth behavior of both the species differs. The growth in case of *sabdariffa* is slow in the initial stages and picks up at later stages, while the growth in *cannabinus* is faster at the initial stages but slows down in later stages. It has been observed that after 60 days of age the growth in *sabdariffa* picks up and surpasses the height of *cannabin us*. It is therefore essential to keep the fields clean up to nearly 40 to 45 days. After that crop grows and covers the weeds. Weeding and thinning is generally done simultaneously. The first weeding is done at the age of three weeks of crop and the second weeding is done after five weeks of age of the crop. It is always advisable to attempt thinning at the time of first weeding. After each weeding hoeing is attempted with the help of wheel hoe to loosen the soil. Weeding with the help of herbicide is also done now-a-days. Although there is no good herbicide recommended for mesta, basalin has given slightly better results. The basalin (Flucholoralin) @ 2 litres per ha as presowing (3 days before sowing) is recommended for mesta weeds.

Irrigation

It is also advisable to irrigate the mesta fields for better yields. However, no irrigation is generally, applied to this crop. The water requirement of mesta is about 50 cm. It is, therefore, desirable to give one or two irrigations at an interval of 15 to 20 'days during sowing time.

Harvesting

Harvesting time is very important in bast fibre crops like jute and mesta. A proper stage of harvesting gives higher yield as well as better quality of fibre. If the plants are harvested prematurely in early stage the quality of fibre is good but the fibre yield is poor. In case the plants are harvested late the fibre yield is better but the quality of fibre goes down. It is therefore necessary to properly understand the course of development of the plant from the seedling stage to the growth stage. Mesta has been found to have two distinct phases of growth like other crops.

These are vegetative phase and reproductive phase. During the vegetative phase the energy of the plant is concentrated in the development of formation and maturation of vegetative parts like stem, leaves and roots. During early phase of growth the plants absorb water at a high rate and often takes in most of the nitrogen, potassium and phosphorus and utilises it for the growth development. The materials are trans located between the leaves, the roots and the stem. This period is also marked with the cambial activity. The cambium deposits xylem inside and phloem outside. This phloem forms fibre of use. It is better to enhance the period during the cambial activity. More is this period, more will be the formation of fibre in the plant and ultimately more will be the fibre yield. The fibre cells formed inside normally do not mature physiologically. They are soft containing nearly 40% of their fibres cells in thin walled state. Since these fibre are soft they are liable to be lost during extraction. This reduces the yield. Therefore, it is necessary to wait for the harvest till these fibres are fully mature.

After the completion of vegetative phase, the carbohydrates start accumulating and the rate of synthesis of protein and the production of new vegetative tissues start declining. The water content in the new vegetative organs also declines. This results in the rapid accumulation of carbohydrates and decline of proteins. The reproductive phase is devoted to the formation of flowers and the development of fruits and seeds on the plant. This phase is marked by three stages namely flower initiation, anthesis and fruiting. In the flower initiation stage, the buds start forming, the growth of internodes is almost checked and the overall activity of the photosynthesis, mineral and water movement is retarted. The cambial activity at the lower part of the stem also stops leaving slight activity of cambium at the upper level. The cambial activity completely stops at the anthesis stage.

The third stage of reproductive phase of the plant goes slowly as the younger flowers continue to bloom even though the older flowers start setting seeds. At this stage the synthesis of proteins and carbohydrates and absorption of water and minerals increase. As the fruits start developing, the nutrients starts accumulating in the fruits and seeds. After this phase, the cambial activity completely stops and clearly becomes differentiated into xylem and phloem elements and since the cambium does not exist at this stage the secondary growth also stops. From the above it is clear that the best time to harvest mesta for fibre is when the plants are left with ten or twelve flowers. The exact time of harvesting will depend on the growth and the accumulation of cellulose and hemicelluloses.

The taller plant will have some of the lower fruits matured without depending much on the reserves in the fibres, while the shorter plant may not have sufficient reserves to mature even a single fruit and should therefore be harvested before the first capsule is mature. Harvesting is done normally by cutting the plants close to the ground with the help of sickle. In some areas the plants of mesta are also uprooted. Such plants take more time to ret and the quality of fibre is adversely affected. The hardy root portion takes more time for retting and the extraction also becomes difficult in such cases. After the harvesting, the plants are sorted out according to the thickness of stems.

This is followed by bundling of plants in convenient sizes of 25 to 30 cm in diameter. These bundles are kept standing- in the field for two or three days for shedding of leaves. The shedding of leaves also simultaneously help in shrinking and rupture of the bark which helps in the entrance of retting micro organisms. In some areas the mesta plants are even kept for 2 to 3 months after harvesting for want of sufficient retting water. These plants are retted when water is available or can be decorticated and the ribbons may be retted in less volume of water.

Varieties of mesta suitable for fibre production

Mesta has many varieties cultivated and suitable for different countries. Most of the countries of the World have bred their own varieties for their local needs. However, some of the important varieties of mesta are given below –

(A) H. cannabinus (kenaf)

Good high yielding varieties of *cannabinus* have been developed in the country. Some of them are

a) *HC-583* : This is a selection from a material collected from Nigeria. It was developed at CRIJAF, Barrackpore. The plants are green with irregular light flush of red pigment. The leaves are entire and cordate. The plants flower in 150-180 days after sowing. The flowers have yellow petal with red stigma. The seed colour is grayish black. The variety is suitable for sowing in the month of April and is suitable for growing in the states of West Bengal, Assam and Orissa. The variety can give fibre yield of 25-30 q/ha.

b) *AMC-108* : This variety is the result of selection of indigenous material and was developed at Agricultural Research Station, Amadalavalasa (A. P.). The plants are light red in pigmentation which even extends to the petiole. The leaves are deeply lobed (5-7 lobes) and slightly red pigmented. It takes 150 days to mature for harvest. The seeds are grayish black in colour. The variety is suitable for April-May sowing in the Southern region of the country, (Bihar and Orissa). The fibre yield is upto 25-30 q/ha.

c) *HC-269* : The variety has been developed at CRIJAF Barrackpore. It is a selection from a local material. The plants are green with irregular flush of red pigment. The leaves are entire and green. The variety matures within 145-160 days for fibre purpose. The seeds are grey and sub remiform in shape. The variety can be sown in April and can yield upto 20-25 q/ha of fibre.

(B) Hibiscus sabdariffa (Roselle)

a) *HS-4288:* This variety has been developed at CRIJAF Barrackpore and is a selection from a cross between variety RT-1 (nonbristled) X RT-2 (bristled). The plants are green with red pigmentation at nodes. The leaves are deeply lobed with at least 5 lobes in each leaf. The upper surface of petiole is red and the lower surface of petiole is green. The variety takes 180 days to flower. The seeds are of coffee colour and reniform. The variety can be sown from mid April to May. It yields 18-20 q/ha of dry fibre. The variety is suitable to grow in West Bengal, Assam, Tripura and Bihar.

b) *HS-7910:* The variety was developed at CRIJAF, Barrackpore. It is a selection from a cross between HS-4288 X RT-1. The plants are green with irregular flush of red pigment which further extends with the maturity of crop. The variety is devoid of bristles on the stem. The leaves are deeply lobed, slightly more reddish than HS-4288. It takes around 180 days to flower. The seeds are reniform and of coffee colour. It can be sown during the month of May and yields upto 18-20 q/ha of dry fibre. Most suitable for the states of West Bengal, Assam, Orissa and Bihar.

c) *AMV-1:* The variety has been developed at Agricultural Research Station, Amadalavalasa (Andhra Pradesh). It is the result of selection made from an indigenous material. The plants are of crimson purple colour with completely green base upto the height of 35-40 cm. The stem is bristled. The leaves are deeply lobed. The petiole is purple in colour. It takes 180-200 days to flower depending much upon the time of sowing. The seeds are of brown colour. The variety is sown normally in between mid April and early May. The fibres yield of 15-17 q/ha is obtained. The variety is suitable for Andhra Pradesh and Orissa.

d) *AMV-2:* This variety has also been developed at Agricultural Research Station, Amadalavalasa (A. P.) It is the result of the selection made from an indigenously collected material. The stem is pink coloured having green base and is fully bristled. The leaves are deeply lobed. The plants take 180-200 days to flower. The seeds are brown and of smaller size. The variety can be sown in early May and is suitable for Andhra Pradesh and Orissa. Fibre yield upto 17 q/ha.

e) *AMV-3:* This is a selection from a cross between AMV-1 X ER-79. The stems are pink with bristles. The leaves are deeply lobed. The plants take 180-200 days to flower. The seeds are brown and smaller in size. Suitable for early May sowings in the states of Andhra Pradesh and Orissa. Fibre yield is upto 18 q/ha.

f) *AMV-4:* This variety is also the result of selection from the cross between AMV-1 and AS-163 made at the Agricultural Research Station, Amadalavalasa (A.P.). The stem is pink in colour with bristles. The leaves are deeply lobed. The variety flowers in between 180-200 days after sowing. The seeds are slightly of bigger size than AMV-1 and AMV-2. The variety is suitable for May sowing in the state of Andhra Pradesh. Fibre yield is upto 20 q/ha.

Cultivation of Mesta for Seed Purpose

Mesta is sown late in the month of July or even in August for seed purpose. The requirement of seeds for raising a good crop of mesta in one hectare of land is nearly 10-12 kg in case of *cannabinus* and 7.5-9.0 kg in case of *sabdariffa.* The land is prepared as usual and the sowing is attempted in rows in a spacing of 40x10 cm. Normally the seed crop of mesta is not favoured with any fertiliser. The crop is raised and harvested when most of the capsules are mature. The seed crop should generally be sown about two months before the day length shortens to about 12 or 121/2 hours as large number of flowers are required for seed production. The seeds start to ripen from the lowest capsule in succession upwards till the time they are fully matured. The capsule shatters when it is fully mature and the harvesting is done before this. The mature plants are left in the field until all the capsules have ripened. The seeds can be thrashed out by beating the capsules on a tarpaulin. The threshing of seeds in mesta is slightly difficult because the bristles on capsules create problems for the workers.

Seed yield in both *cannabinus* and *sabdariffa* are highly variable. Generally in *cannabinus* the light insensitive and early maturing varieties give higher seed yields (upto 1200 kg/ha) while in *sabdariffa* types the seed yield is low. Since the seed size in *sabdariffa* is smaller the proportionate yield are more or less same. Some of the Indonesian types give higher seed yield of even 1300 kg/ha. However, in normal cases the seed yield is 700800 kg/ha. According to the basic fundamental rule, a seed crop of mesta covering an area of one hectare will give sufficient seeds to cover 20 ha of land.

It has been estimated that one kg seeds of *cannabinus* varieties if counted comes to nearly 35, 000 to 40,000 in number. A plant of *cannabin us* variety contains 40 to 60 capsules and each capsule have 15 to 25 seeds in it. The varieties of *sabdariffa* have 50 to 100 capsules per plant and each capsules possesses 15 to 20 seeds in it.

The seeds of mesta should be stored in perfect dehumidified conditions. Seeds with 8 percent moisture content can retain viability for more than five to six years if stored upto 10°C temperature. The mesta seeds are generally affected by weevil's attack. It is, therefore, recommended that the seeds should be treated with Malathion or fumigated by Carbon disulphide before they are stored. The seeds of *Hibiscus cannabiuus* also produce good edible oil. Nearly 21% oil can be extracted. The oil has been found to be good for culinary purposes. It has already become popular in many countries of Africa and Asia including India. After the seeds are harvested, they are dried properly under the sun for a day or two depending on the sunlight. The dried seeds are then sent to the crushing mills. The oil so extracted is refined properly and tinned for general market sale.

Mesta oil is already in use by the soap industries in almost all parts of the world. After the extraction of oil, sufficient quantity of oil cake is left behind. This oil cake is used as a cattle feed. The oil cake contains following ingredients -

(i) Moisture - 9.26%

(ii) Proteins - 29.18%

(iii) Oil - 9.76%

(iv) Crude materials - 15.68%

The oil of mesta is also rich in certain physical and chemical properties. The analysis of oil is given below -

(A) Physical properties:

(i) Acid value = 2.10% by weight

(ii) Saponification value = 187-190% by weight

(iii) Iodine value = 90-94 by weight

(iv) Refractive index at 40°C = 1.465 q

(B) Chemical properties:

(i) Palmitic acid = 15.8%

(ii) Stearic acid = 6.8%

(iii) Oleic acid = 51.0%

(iv) Linoleic acid = 26.4%

Higher value of iodine value clearly indicates that the mesta oil can easily be mixed with any other edible oil. Higher value of oleic and linoleic acid also provide advantage of this oil over other edible oils.

SISAL

Sisal (agave sisalana perrine), introduced from Mexico, is a semi-perennial plant and yields cream-white fibre from its leaves. The fibre is used for making ropes, matresses, brushes, doormats and fancy articles of various kinds. Moribund plantations of sisal are found in southern Bihar, Orissa, Andhra Pradesh, Karnataka, Maharashtra, Madhya Pradesh and West Bengal, raised in the early part of this century.

The technology for raising and maintaining a sisal plantation in India has been developed (sisal in India-JARI). As a result old plantations are being renovated and new ones are being planted on scientific lines to cater for commercial needs. The planting material can be obtained from the Sisal Research Station, Bamra, and from other plantations, particularly in north-western Orrisa. The plant popularly known as century plant, has a short thick stem (bole) which is almost concealed in early stages by the surrounding sessile leaves which are thick, long and broad. From slender rhizomes, buds grow and sprout above-ground to form baby-plants or suckers. After 8-10 years the plants flowers. The stout floral axis develop slowly and produces flowers which drop off and bulbils are small baby-plants develop on the seat of flowers. Both bulbils and suckers furnish the propogation material for plantations.

agave sisalana perrine

Bulbils and suckers are nurtured with copious nutrients in nutrients for a year in close spacings and later transplanted when 10 to 18 months old. Depending on growth the harvesting of the leaves starts from 2 to 3 years old plants and such annual harvests last for 7 to 8 years comprising a full cycle. The fibre is extracted with power-driven decorticators. Raspador decorticators are being manufactured in India. After a cycle is completed, the plantation is cleared, heavily manured and is preferably put for a year under a green manure crop and then replanted.

The yield of two tonnes per hectare per year may be considered good in the dry, infertile areas of north-western Orissa or Chhota Nagpur. With good management, the yield can be much improved, more so in better types of soil, with a rainfall between 650 and 1300 mm. Good drainage is essential. Gravely or highly acidic soils, poor in nutrients, are unfit for fibre production.

When exposd to rainfall above 1500 mm, sisal leaves become susceptible to a number of fungal pathogens, including colletotrichun, diplodia and alternaria spp. in the wilting of plants has been observed, but not frequently.

Sisal suffers most from imbalance in nutrition. It requires large quantities of calcium, magnesium and potassium, besides nitrogen and phosphorus. In acid soils, with pH below 6, this imbalance begins to be reflected in various kinds of symptons and the remedy lies in the application of soil correctives, organic manures and fertilisers. Phytophthora and pythium spp. have been reported on exotic varieties introduced recently. This is a serious disease and planters must use only indigenous resistant types. With assured internal consumption, the present target is 8000 hectares. There is scope for many more plantations.

The other two species found to have been used in India for plantations are Agave cantala Roxb. and Agave veracruz Mill. The fibre of these two species is often marketed as sisal fibre.

SUNNHEMP

Sunnhemp *(Crotalaria aria juncea,* L), a plant of sub-order *Paplionaceae* of order *Leguminoseae* is an annual shrub cultivated as multipurpose legume especially for its fine fibre in many countries including India. The crop is grown also for legume or as a fodder. The genus name *"Crotalaria"* means 'rattle' and refers to the noise made by the seeds shaken in the mature pods. Species of this genus are wide spread throughout tropical, sub-tropical and to a lesser extent temperate countries. The sunnhemp crop is native to India. The crop is known in India by various names like 'Sonai' or 'San' (Hindi), 'Sanpat' (Bengali), 'Tag' (Marathi), 'Vakku' (Malayalam) 'Sanpat' (Oriya), 'Janumu' (Telegu), 'Saab' (Kannada).

Early History

It is one of the earliest of the distinctly named fibre of India, being mentioned in the early Sanskrit under the name of Sana. It is enjoined in the laws of Manu that the sacred threads used by 'Khatriyas' or Rajput (warrior class) should be made of 'Sana'. The 'Ain-IAkbari' has clearly described two types of fibre plants one with flowers like cotton and the other with yellow flowers. The former can no doubt be *Hibiscus* and the later can only be 'Crotalaria'.

The fibre from the Indian Sunnhemp was first introduced into Europe in 1791 and 1792 by the East India Company. During those early years, jute was considered as fibre plant of less importance. It was Roxburgh who proved that the quality of fibre of sunnhemp is excellent and not at all inferior to European Hemp. Later on Watt (1908) confirmed that sunnhemp was a far superior fibre than jute.

Sunnhemp as a fibre and green manure crop is cultivated in all states of India. It also serves as a good fodder crop in many parts of the country and is the largest producer of sunnhemp fibre followed by Bangladesh and Brazil. Of late, China is the highest yielder of sunnhemp fibre in the world.

Area and Production

As stated earlier, sunnhemp is grown in almost all states of India either as a fibre crop, green manure or fodder crop. The states of Bihar, Madhya Pradesh, Maharastra, Rajasthan, Orissa and Uttar Pradesh grow this crop mainly for fibre. These states cover nearly 87 per cent of the total area under cultivation of sunnhemp crop. Among these states, Orissa alone produces 26 per cent of the total sunnhemp produced in the country. Average productivity in the country is around 600 kg ha-1. The most concentrated sunnhemp growing districts of U.P. are Varanasi and Jaunpur, Aurangabad and Jalana are the two main districts of Maharastra, which cultivate this crop in large scale. Besides, Ganjam District of Orissa also grows this crop. Above five districts accounts for 20 percent of the country's area and 23 per cent of production. The fibre yield in the Balasore and Sundargarh districts of Orissa and Chittaur districts of Rajasthan is the highest in the country followed by Ganjam and Kalahandi districts of Orissa.

Besides India, the crop is cultivated in many other countries like China, Korea (DPR), Pakistan, Bangladesh, Romania, Russia and CIS countries. These countries cover nearly 320 thousand hectare producing approximately 200 thousand metric tons of fibre. The six countries mentioned above cover nearly 92 percent of World's coverage with 83 percent of total production. India contributes 23 percent of production with 27 percent of world's area under cultivation.

Germplasm

Out of several well recognised types 'Madaripur' and 'Seraganj' are best known in Pakistan. The first one is considered best having creamy white fibres of good strength, relatively free from dirt. 'Somerset' is a good cultivar in South Africa. Similarly, KRC-1 in Brazil and 'Tropical sunn' in U. S. A. are good cultivars. In India, 'T-6' is a day neutral. The variety 'K-12 yellow' which is the result of mass selection is a proven cultivar of 'Kharif-sunn' superior in yield to all other cultivars. It is a longer growing cultivar, resistant to stem break disease exhibits tolerance to disease, draught, and insect and can grow well on lateritic poor soil in slopes and with weeds.

A new selection of Sunnhemp (SS 11) has been found to have clear edge over K-12 yellow. The varieties K-12 (black), SJ-67-42 and Chhindwara in terms of fibre yield at 90 days cropage have proved the best for the last five years at Sunnhemp Research Station, Pratapgarh (U.P). A total of 122 sunnhemp germplasms are being maintained at

Sunnhemp Research Station, Pratapgarh. Germplasm from Russia is found day neutral and early flowering type.

Botanical Description

The sunnhemp is a herbaceous shrub, annual, 1-3 m. tall, vegetative parts covered with short downy hairs, tap root long, strong with many well developed lateral roots and numerously branched, lobbed nodules. The stem is approximately of 2 cm. in diameter, leaves are simple with minute pointed stipules, petiole short, about 5 cm long with pulverous, blades linear elliptical to oblong, entire 4-12 cm long, 0.5-3 cm broad and bright green in colour. Inflorescence is a terminal open raceme, 25 cm long with very small linear bracts.

Sunnhemp

The flowers, are showy small with five hairy sepals, shortly united at base, the lobes are pointed with 3 lower sepals united at base. The petals are deep yellow, the standard erect is about 2.5 cm in diameter, rounded, sometimes streaked purple on dorsal surface, the wings are shorter and keel is twisted, stamens are 10, almost free at base, 5 with short filaments and long narrow anthers and 5 with long filaments and small rounded anthers; fruit-an inflated pod about 3cm long, 1cm wide groved along the upper surface, with a short pointed beak, light brown when ripe, several seeded softy hairs; seeds numerous, small flattened, dark grey to black loosened in the pod at maturity. Seed weight is variable, ranging from 18000-30000 per kg depending upon variety. The chromosome number of this cross - pollinated crop is sixteen. The treatment with indole butyric acid partially overcome self-incompatibility and tetraploids have been induced by treating young root tips with Colchicine.

Ecology

Sunnhemp is the fastest growing species of the genus and is very effective in smothering out weeds. Almost any well drained soil is suitable for 'kharif' crop. Sunnhemp grown during rainy season is mainly utilised as a green manure crop. For fibre, it is grown on fairly light well drained soil (alluvium soil-old or new) having sandy loam or loamy textured soil that retains sufficient moisture. Soil moisture per cent of 30 helps in good and effective germination.

Sunnhemp is a short day crop, but vegetative growth is favoured by long days. It can tolerate an annual precipitation of 49-429 mm, annual mean temperature of 15-37.5° C and pH of 5-8.4. However, the near neutral pH range (6-7) where phosphate availability high is found good for the crop.

The sunnhemp growing areas of India range from latitude 17 - 30° N with relatively high humidity. in the Northern Provinces of India, sunnhemp is grown as 'kharif' crop whereas in South India it is raised as 'rabi' crop where winter is not pronounced.

The sunnhemp survives in both tropical and sub-tropical climate. In Brazil, the crop is grown in Amazon belt under almost equatorial region to latitude of 22.5° in the States of Sau Paulo under humidity and temperature having moderate to heavy rainfall.

Photoperiodism

The date of sowing has direct bearing on the vegetative growth and development of sunnhemp. The crop sown immediately after the break of the monsoon gives significantly higher yields than those obtained from crops sown later. Sengupta and Talukdar found the planting if made during the month of April to September, flower between September 1 to October 14. The best growth is obtained from plants sown in April and May. As the vegetative growth period become progressively shorter with later plantings, the plant height becomes progressively shorter. The number of internodes are fewer and the stem diameter also becomes smaller. The response to later plantings is typical of other fibre bearing species such as jute, mesta, roselle and urena.

Pollination

The lowest flowers of the inflorescence open first and remain open for 2 days. The flowers open between 2 and 3 am. The dehiscence of anthers is complete in between 12 noon to 1 pm in about one hour and a half. The stigma remains receptive on the day of the opening of flower. In the bud the stamens with the linear anthers are longest and dehisce with bud. The filaments of stamens with ovate anthers then elongate, pushing the liberated pollen to the orifice of the keel. The stigma lies at the orifice above the pollen mass. The wings and the keel are articulated by a ball and socket joint. Large bees such as *Xylocopa* spp. and *Vegactile* spp. alight on the wings and set ball and piston mechanism into operation. The ovate anthers dehisce at the top of the keel at the end of the first day or following morning and the pollen from them is liberated with the help of the bees. Extensive crosspollination occurs and self-pollination only takes place after the stigmatic surface is stimulated by the bee or mechanically. According to Kundu, seed setting is practically nil when the flowers are kept inside muslin bags. Slight scratching of stigma with the help of a camel hair brush before pollination also helps in getting good seed set in this crop. A normal pod contains nearly 10 to 15 seeds. Thus sunnhemp may be considered as a highly cross-pollinated crop, and the flower must be tripped as in alfalfa to be fertilised.

Varieties Cultivated

A good number of varieties of Sunnhemp have been developed and released in the country. However, the variety which has become most popular is K-12 (yellow). This is a selection made from an old variety K-12 (black) developed by Prof. T.S. Sabnis of Government Agricultural College, Kanpur (U.P). The variety K-12 (yellow) was developed by the Central Research Institute for Jute and Allied Fibres, Barrackpore. The variety was officially released in 1971. It has the potentiality of yielding 15 q of fibre per hectare. No variety after that has been able to yield better than this variety till now at national level. Besides this variety, some other varieties have been developed for specific areas with very limited adaptability. Some of the important varieties are described below -

i) *K-12 (black):* This variety was developed by Prof. Sabnis of Kanpur during 1923-26 following mass selection from a local type. The seeds of this variety are black in colour, yield upto 10-12 q fibre per hectare in 100-120 days maturity. It possesses fibre of good quality. It also shows some resistance to wilt disease and shoot borer Recommended for U.P.

ii) *K-12 (yellow):* The variety is a selection from the old variety K-12 (black). The variety has been developed by the Central Research Institute for Jute and Allied Fibres, Barrackpore. The selection was made at Sunnhemp Research Station, Pratapgarh. (U.P.) and was released in 1971 for general cultivation. The fibre yield potentiality of the variety is 15 q ha^{-1}. It has shown some resistance to wilt disease. The variety has wider adaptability.

iii) *M-18:* The variety has been developed by the Madhya Pradesh Department of Agriculture. It is a early maturing type with high yield potential. It is resistant to draught, suitable for light soils with low rainfall areas of M.P.

iv) *M-35:* This is more or less similar to M-18. It shows some resistance to shoot borer.

v) *BE - 1:* The variety was evolved at Sabour (Bihar) through selection from the local material. This is a popular variety of Bihar and Orissa. Both yield and fibre qualities are good.

vi) *Belgoan:* The variety was developed by Madhya Pradesh Department of Agriculture. This is suitable for heavy soil and is grown in M.P.

vii) *Chindwara:* Developed from material collected from Chhindwara (M.P). Suitable for M.P. and other states.

viii) *Jabalpur:* This is also a selection from a type collected from Jabalpur region of M.P. Cultivated in M.P. and U.P.

ix) *Bellary:* Developed through selection from a local type. It is a high yielder and most popular in Tamil Nadu.

x) *D - IX:* The variety has been selected from a local type collected from eastern part of Maharashtra state. It shows resistance to wilt disease and is popular in Maharastra.

xi) *T - 6:* The variety has been developed by the Central Research Institute for Jute and Allied Fibres through selection from a local type of Midnapore district of West Bengal. Most popular in W.B. as green manure crop.

xii) *ST - 95 :* This is a selection from a type introduced from Taiwan. It has also showed slight resistance to wilt and shoot-borer.

xiii) *ST-55:* This variety is a selection from a local indigenous type. Gives high yield of 14 -15 q ha -1. It has shown moderate resistance to wilt diseases. A new cultivar of sunnhemp (SS II) has been developed at Sunnhemp Research Station, Pratapgarh which is a better yielder as compared to K-12 (black and yellow), SJ- 67-42 and Chhindwara at 90 days cropage. This is in pipeline and awaiting release soon.

Cultivation

The crop is cultivated in a wide range of climate in all parts of the country. It grows almost in every type of soil but cannot stand water logging. Growth period is favoured by the atmospheric temperature of 23-30°C and rainfall of 400mm. Land is required to be ploughed properly. Good ploughing and laddering is done to make a fine tilth which is most suitable for the germination of seeds. The moisture percentage of soil should not be below 30 per cent as the big size seeds require high moisture for good germination.

a) *Sowing time:* It is cultivated mostly as a 'kharif' crop and is sown with the commencement of South-West monsoon in the month of May and June and harvested in the month of October. Kundu writes about the above sowing time. However, in Ratnagiri and Panchamahal districts of Maharashtra, 24-Parganas and Maida of West Bengal and Purnea region of Bihar, sunnhemp is grown as 'rabi' crop where it is sown in October-November and harvested in February-March. For green manure purpose, the crop can be grown in any month where it is ploughed down normally within 30-45 days of crop age.

Sunnhemp is sown as broadcast crop in 90% cases. Sowing in lines or rows are advocated. The sowing is done in line by dropping seeds behind the plough keeping rows at 2530 cm apart. Normally seeds are put nearly 2-3 cm deep in the soil. The plant to plant distance is kept at 5-7 cm preferably through thinning. The germination is epigeal and seed germinates rapidly in about 3 days time.

The growth of the plant is very much influenced by the date of the sowing and the nature of rainy season in the early part of crop growth. Generally, the crop sown immediately after monsoon gives always higher yield than the crop sown at a later date.

b) *Seed rate:* Seed-rate is dependant on the germinability and soil moisture content. In row-cropping, 25 kg seeds per hactare is considered sufficient if germinability is 80% or above. For broadcasting method, about 35 kg seeds per hactare is required. The seeds are casted once from north to south and next from west to east to effect uniform distribution. After casting the seeds the soil is raked and laddered to bring the seed 2-3cm below soil surface.

c) *Weeds:* In broadcast, heavy seed rate ensures upright erect stems which helps to smoother the weeds. It is valuable for controlling Kans grass *(Saccharum spontaneum,* L). The crop, however, cannot control nut-sedge. Generally, weeding is not done. Care should be taken in weeding out *Ipomoea* spp. whose seeds germinate, grow, flowers and fruits at the same time as that of sunnhemp.

Rotations

Sunnhemp as a rotational crop has a major advantage and being a legume it has potential to meet its requirement of nitrogen by fixation of atmospheric nitrogen by bacteriods located in the nodules in the root and good for soil renovation. The crop is useful in rotation with potato, tobacco and other crops that are subjected to nematode infection. The crop is also rotated with sugarcane, rice, wheat and jute. It is also favoured as a cover crop in citrus, tea, cocoanut and rubber estates. It has proved as a special rotational crop with cotton in U.S.A. and with Kenaf in Cuba. In Brazil, sunnhemp is rotated with wheat, sorghum, sugarcane and beans. In dry production, yield increased on an average of 41 % over control in eight locations when bean crop was followed by sunnhemp..

In India, cropping schedules found suitable are: Sunnhemp-Wheat; Sunnhemp-Potato; Sunnhemp-Mustard; Sunnhemp-Paddy-Wheat; Sunnhemp:-Mustard-Wheat. Among the various rotation tried, the Sunnhemp -.Wheat rotation was found to be best where the yield of wheat was found to increase by 10 -15 percent.

Manures and Fertilisers

The mineral nutrition of legume is directly related to the mineral nutrition of Rhizobium.

The mineral nutrient status can affect Rhizobium at a number of different stages of nodulation.

pH : Legumes usually need neutral to slightly alkaline soil for nodulation. Root hair infection and deformation and consequently the formulation of nodules are inhibited by soil acidity of around pH 4.5.

Organics: Greater population of *Rhizobium leguminosarum* has been found in soils treated with farm yard manure than in control unamended soil. The humic and fulvic acid content of soil organic matter may have an ameliorating effect as both appreciably improve rhizobial growth. It is advisable to apply organic manure @ 3 to 5 tonnes per hactare at the time of land preparation. The organic matter serves carbon and energy source to rhizobia.

Inorganics: Like all leguminous crop, sunnhemp is also not supplied with any nitrogenous fertiliser. The role of legume in enriching the fertility of soil by controlling nitrogen through symbiotic nitrogen fixation by Rhizobium is well established. However, for the initial boosting up of the crop sunnhemp is also sometimes supplied with 5 to 10 kg nitrogen per hactare. At CRIJAF, it did not produce any significant change in the yield of fibre.

Phosphorus is considered to be the most important nutrient element for leguminous crop. Application of phosphatic fertiliser resulted in a large population of rhizobia occurring in cultivated soil than in virgin soil. Phosphate increases the amounts of nitrogen fixed by increasing overall plant growth. Phosphorus @20kg P 2°5 ha^{-1} is recommended for sunnhemp.

Potassium fertiliser was also applied @40 kg K20 per hactare as a safe dose. The response from the combination of P and K (P 2oK4o) produced the highest response and yield (10.6 q ha^{-1}) of sunnhemp fibre at CRIJAF. A high rate of potassium uptake was observed all throughout the growth period. Potassium levels of soil have been shown to have a reflection on the occurrence of rhizobium. *Calcium* has an important effect on the rhizobial population in the soil. It is essential for pod filling of leguminous plants. It is rendered in the presence of calcium and magnesium as carbonate and hydroxides. Calcium neutralises the acidity which chiefly increases the rhizobial growth and survival. The requirement for Mg is more than Ca. The total need of the rhizobia for Ca and Mg ions (0.5 mM) has been found to be approximately 20 times the specific requirement in respect of calcium ion and 5 times for magnesium. Ca concentration below 0.2 mM inhibits nodulation. Mg salts have been found to stimulate nodulation in legumes.

Micronutrients are essential for the growth and nutrition of leguminous crops. The low concentration of zinc and manganese are required for the rhizobial species. Molybdenum has been reported to have a salutary effect on the growth and nodulation of leguminous plants. A small amount of molybdeunm (50 g ha^{-1}) added to the soil or as foliar spray, markedly improve growth and nitrogen fixation.

Molybdenum is required in small quantities for N_2 fixation process in the symbiotic system. It is required mainly for nitrogenous synthesis which help reduction of N_2 to NH_3' In acid soil, molybdenum availability is affected although present in ample quantity. Liming of soil to increase the pH can overcome the molybdenum deficiency. The synthesis of vitamin B has been recorded by the rhizobia and is the reason for their cobalt requirement. Cobalt does not exert effects on the rhizobial multiplication in the soil or in the rhizosphere. The primary effect of deficiency is the lowering of number of nodule initials and such nodules as develop have very much lowered concentration of bacteriods. The load bacteriods number appears to be associated with defective synthesis of vitamin B_{12} and consequent effect on bacterial multiplication. Bacteriods are the seats of N_2 fixation. At CRIJAF, cobalt (50 g ha^{-1}) individually and in combination with phosphorus (@ 20 kg P_2OS ha^{-1}) and molybdenum (@ 50 g ha^{-1}) brought about significant performance in fibre yield. Like molybdenum, iron is another essential element required for the healthy growth of symbiotic nitrogen fixing system whether they are utilising elemental or combined nitrogen. Iron is necessary for the nitrogenous activity and leghaemoglobin production. Boron is also essential for nodulation which starts almost at two leaf stage of the seedlings.

Harvest

Harvesting of the crop at proper time is very important for ensuring not only better yield but also better quality of fibre. The proper time of harvest, however, differs from place to place. The crop is still harvested at pod formation stage in central parts of the country like U.P. and M.P. In the southern parts of the country, the crop is harvested at the pre-flower or full blossom stage. In Bihar, harvesting at green pod stage is preferred. The case is entirely different in Maharastra and West Bengal where the harvesting is done very late at fully matured pod-stage. Kundu states that there is little difference in strength and fibre quality at flowering or when the seed pods are fully mature. Since some income can be obtained from the sale of seeds, he recommended harvesting at matured pod stage. Various experiments conducted have shown that it is better to harvest the crop at 50% flowering or early pod-stage to compromise between yield and quality of fibre. In any case, the fibre crops get ready for harvest in 90-100 days or 120-150 days (50% flowering onwards). Flowering normally starts in September. Harvesting is normally done manually with the help of sickles at the ground level. In U.P. and Bihar the tops of the plants (approximately 30 cm) are cut at harvesting for use as a fodder. The thick and thin plants of equal sizes are sorted out and tied in the bundles of 20-25 cm diameter. The bundles are kept in the field for 2-3 days for shedding of leaves. In some areas, the leaves are removed by hand and plants are sent for retting immediately. Areas where sufficient water for retting is not available, the plants are kept in bundles till next monsoon when water is available for retting. In the Ratnagiri district of Maharashtra, the plants are simply spread on the ground for 10-15 days. The retting takes place due to sufficient dew which is quite heavy during nights.

Reting

Retting is the process in which the fibre in the bark is separated from the woody portion by the action of microorganisms present in the water. The process involves steeping and keeping the stems submerged in water for certain periods. The number of days required for retting depends on water temperature, locality, time of year, weather conditions, depth and source of water, thickness of stalks and quantity of straw in relation to volume of water. Cement tanks are good for retting. But earth pits, drains, streams~ and back water pools of river are also used. If more than one ret is to be carried out in a pool, sufficient flow of water must be ensured to avoid fouling the water which discolours the fibre.

Shaliow water of about 1 m to 1.5m deep can serve the purpose. In the process of steeping, the bundles are arranged side by side preferably in single layer in the water. The bundles are weighted down with the help of some cement blocks to a depth of 10 to 15cm in the water. The weighted materials which release tannins should not be used. The retting period in India during September is 3 to 5 days. In cold water, retting takes about 7-9 days. If on testing of the retted stalks fibre is found loosening, it is considered ready for extraction. The softening of fibre takes place due to action of the enzyme released by the bacteria acting on the stem. The fibre is removed from the stem by hand.

Precautions must be taken against under-retting or over-retting of fibre. In underrated condition, removal of fibres will be a difficult task as bark will adhere to the extracted fibre. While in over-retted condition, micro-organisms will break down the tissues surrounding the fibre and this will result in weakening of the fibre. The methods of fibre extraction differ from place to place. The bark is partially removed by pulling a handful of the retted stalks between the thumb and the first finger. The stems are then stripped one by one.

The person after breaking the lower end of the stalk gets free end of the fibre. He grasps this end of the stalk with his left hand and removes the fibre in strips by running up the thumb and first finger of the right hand between it and the stalk. However, in some areas of U.P. and Bihar, the retted stalks are dried after they are removed from water. The fibre is then extracted from the dried stalks by breaking the stalks at different places and pulling it away from the wood portion. The fiber extracted in this way from dried stalks is of better quality than the fibre extracted from the wet stalks. One person can strip only the equivalent of 2.7-3.6 kg of dry fibre per day. After extraction, the fibre is washed in water and dried in sun. The adequate washing leaves lot of gums which binds the finer component of fibre into coarse strands. The bundles of fibre are made by twisting the fibres in convenient sizes before sending it to market. In most areas of Andhra Pradesh, the fibre strands are not twisted but tied at the lower ends. In sunnhemp (K-12 yellow) with good retting, high grade white fibre yield of nearly 10^{-12}q ha^{-1} can be obtained in farmers' fields.

Mechanical Properties of Sunnhemp Fibre

Strength is one of the most important properties which largely determine the quality of yarn. This is usually expressed as tenacity, a ratio of breaking load of fibre (in g), to the fineness (in tex). Samples are taken either in the form of bundles of fibres to get bundle strength or in the form of single filament. Extensibility, fatigue property, torsional rigidity, fineness and density are other macroscopic properties of fibre.

Physical properties of sunnhemp fibres.

1. Ultimate Cells
   a) Length (mm) 5-20
   b) breadth (10-3mm) 12-35
   c) Length/Breadth 450
2. Filaments
   a) Gravimetric fineness (tex) b) Tenacity (g / tex) 5.5-17.0
   b) Tenacity (g/tax) 30-40
   c) Extension at break (%) 2.5-3.5
   d) Modules of torsional rigidity (X 1010 dynes cm-2) 1-2
   e) Flexural rigidity (dynes cm-2) 125-175
   f) Transverse swelling in water (%) 18-20
3. Bundles
   a) Tenacity (g/tex) 15-35
   b) Density True (g/cc) 1.53 Apparent (g/cc) 1.34c) Moisture regain (%) at 65% RH 10.5 & at 100% RH 28.5
4. Degree of crystallinity (X-ray, quantitative) High
5. Herman's angle of orientation (X-ray in degrees) 9 – 10

Fineness: Fibres are entangled in mesh structure. Single filaments are obtained by combing- and splitting the mesh structure. The length and breadth ratio (LIB) of ultimate cell is a measure of fineness. The length and breadth ratio of sunnhemp (450) is much more than jute, mesta, roselle, sisal, coir, aloe, banana, manila hemp, bhindi, dhaincha and areca nut (75-140) indicating its finer nature. However, ramie, flax, true hemp and cotton (900 - 3500) are finer than sunnhemp on this count.

Tenacity: In terms of bundle tenacity, toss a jute and sunnhemp fall in the same group and have edge over white jute, mesta, roselle, bhindi, dhaincha, coir and aloe. Ramie, flax, true hemp, sisal and Manila hemp are stronger than sunnhemp fibre. This property of fibre is influenced mostly by retting conditions. Hence the above statements may not al ways hold true.

Density: The bulk density may be evaluated in terms of total volume of certain weight of the fibre, cut to a definite length, arranged in a particular way, and put under a constant pressure. Any fibrous substance is obtained best by immerging it in some inert, non-swelling liquid. The liquid is allowed to permeate all air pores and the density thus obtained is the 'true' density of the fibre. If the liquid used for immersion is a light mineral oil of low density, it enters the inter fibre spaces and envelops the fibre elements for a few minutes before penetratng into the micropores and lumen of the fibre. Density obtained before penetration occurs, has been termed as *'apparent'* density. The co-efficient of correlation between yarn tenacity and apparent density is + 0.633 and between yarn tenacity and bulk density is + 0.559. Heavy bodied samples have a higher apparent or bulk density which in turn is associated with higher yarn tenacity.

These two values of sunnhemp fibre bundles are more in comparison to tossa and white jute, mesta, roselle, bhindi, dhaincha, sisal, manila hemp, coir, aloe, banana and areca nut but less than that of ramie, flax and true hemp.

Friction: Friction has an important role in yarn spinning. Both static and dynamic friction properties of all bast fibre crops have been studied. The co-efficient of friction of sunnhemp fibre are 0.50+0.02 r (parallel) and 0.40+0.004 r (perpendicular). These values are less than that of white jute, roselle, urena and ramie but more then tossa jute and flax.

FLAX

Flax or linwed (*Linum usitatimsutm* L.) fibre con from the stems and the oil from the seeds. The fibre obtained from the linmwda is inferior in respect of length and quality. There are some types exclusively for fibre. The flax fibre is valued ort its strength and durability which are superior to those of cotton. It is soft, shining and poeon high water absorbency. It has low elasticity and is stronger wheis wet than when dry. It is resistant to high temperatures, moiature and mudow. ltd not readily reaettomordants and dym, unlike cotton. B'abrim made of flarlaunder well. Max contains 80 to 90 per cent cellulose

Flax in woven into fine fabrics, Much an lawns, cambric and drills, canvas and buckrarns. It is used in linen. stitching, making twines and nets for fishing, ropes, carpet-backing. wrapping cloth and house furnishings. It is a good raw material for tissue paper, firefi htirig hose and water-baga.

Although in the Gangetic flax types introduced Italy, were not found the bre amp; on the contrary, they deteriorated into the linaced types. The attempts to produce a dualpur, i.e., strain, have so far failed. We are bout SW tonnes of flax to meet the demands, chiefly from the Defence Organisation. India's own efforts to cultivate flax have not been successful because of the climate.

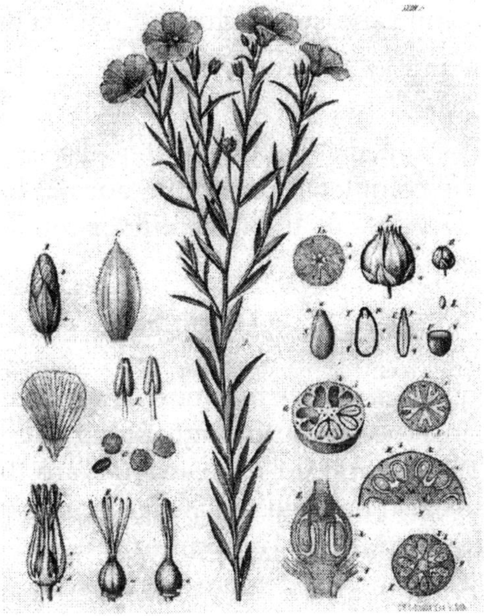

Linum usitatimsutm

Climate

Flax is a crop of the temperate zone. A cool humid climate with temperature ranging from IT to 27'0 and a diurnal fluctuation of relative humidity between 30 and 95 per cent, cloudy weather during the growing period is preferred. In the hot and dry climate, flax types tend to branch and grow as linseed types. The ideal geographical regions envisaged for successful flax production in India are the valleys of the Himalayan ranges where the winter iinulates a European summer; this possibility bm been proved by some trials.

Soils

Heavy clay and light soils are unsuitable. Rich loams or clay loams are considered best. The 'char' lands of the Ganges and the Brahmaputra which receive silt from the annual floods are suitable in winter.

Cultivation

For the Gangetic plains, the sowing time is early November. For the high rangu, it will be early October. The seed is drilled in rows 20 em apart. After germination, the seedlings are thinned to keep a plant-to-plant distance of 1 to, 2 em. The wed-rate is 40 kg, the expected germination heirs 90 per cent. One or two irrigations will be needed. The crop matures in 100 days, one month after the initiation of the first flower. Harvesting is done by cutting the stems close to the ground, or by pulling them out.

Retting

Bundles of stalks (straw) steeped in retting water just enough to cover. It take about 3 days for the retting to be completed under the normal temperature prevailing at the harvest time. Over- retting weakens the fibre; under-retting makes extraction difficult Betted straw is then dried.

Extraction

The extraction of fibre requires a simple machine. The machine is indigenously designed on the, principle of passing the straw through fluted rollers to break the woody core and separate the fibre. The eeparat ed fibre stranda are then brushed and rolled into bundles. This machine can also he used to salvage fibre from the linseed straw that go or at best can he ntmed as fodder or can be burnt a fuel. The fibre obtained way he inferior in quality, but it can be used for paper-Making.

Varieties and Yield

Flax Mtivation of the put depended on varieties straightaway impot from abroad. The recurring imports of wed in large quantities (owing to the deterioration of the crop) was not economical.

CONGO JUTE

Cotton is one of the most important commercial crops playing a key role in economic political and social affairs of the world. Some new selections ('F 895' and 'F 896') have been made and they have become acclimatised.

Cotton is one, of the most important commercial crops playing a key role in economic, political and social affairs of the world. Chiefly as a fibre crop cottt)ll is cultivated in about 60 countries of the world but ten countries, viz. the USST, the USA, China, India, Brazil, Pakistan, Turkey, Egypt, Mexico and Sudan account for nearly 85 per cent of the total production.

The organised sector of the Indian. textile industry constitutes the largest single industrial segment in the country in terms of the annual value of output and labour employed, both direct and indirect. There are 688 cotton textile mills in India, as enumerated on 1 January 1973, having an installed capacity of 18.4 million spindles and

207 thousand looms. The industry provides di.ect employment for nearly 900,000 workers and indi, rect employment for several millions. The decentralised sector comprising the power-loom, hand-looms, charkha, etc. is reported to provide - employment for over 2.5 million.

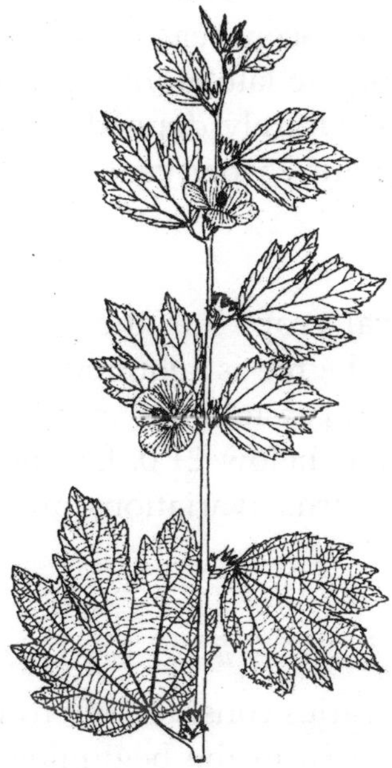

Cotton plant

With nearly 8 million hectares under the cotton crop, India ranks first in the *ii,*orld in respect ot area and tourth ın total production which reached the level of 63.3 lakhs of bales during 1971-72. The other major cotton-growing countries of the world are the United States of America, the USSR and China.

Species Composition

India grows on a commercial scale varieties falling under all the four cultivated species of Gossypium, viz. G. hirauguin, G. barbadense, G. arboieum and G. herbacin. The predominant species cultivated is G. hirsulunt, which covers about 50 per oent of the area of 8 million hectares, followed by G. ambwmm with 29 per cent and G. Aeum with 21 per cent. The area under G. barba nse is negligible and covers only a few thousand hectares. Cotton research and development efforts are directed towards progressively increasing the area under G. hirmaum and replacing, as far as possible, the area under the old world species. The cotton crop in the United States is now approximately 99 per cent G. hirsutum and one per cent G. barbadmst In the USSIT, G. hirmauits is grown over 92 per cent of the area and G. barbadenje over the rest of the area.

Area under Irrigation

Cotton in India is largely cultivated under rainfed or barani conditions. lit is mainly raised during the tropical monscon season, although in the southern parts of the country, it in cultivated during the late monswn season in winter. In nearly 75 per cent of the area, the cotton crop is entirely dependent on rainfall, whereas supplementary irrigation facilities exist in about 25 per cent of the area.

Climate and Soil

Cotton is tropical subtropical crop. For the successful germination of its seeds a minimum temperature of 15 degree is required. The optimum temperature range for vegetative growth is 21-27°C. It can tolerate temperatures as high as 43°C, but does not do well if the temperature falls below 21'0. During the period of fruiting warm days and cool nights with large diurnal variations are conductive to goog boll and fibre development.

Crop Season and Rotation

The sowing season of cotton varies considerably from tract to tractand is early sowings done in the Dharwar-Gadag area in the beginning of September have been found to give better yields. In addition, summer sowings in Tamil Nadu ire done during February-March.

The sowings of cotton in the rice fallows of Andhra Pradesh and Tamil Nadu extend from the second half of December to the, middle of January. The honest rotations for cotton in the irrigated .areas of the Punjab, Haryana, northern Rajasthan and Uttar Pradesh are cotton-wheat, cotton-jowar and cotton-wheat-toria. The growing of' berseen and guar has been found to have a beneficial effect on the succeeding cotton crop.

In the central and western India, cotton- jowar, cotton-bajra, cotton-wheat, cotton-gram and cotton-sesamum are the usual rotations followed. The raising of the cotton crop after a legume in rotation and preceded by a grain crop has been found to give a better yield. Cotton following groundnut has been found to give 15-20 per cent higher yield than that following wheat or jowar. Another common practice in cotton cultivation is mixed cropping with maize, jowar, ambadi, sesamum, pulses or vegetables as practised in many parts of central India. Intercropping with ragi, other millets or ground. nut is also quite common in parts of Tamil Nadu, Karnataka and Andhra Pradesh. Intercropping and mixed cropping under rainfed conditions serve both as an insurance against crop failures and as a preventive against soil erosion.

Preparation

For the irrigated cotton crop, following wheat in the northern states of India, the preparation of land is usually hurried through, since the interval between the harvesting of the grain crop in March-April and the sowing of the irrigated cotton in April-May is very short. The soil is first given a heavy irriga- tion and one or two ploughings. Thereafter, a light irrigation is given. followed by one or two ploughings and then the soil is planked with a wooden plank (sohwa) before cotton is sown. The black-soil areas of central and southern. India are prepared for sowing the rainy cotton by harroing-field 3 or 4 times with the blade harroing. The land is given a deep ploughing only once in 4-5 YCAM to remove weeds.

In the red and lateritic soils of southern India, which are poor in retaining soil moistures, the field is usually given two or three light ploughings or hamwings for raising cotton as a rabi crop. The irrigated crop in these soils is given one watering, ploughed once and the land is well for aowiiig the irrigated winter crop.

Seed-rate and Spacing

For obtaining mali. yields, proper seed-rate and spacing are essential. Depending on the variety, the level of soil fertility and the cultivation practices, smd-rates and spacing & have bets recommended for different improved varieties.

A rate of 15 to 25 kg per hectare and a of 75 to 90 em between the rows are generally recommended for cotton grown under irrigated conditions. For the raiu-fed deei cotton, a seed-rate of 12 to 16 kg per hectare and a spacing of 45 to 60 em between the rows are adopted. For the rainfed American cotton, the seed-rate is 12 to 16 kg and the apacing is 60 to 75 em between the rows. B,roadeasting the or sowing it in lines either with a Mechanical drill or by hand, are the two common methods. Line-gowing with wod.drills is recommended in order to ensure uniform germination, better stand and easy inter-cultivation.

In the case of the ridge-sowxk irrigated crop, hand dibbling of the seed at the spacing is commonly followed. This method is also followed in dryland areas of parts of Madhya Pradesh and Andhra Pradesh, adopting the square-pocket system of cultivation.

The Cotton crop in most of the areas in India is usually not adequately manured. The soils in which cotton is grown are generally sufficient in phosphate and potash, but since they are deficient in nitrogen and matter, it is necessary to apply them for better Yields. Any leguminous crop. cithe grecli-manure crop or as a crop in rotation preceding cotton, increased the yield of cotton.

Farmyard manure is rarely applied to the cotton crop in the states of Punjab, Haryana, Rajasthan and Gujarat. Of late, sonic of the progreesive farmer, states apiply 6 to 12 tonnes farmyard manure per hectare to the rain-fed crop and 15 to 28 tonnes of it to the irrigated crop. In Maharashtra, the cotton crop is manured with farmyard manure once in 3 or 4 years at the rate of 12 to 15 tonnes per hectare. In Tamil Nadu, the irrigated American cotton crop is usually manured heavily and auppleinente with fertiliser.

The need for applling chemical fertilisers for increas. ing cotton yield ham been amply proved. The usual dose recommended is 40 kg of nitrogen per hectare in the case of the irrigated crop and 20 kg in the case of the rainfed crop. In northern India, a higher dose of 60 to 100 kgN/ha has been found to he economical. Good response to high doses of 80 to 100 kg Niba has obtained in parts of Maharashtra, in the fallown of Andhra Pradesh and in the case of winter-sown cotton in Tamil Nadu.

The application of nitrogen by band placement has been found to give better yield broadcasting. For the irrigated cotton crop, it has been found to be economical to apply half of the basal dressing at the tigne of sowing and the other half as top-dressing during thinning or just before flowering.

In the case of the irrigated and row-gown cotton crop, intercultivation is done fairly regularly, with either a blade harrow or with a three tined however with a dui plough. In the case of the crop that is gown broadcaat, one or two hand-booings are given to remove weeds. Intercillitivation not only cheeks the growth of weeds, but also lead to better soil aeration and soil-moisture conservation.

The thinning of the cotton crop i,3 a special feature of the irrigated crop sown on ridges in Peninsular and in parts, of southern Gujarat. Manuring is desirable for maintaining the optimum population of plants to obtain a high yield. During this operation, the vigorous things are retained and weak and oil-type ones are removed.

In the case o irrigated cotton, the optiown moisture supply during the critical stages of growth and development is the most important aspect of cultivation. A proper irrigation schedule in the case of this crop in relation to soil type, eather conditions, cultivation practices and the variety is the dessisive factor in getting the high yield.

The cotton is grown in variety of soils it require to amenable to good drainage as it does not tolarate water logging. It is grown mainly as a dry crop in the black cotton and and medium black soils as an irrigated crop in the alluvial soils

In Tamil Nadu, the major portion of irrigated and rainfeed crop is planted in August-September. Early sowings done in the Dharwar-Gadag area in the beginning of September have been found to give better yields. In addition, summer sowings in Tamil Nadu are done during February-March. The sowings of cotton in the rice fallows of Andhra Pradesh and Tamil Nadu extend from the second half of December to the middle of January.

The commonest retations for cotton in the irrigated areas of the Punjab, Haryana northern Rajasthan and Uttar Pradesh are cottori-wheat, cotton-jowar and cottonʋheat-toria. The growing of berseem and guar has been found to have a beneficial effect n the succeeding cotton crop. In the central and western India, cotton, cotton-bajra, otton-wheat, cotton-gram and cotton-sesamum are the usual rotations followed. The aising of the cotton crop after a legume in rotation and preceded by a grain crop has een found to give a better yield. Cotton following groundnut has been found to give 5-20 per cent higher yield than that follwoing wheat or jowar.

Another common practice in cotton cultivation is mixed cropping with maize, jowar, mbadi, seasamum, pulses or vegetables as practised in many parts of central India. ntercropping with ragi, other millets or ground nut is also quite common in parts of 'amil Nadu, Karnataka and Andhra Pradesh. Intercropping and mixed cropping. under ainfed conditions serve both as an insurance against crop failures and as a preventive gainst soil erosion.

eed rate Spacing and Sowing

or obtaining mali. mtim yields, proper seed-rate and spacing are essential. Depending n the variety, the level of soil fertility and the cultivation practices, smd-rates and pacing have bets recommended for different improved varieties.

A seed rate of 15 to 25 kg per hectare and a ng of 75 to 90 em between the rows re generally recommended for cotton grown under irrigated conditions. For the rainfed eei cotton, a seed-rate of 12 to 16 kg per hectare and a spacing of 45 to 60 em between he rows are adopted. For the rainfed American cotton, the seed-rate is 12 to 16 kg and he apacing is 60 to 75 em between the rows.

Broadeasting the seed or sowing it in lines either with a mechanical drill or by hand, re the two common methods. Line-sowing with woddrills is recommended in order o ensure uniform germination, better stand and easy inter-cultivation. In the case of he ridge-sowxk irrigated crop, hand-dibbling of the seed at the spacing is commonly ollowed. Thia method is also followed in dryland ar as of parts of Madhya Pradesh and ndhra Pradesh, adopting the square pocket systenl of cultivation.

Ianuring

'he cotton crop in niost of the areas in India is usually not adequately manured. The oilsin which cotton is grown are generally sufficient in phosphate and potash, but since hey are deficient in nitrogen an matter, it is neccssary to apply them for get Yields. ny leguminous crop either as greenmanure crop or as a crop in rotation preceding otton, increased the yield of cotton.

Farmyard manure is rarely applied to the ootton crop in the states of Punjab, Iaryana, Rajasthan, and Gujarat. Of late, sonic of the rogreegive farmexr, ilk tlicee states piply 6 to 12 tonnes of farmyard manure per hectare to the rainfed crop and 15 to

28 tonnes of it to the irrigated crop. In Maharashtra, the cotton crop is manured with farmyard manure once in 3 or 4 years at the rate of 12 to 15 tonnes per hectare. In Tamil Nadu, the irrigated American cotton crop is usually manured heavily and aupplleinente it with fertiliser. The need for appling chemical fertilisers for increasing cotton yield ham been amply proved. The usual dose recommended is 40 kg of nitrogen per hectare in the case of the irrigated crop and 20 kg in the case of the rainfed crop In northeria India, a higher dose of 60 to 100 kgN/ha has been found to he economical Good response to hiah doses of 80 to 100 kg Niba has I) Ceii obtained in parts of Maharashtra, in the fallowing of Andhra Pradesh and in the case of winter-sown cotton in Tamg Nadu. The a lication of nitrogen by band placement ban, PI been foun to give better yield an broadcasting. For the irrigated cotton crop, it has been found to be economi. cal to apply half of the a basal dressing at the tigne of flowing and the other half as top-dressing during thinning or just before flowering.

Pest and Diseases

The cotton crop is subject to damage by a number of pests right from the, time of germination till the final picking. Very intensive work has been carried out on the cotton pests to evolve measures to control them and obtain economic yields. The pests of major significance in most of the cotton-growing tracts include aphids, jamids, thrips bollworms and mites, which affect the yield and quality of cotton. Ephylactic control schedules have been developed for checking the damage from pests and they have formed the basis for the crop-protection programmes in cotton cultivation.

Diseases of cotton cause reduction in Yields by affecting germination, by killing the seedlings, by lowering the productivity of plants and also by affecting the quality of lint. Root-rot and wilt, caused by several fungi, bacterial blight and anthraenose are the major diseases of cotton in India. Control measures, including seed treatment, spraying the foliage with fungicides and the use of resistant varieties have been developed for the major diseases.

REFERENCES

Collinson, M. (editor): *A History of Farming Systems Research*. CABI Publishing, 2000.

Crosby, Alfred W.: *The Columbian Exchange: Biological and Cultural Consequences of 1492*. Praeger Publishers, 2003 (30th Anniversary Edition).

Davis, Donald R., and Hugh D. Riordan (2004) Changes in USDA Food Composition Data for 43 Garden Crops, 1950 to 1999. Journal of the American College of Nutrition, Vol. 23, No. 6, 669-682.

Friedland, William H. and Amy Barton (1975) Destalking the Wily Tomato: A Case Study of Social Consequences in California Agricultural Research. Univ. California at Sta. Cruz, Research Monograph 15.

Mazoyer, Marcel; Roudart, Laurence (2006): *A history of world agriculture : from the Neolithic Age to the current crisis*, New York, NY : Monthly Review Press.

5

Tea

Tea *(Camellia spp.)* is an important beverage and the world drinks more of it than any other beverage. It is made from the tender or young leaves and unopened buds of the evergreen tea-plant, popular as a 'healthful herb'. The tea plant is in the polymorphic species *Camellia sinensis* Kuntz, but recent findings show that this plant of commerce is derived from more than one species. Two distinct varieties of tea-plant are generally recognised, the small-leaved China *(sinensis)* and the large-leaved Assam *(assamica)* which have been raised to a specific rank by a well-known tea botanist. Careful field observations, however, reveal that more than one or two species are involved in the evolution of the present-day tea-plant of commerce. Considerable interspecific hybridisation has taken place in nature. Thus, the taxonomy of the tea-complex is confounded.

The tea-plant, in the natural state, grows into a small or medium-sized tree, but in commercial plantations it is pruned and trained to form a many-branched low bush and is encouraged to produce vigorous vegetative growth by adopting an appropriate schedule of fertiliser applications. The important tree growing countries are India, Sri Lanka and East Africa; Japan and Indonesia also produce sizeable quanties of tea. It is also grown in Bangladesh, China, Georgia, Argentina and some other countries. In India, the crop is grown in Assam, West Bengal, Kerala, Karnataka and Tamil Nadu, and to some extent in Tripura and Himachal Pradesh. The total area under tea in our country is about 3,58,000 hectares and over 468 million kilograms of product is obtained annually. Over one million workers are employed by the tea industries. Yields have steadily increased during two decades. Tremendous progress has been made on the selection of clones and those with high yields and improved quality have been released to the industry. The tea planters are encouraged to uproot poor section of their plantation and replant them with improved planting material.

CLIMATE

Tea is grown on plantation scale in many areas of the world in warm and moist climate.

These range from Georgia (USSR) to Corrientes (Argentina). It is grown from sea leve to about 2460 m. A well distributed rainfall is essential for its cultivation. The rainfa varies from 125 cm to 750 cm in the tea-growing regions in this country.

Rainfall

The average annual rainfall in North East India ranges from 2000-4000 mm. Howeve more than the total amount, the distribution of rainfall matters a lot for sustained hig yield of tea throughout the season. In the North East India, the rainfall distribution i not even. The excess rainfall in the monsoon months of June-September causes drainag problems. The average monthly rainfall during November to March is less than th evapotranspiration loss and the resulting soil moisture deficit affects tea bushes. If th dry spell persists for a longer period, tea plants suffer heavily and crop goes down i spite of having sufficient rainfall in the monsoons. Thus, adequate rainfall during wint and early spring is crucial for high yield. Seasonal variation of rainfall (long-ter average) is given in the following table:

Average rainfall distribution in the North East India's tea growing regions in mm

Season	*UA*	*CA*	*NB*	*BV*	*Dooars*	*Terai*	*Darjeeling*
Winter (Dec-Feb)	4	3	3	3	1	1	2
Pre-monsoon (Mar-May)	24	27	19	28	14	13	15
Monsoon (June-Sep)	66	63	72	62	79	82	77
Post-monsoon (Oct-Nov)	6	7	6	7	6	4	6
Total	2720	2065	2185	3000	3990	2965	2345

Temperature and RH

Temperature affects tea yield by influencing rate of photosynthesis and controlli growth and dormancy. In general, the ambient temperature within 13°C and 28-32°C conducive for growth of tea. Maximum ambient temperature above 32°C is unfavourab for optimum photosynthesis more so if it is accompanied by low humidity. In the t belts of this region, the average winter minimum temperature (Dec-Feb) remains belo 12°C and there is hardly any growth during this period. Flushing commences from Mar with the rise in temperature. Winter dormancy however is the result of interaction short day length and low temperature. Low temperature causes slower growth and lo yield in the hill district of Darjeeling in comparison to the plains of Dooars and Assa A humid climate and high RH favours growth of tea.

Day Length

Day length influences growth and dormancy in tea bushes. When days of less than 11 15 min duration last for at least six weeks tea bushes become dormant. Hence the leng of growing season decreases with increasing distance from the equator. Seasonal dormancy appears from around 18° North and South latitudes. In the Northeast India (25°-27°N latitude), the tea bushes remain dormant during the winter season for about 3 months on account of the combined effects of short days and low temperature.

Soil

Tea is planted on soils of widely different geological origin and of almost all physical types, a well drained, deep and friable loam or forest land rich in organic matter is ideal. Tea soils are acidic (pH 4 to 6), low in calcium and generally rich in iron and manganese.

Tea grows well on high land well drained soils having a good depth, acidic pH in the range 4.5 to 5.5 and more than 2% organic matter. Shallow and compacted sub-soils limit root growth. Tea plants growing on such soils are liable to suffer from draught during dry period and water logging during the rainy months. There should not be any hard pan or concretions in the subsoil within 2m depths. The depth of ground water table should not be less than 90 cm for good growth of tea. Catchment planning is required for improved soil and water management practices in a tea estate for which land survey designed to identify all major and minor topographical features needs to be carried out.

Land Preparation

The vegetation is cleared one season ahead by ring barking trees followed by uprooting the stumps along with roots and leaving the litters to decompose in situ. The pits so formed are filled up with excavated soil and the ground is levelled with minimum soil disturbance. The drains are then laid out and excavated soil spread thinly over the ground. Following this, staking for tea plants is made and seeds of green crop like Crotolaria anagyroides are sown between the tea lines for providing quick ground cover and temporary shade to the young tea plants. Tea is planted as soon as land preparation is completed. In general, planting is completed by end-May/June in draughty areas and by September to early November in other places. The newly planted tea fields are then adequately mulched. Between clearing the land and planting, there should be minimum time interval. The drains and the tea planting should be along the contour in hilly areas or in land with more than 2% slope.

Land preparation for replanting on uprooted land

The pits and the subsidiary drains should be filled up with soil and the land is levelled by hand hoeing or mechanically. The drainage lines should be marked according to topography of the land. Ploughing and cross ploughing should be undertaken when the soil is moist and sub-soiled where necessary, taking care not to plough the lines of drains. If the soil is highly acidic, 2-3 tons of powdered slake lime or dolomite is broadcast before ploughing. After levelling the land, drains are dug and the area put under rehabilitation crops for at least two years. After rehabilitation period is over, the grass is cut at ground level leaving the lopping in situ. The stumps are removed with minimum soil disturbance. The drains are deepened or re-graded and the area is replanted with vigorous clones or seedlings.

After a prolonged period of monoculture under tea, the physico-chemical as well as biological properties of the soil deteriorate considerably. Old tea fields due for uprooting usually carry a lot of vacancies. The death and debility of the bushes are caused by diseases and pests and faulty agronomic practices, deterioration of soil fertility and physical conditions due to inadequate fertilisation, erosion, poor drainage compaction, etc., and old age is the cause of low yield of fields due for uprooting. Proper rehabilitation of the uprooted soil is a pre-condition for the success of any replanted operation.

The old tea fields, after uprooting, are put under rehabilitation crops for a period of 18-24 months. The rehabilitation crops used in Northeast India are Guatemala grass (*Tripsacum laxum*), Pusa Giant Hybrid Napier (*Pennisetum purpureum*), Citronella grass (*Cymbopogon wintarianus*).

Characteristics of a good rehabilitation crops are: (1) it should yield a large quantity of green material in a short period of time, (2) should be quick growing to be able to suppress the weeds, (3) should have more leafy growth so that its decomposition will be rapid and (4) should have a deep and fibrous roots system to penetrate deeper into the soil and improve soil structure. Guatemala grass has been mostly used in the tea estates of North East India. Benefits of rehabilitation can be summarised as:

1. it adds organic matter and nutrients to the soil,
2. helps in improving soil structure and thereby better aeration and moisture storage,
3. draws nutrients from deeper soil layers and added to the top soil through the lopping,
4. increases the activity of micro organisms
5. conserves the top soil, and
6. breaks the food chain of the primary root diseases.

Rehabilitation crops should be planted in spring in moist soil and manured with 15-20 tonnes of cattle manure per hectare or N:P:K at the rate of 60:30:60 kg/ha (10:5:10) mixtures @ 600 kg/ha initially, and @ 300 kg/ha after each lopping. It is strongly recommended that the lopping be left in situ. The rehabilitation period will depend on the build-up of organic carbon in the soil preferably above 1%. For achieving this normally a period of 1.5 to 2 years is required. If the growth of the rehabilitation crop is not satisfactory, the period may be extended by one more year.

Effect of rehabilitation crops on soil properties was studied at Tocklai. The results showed that roots of Guatemala grass proliferated down to a depth of 120cm and in the process reduced compaction (2.8-7.7%), increased water stable aggregates (24.4-162.5%), hydraulic conductivity (16.0-81.0%), porosity (0.0-15.4%) and organic matter status (5.4-119.6%). It is clear from this study that rehabilitation with Guatemala grass improved the soil health.

CULTIVATION

Camellia sinensis is an evergreen plant that grows mainly in tropical and sub-tropical climates. Nevertheless, some varieties can also tolerate marine climates and are cultivated as far north as Cornwall on the UK mainland and Seattle in the United States.

In addition to a zone 8 climate or warmer, tea plants require at least 50 inches of rainfall a year and prefer acidic soils. Many high-quality tea plants are cultivated at elevations of up to 1500 meters (5,000 ft): at these heights, the plants grow more slowly and acquire a better flavour.

Only the top 1-2 inches of the mature plant are picked. These buds and leaves are called *flushes*. A plant will grow a new flush every seven to ten days during the growing season.

A tea plant will grow into a tree if left undisturbed, but cultivated plants are pruned to waist height for ease of plucking.

Two principal varieties are used: the small-leaved China plant (*C. sinensis sinensis*), used for most Chinese, Formosan and Japanese teas (but not Pu-erh); and the large-leaved Assam plant (*C. sinensis assamica*), used in most Indian and other teas. Within these botanical varieties, especially the small-leaved China plant, there are many traditional strains and modern clonal varieties. Leaf size is the chief criterion for the classification of tea plants: tea is classified into (1) Assam type, characterised by the largest leaves; (2) China type, characterised by the smallest leaves; and (3) Cambod, characterised by leaves of intermediate size.

Tea is propagated from seed, but high yielding clonal material has also become popular. The plants are initially raised in nurseries to produce healthy plants. 1 to 1 and 1/2 year old nursery seedlings are used for planting in the field. The vegetative propogation of selected tea-plant cuttings, each with a leaf and an internode has been standardised and progressive planters now raise colonial nurseries on a large scale.

The area intended for planting is first cleared of forest growth and adequate steps are taken to prevent soilerosion. Planting lines are traced and pits of a convenient size are dug. The normal accepted spacing is 120 X 75 cm, accommodating about 10,000 plants in a hectare. The soil is heavily mulched during the initial one or two years.

Camellia sinensis

In most of the tea areas in our country, a system of shading the tea plants is practised. In southern India, the silver oak (Grevilla robusta) is a shade tree of choice (planted 600 by 600cm, later thinned to 600cm by 1200cm or 1260 cm by 1200 cm as required). Other trees used for shading are *Erythrina lithosperma (dadap), Acacia and Albizia.* Albizia species are used in tea gardens in northern India.

Weeds compete with the tea plant for nutrients, and weed control is important in tea plantations. In recent years, the control of weeds with chemicals has gained popularity. Grasses and broad-leaf dicot weeds are kept under check. Pests and diseases are effectively controlled with chemicals.

A manuring scheme is followed from the nursery stage upto maturity. A special 1:1:1 NPK soluble fertiliser mixture is used in the nursery; a 4:3:3 NPK mixture in the first 3 years in the clearing and thereafter the dosage of the fertiliser applied is varied, depending on the type of tea, its performance, soil fertility and the yield potential of the crop. On an average 10 kg of nitrogen is applied for every 100 kg of the crop.

Tea readily responds to the application of nitrogen and potassium. Potassium is applied at the rate of 40 to 50 percent of the amount of nitrogen applied, depending on the soil type and weather conditions. In southern India tea responds to nitrogen upto

300 kg per hectare applied to some high-yielding varieties of tea, whereas in northern India, the application of nitrogen is limited to about 120 kg whereas 30 to 40 kg of phosphate per hectare is adequate. Zinc deficiency is a limiting factor in crop production in most areas and a schedule of foliar application of zinc sulphate usually 11 kg per hectare in alternate years is recommended.

SEED PROPAGATION

Sandy loam soil is the best for raising nursery on orthodox beds. Wherever possible, the nursery site should be on virgin soil. In absence of virgin soil, uprooted tea area or area with poor soil may be used after the site has been put under green crops like Crotolaria anagyroides, Mimosa invisa or grasses like Tripsecum laxum (Guatemala), Cymbopogon (Citronella), etc., for a period of about two years. The plants need lopping 2-3 times a year and the loppings should be left in situ for decomposition. Alternatively, the area may be heavily dressed with cattle manure @ 25 tons/ha. The area should be criss-crossed ploughed and harrowed about a month before the beds are ready for sowing. Beds should not be prepared during rains. If the green crop or Guatemala grass was raised, the land should preferable be deep hoed only. The seedbeds may be of any convenient length but should not be wider than 1.5m for convenience of cultural operations. In North East India, the beds are generally prepared flat 10-15 cm above the surrounding ground level. Each nursery bed is enclosed by a drain 20 cm deep and 20 cm wide, which is linked up with outlet drain. Single super phosphate (SSP) should be applied to the soil @ 50g/m^2 before sowing the seeds. A nursery bed measuring 30m x 1.5 m or 37.5m x 1.2 m will accommodate about 1300 seeds when sown at 20cm x 20cm (triangular) spacing.

Tea seeds are planted on nursery beds in rows at a depth of 1.5 cm with the "eye" facing downwards or parallel to the ground surface. If the seedlings are to be lifted with the bheti for transplanting in the field later, the seeds are sown with a spacing of 20 cm. This spacing can be reduced to 15 cm or so for preparing the seedling for stump or carrot planting where the lifting the plants with the bheti is not required. The nursery beds can be of any convenient length and up to 120 cm wide.

Polythene Sleeve Nursery

When land is insufficient, seedlings are raised in polythene sleeves. Soil can be brought from any convenient site but must be tested for acidity and eelworm population. It should be mixed with superphosphate @500 g/m^3 of soil and the sleeves filled with the soil 3-4 weeks before sowing. Seeds are sown at of 0.5-1cm depths with the "eye" facing downwards and covering the seeds with a thin layer of soil. The sleeves can be placed in a permanent nursery site where irrigation facilities exist. If irrigation facilities are not available, sleeves can be placed under usual overhead shade or green crop shade. Vigorous seedlings can be planted in the field in April-May when they are 5-6 months old. For planting 1-year old seedlings, seeds should be sown in 30 cm long sleeves.

However, it will be necessary then to shift the sleeves once or twice after mid-July to prevent the taproot from penetrating the ground. For planting one hectare one would require 40-60 kg tea seeds depending upon the spacing.

Pre-germination of the seeds in moist sand kept in germinating pits with mulch cover is necessary in order to raise good nursery. They are ready to be transferred to the sleeves as soon as the seeds crack. Seeds that do not germinate even after 2 months should be discarded.

After sowing, the bed or the sleeves can be kept under mulch when the nursery is raised under green crop shade. Frequent examination is required to ensure scrapping of the mulch around the germinating seeds so that it does not cause any obstruction. Excess watering should be avoided. Only hand weeding should be done carefully and proper care against pests and diseases should be ensured to protect the plants.

VEGETATIVE PROPAGATION

The nursery is raised on a high land with adequate drainage and sunlight which is convenient for irrigation and frequent inspection. The soil should be neither too sandy nor too clayey with pH 4.5-5.0 and free from eelworm. If the soil is too acidic, it requires a treatment with slaked lime @ 2t/ha. Sub-acidic soils can be amended with 2% aluminium sulphate solution 2-3 times. Soil should be tested before planting cuttings.

The nursery beds should be arranged east-west direction under overhead shade (bamboo lath and thatch) and keeping the north side open for better air circulation and light penetration. The northern height of overhead should not be more than 180 cm and should slope down towards south by 60 cm.

Beds under low shade should be north south oriented. Shade is provided with the help of bamboo lath frame about 22 cm over the propagation beds, supported by sidewalls. The bamboo lath frame should be raised gradually with increase in height of the plants. It is necessary to prepare the beds at least 6-8 weeks before planting cuttings. The period of heavy rain should be avoided for preparation of beds as clods are difficult to break when the soil is wet. The bed is generally 120 cm wide and of any convenient length. Each bed is enclosed by a 45 cm deep and 30 cm wide drain, which should run into one outlet drain. The surface of the beds should be slightly cambered for quick run-off of rain or irrigation water. Application of superphosphate @500g/m^3 of soil used for filling sleeves or @50g/m^2 of bed by incorporating in the soil to 15-20 cm depth helps in root development.

Polythene Sleeve Nursery

Clones can also be planted directly into sleeves. Smaller sleeves produce weaker plants. Before filling sleeves, the soil should be sieved through a No.4 wire mesh to eliminate undecomposed organic matter and pebbles. The sleeves should be filled with the sieved soil without much ramming and the filled sleeves should be kept covered preferably under shade for 6-8 weeks prior to planting cuttings.

Type of Cuttings

Cuttings should be taken from the primaries with dormant apical buds (banji state) for higher success. In some clones, lateral shoots are produced from the lower portion of the primaries quite early. The unbranched portion of the primaries in such cases should be taken first followed by the laterals when they mature after a few weeks. A good cutting should have a fully mature healthy undamaged leaf, dormant or just swelling bud and a hard green stem of about 2.5 cm length below and about 0.5 cm above the leaf. A basal cut parallel to the slope of the leaf is more convenient for handling and for quick propagation of cuttings. In large leaf clones, reducing the leaf area by cutting away top-half of the leaf blade results in better rooting.

Planting of cuttings

With a dibber (small bamboo sticks slightly thinner than the stem) 2cm deep holes are made 5 cm apart on the bed. Cuttings are then planted firmly by holding the cutting at the node with the leaf pointing away from the hand in a semi erect position and inserting the stem on the holes upright. In north south oriented beds, the tip of the leaf should point towards north and in east west aligned beds, the tip should point towards west. Care should be taken to keep the petiole above the soil surface to prevent rotting of the mother leaf. The mother leaf should be more or less in an upright position after insertion into the holes. There should be no air pockets in the interspaces of cutting and soil. Dipping the cuttings in 0.1 per cent zinc sulphate solution before planting helps root initiation. The soil moisture should be maintained at field capacity. The cuttings should be transferred to sleeves just after initiation of roots. Over shading leading to excess moisture in soil and pH higher than 5.0 may cause excessive callus formation at the expense of root growth. Shade should be such that cuts about 50% of the mid-day sun light falling on the nursery bed. In the low lath frame shade, the bamboo frames should be raised on the east after the cuttings have produced 4-5 new leaves. Where thatch is used in overhead shade automatic thinning occurs in time. When bamboo lath is used, gradual removal of the roof (alternate one) ensures adequate hardening of plants. Young plants need protection from pests and diseases by high volume spray of suitable pesticide. Weeds are to be removed by hand. The plants should be kept outside without shade for a couple of weeks before taking them to the field.

Cutting nursery in Darjeeling

In Darjeeling, the nursery sites should preferably be located at low or mid elevation and on the northern aspect. The nursery site should be protected by contour drains and the beds (about 1.2 m wide and 15-20 cm high) must be aligned along the contour to guard against soil erosion. The sleeves should be arranged on raised beds (about 15 cm high) that have a slight camber as in orthodox propagation beds. Up to an elevation of about 1500m, bamboo lath frames or overhead shade of bamboo or thatch may be used. At higher elevation overhead shade of bamboo or thatch is generally more useful. Below the shade, the beds may be completely covered by a polythene tent (45 cm above the cuttings) with semi-circular bamboo frames to conserve moisture and heat to help quick growth of the cuttings. The polythene tents should be gradually removed when the cuttings have rooted and most of them have produced 2-3 leaves. With further growth, the overhead shade should be gradually thinned.

Manuring of Nursery Plants

When the plants have produced 3-4 leaves, they are manured with a mixture of NPK (2:1:2) and sand at 1:9 ratio. One teaspoonful of this mixture should be applied per plant at 3-4 weeks interval during the growing season. If required, foliar application of nutrients can be started when the cuttings have produced 6-8 leaves during April-May and August-November.

PLANTING MATERIALS

Seed

Commercial planting of tea in Northeast India began in the fourth decade of the 19th century with seeds of small dark leaf race of tea imported from China. Meanwhile the merit of the large, light leaf Assam race of tea was demonstrated by the auction of a small consignment of this tea in the London tea market on 10th January 1839. This was a momentous occasion because besides establishing the worth of Assam race of tea, it determined the further course of tea cultivation throughout the world. Much more tea is commercially made from the Assam race of tea than from the China race. The China race is grown in areas where the temperature drops very low during a part of the year.

Discovery of the value of the Assam race of tea induced the early planters in Assam to explore the tea tracts abandoned by the migratory hill tribes inhabitating the Borail mountain range along the South and Southeast border of Assam. The limited quantity of collected seeds of the Assam race planted in nurseries provided some material for commercial planting and some seedlings were used for planting seed orchards. To meet the requirement of seed for the rapidly expanding tea areas, a fairly large number of seed orchards were established primarily in Assam during the second half of the 19th century.

Tea seed production became a lucrative trade since these seeds are in large demand not only in North East India but also in South India, Indonesia, Africa and Sri Lanka to which large consignments of seeds are exported.

It may be mentioned in this connection that the three races of tea hybridise freely which must have happened to some extent under natural conditions in Southeast Asia, which is the home of tea, and other Camellias. Hence the races of tea are not genetically pure but are assigned to the three types on subjective judgement based mainly on shape, size and colour of leaf and nature of bush frame.

Clone

Until the middle of the last century tea was grown essentially from seed throughout the world. Development of a simple, cheap and rapid method of vegetative propagation of the tea plant in the third and fourth decades of the last century at Tocklai made it possible to release elite plants and multiply them vegetatively as clones for economical planting. Tocklai released the first tea clone in the world in 1949.

More clones had since been released with the progress of time, the total number at the end of 2002 being 30 excluding 151 clones selected in commercial tea estates. Two new TV clones are in the pipeline, which are under multilocational trials. Following are the clones so far released by Tocklai for commercial cultivation:

Standard clones: TV2, TV3, TV4, TV5, TV6, TV7, TV10, TV11, TV12, TV13, TV14, TV16, TV17, TV20, TV24, TV27, AND TV28 have above average yield & quality with high yield potential of about 3000-3500 kg made tea/ha.

Yield clones: TV8, TV9, TV15, TV18, TV19, TV22, TV23, TV25, TV26, TV29 AND TV30 have average quality but high yield with yield potential of 4000 kg made tea/ha and above.

Quality clones: TV1 and TV21 have high quality but average yield with yield potential of about 2500-2800 kg made tea/ha.

Seed and clone

Being the progeny of a hybrid parents, no two bushes in a population of tea raised from seed are alike. Hence, seed populations can be fitted into a wide range of environmental and cultural conditions without much change in their overall performance. On the otherhand all bushes of a clone widely separated in space and time behave in most ways as a single bush. Being genetically different all bushes of a seed population are unlikely to suffer equally from the attack of a pest or a disease while all bushes of a clone will be equally susceptible. Hence absolute dependence on clones for commercial cultivation of a long duration crop like tea is a risky proposition. Realising this danger Tocklai started breeding clonal seed stocks (varieties) along with the selection of clones.

Tocklai has already produced and released 14 biclonal seed varieties. Then the clonal seed stocks are of equal or nearly equal merit as the selected clones. The seed stocks are produced to serve the needs of the different soil climatic regions of North East India.

The tea industry is now using these seed stocks for replacement or extension of tea areas along with clones. The biclonal seed stocks released by Tocklai till the end of 2002 are TS 378, TS 379, TS 397, TS 449, TS 450, TS 462, TS 463, TS 464, TS 491, TS 506,TS 520, TS 557, TS 569 and TS 589.

Guidelines for Use of Seed and Clone

1. No single clone should occupy more than 10% of the total area of a tea estate.
2. A minimum of 5 clones should be grown in an estate.
3. Ideal clone to seed ratio is 1:1.

PLANTING

This is a very delicate operation and needs adequate planning and proper supervision. Correctly planted tea plants establish in the field quickly, grow vigorously and come into full bearing earlier. On the other hand, a slight error during planting can cause high percentage of mortality or permanent setback to the plants.

Time of Planting

Planting can be done in April-June and September-October or October-November with adequate irrigation. Periods of heavy rains should be avoided.

Type of Plants used for Planting

Only healthy plants 40-60 cm high with at least 12 good mature leaves and of pencil (0.5 cm) thickness (at collar) should be taken for planting in field. In general, 9 to 12 month old plants attain this stage. Sub-standard plants should be discarded. Before plants are removed from nursery, they should be hardened by gradual exposure to full sun. Transportation of the plants to the planting site in field should be done with utmost care and only after proper labelling.

Type of Planting

There are two types of planting, i.e., pit planting and trench planting.

Pit planting

This method is followed when spacing between plants is wide enough to allow digging of individual pits of proper size and without much difficulty. Pits should be about 45 cm wide and 45 cm deep, circular and straight walled. Smaller pits restrict root growth and retard shoot growth and development. The excavated soil is conditioned by mixing with 4-5 kg well-decomposed cattle manure or 150-200g well-decomposed oil cakes and returning the soil into the pits. No other manure is used except 30 g rock phosphate and 30g SSP at the time of planting.

Trench planting

This method is adopted for closer spacing and in heavy soils. Trenches 30 cm wide and 45 cm deep are dug along the rows. The excavated soil is conditioned and returned back as in case of pits and tea is planted directly on the trenches.

Method of Planting

There are two methods of planting, for plants raised in nursery beds. They are bheti planting and stump planting.

Bheti planting

Here, the plants are lifted along with the bheti and the roots intact from the nursery bed. This is convenient with plants grown in ploythene sleeves, which reduces difficulty in transportation, reduce root damage and gives a very high percentage of survival. The polythene is removed carefully by slitting the tube and the bheti is held in the pit half-filled with the conditioned soil in such a manner so that the top of the bheti is flushed with the ground surface. 30g rock phosphate is added at the bottom of the bheti and the pit is filled with soil with adequate ramming. At about 5 cm depth 30g SSP is added around the bheti and the pit is filled up to the collar of the plant with soil. Adequate ramming is necessary to prevent sinking of the pit level later, which will cause localised waterlogging. The same method can be used for plants grown in sleeves.

Stump planting

Plants are lifted from the nursery bed without having any soil around the roots. The shoot portion is cut off 15-20 cm from the collar and the excess roots trimmed off before putting them into the pits. This method is generally followed with overgrown nursery plants and has the advantage of easy transport and reduced chances of withering after planting. However, the percentage of survival is much less than bheti planting.

Spacing

About 14000-16000 (up to 17000 in hilly areas) plants per hectare have been found to be ideal bush population with spacing of 105-110 cm between rows and 60-75 cm between plants. The planting can be done either as single or double hedge.

Shade to tea areas

In N.E. India where ambient temperature rises to 30-32°C during the main tea growing season of June-October, provision of moderate shade to tea areas by planting a number of leguminous trees has been found beneficial for health and productivity of tea bushes. The light, broad leaf bushes of the Assam race of tea benefit more from shading than bushes of the dark leaf China race. Reduction of excessive heat and light radiation in the main causes of benefit from shading. Addition of organic matter by leaf fall, prevention of soil moisture during the dry winter months, reduction of red spider incidence, etc., are the other benefits of shade. The popular shade tree species in N.E. India are :

- *Albizzia odoratissima*
- *Albbizzia chinensis*
- *Albizzia procera*
- *Accacia lenticularis*
- *Derris robusta*
- *Albizzia lucida*
- *Delberzia sericea*
- *Albizzia lebek*
- *Adenanthera pavonina*
- *Albizzia moluccana*

The beneficial effects of shade trees in tea plantations were realised long ago and organised plantation of shade trees in Northeast India began towards the close of the last century. The "Sau" (*Albizzia chinensis*) was the first shade tree used extensively for planting in tea plantation. Gradually other leguminous species were introduced as shade trees in tea. Since early 1940s, extensive investigations were carried out at Tocklai on various effects of shade trees on tea plants related to leaf temperature, light interception, photosynthesis, portioning of assimilates, incidence of pests/diseases, moisture conservation, addition of nutrients to soil, etc. The information generated showed the following:

- A shade tree sp. with single layer canopy composed of small leaflets which intercept 30-50% of the available sunlight in the form of sunflecks is considered an ideal shade tree. The shade tree Albizzia chinensis meets these requirement to the best.
- Optimum population of shade tree helps to conserve soil moisture by 1-3 % during the dry season and lowers the leaf temperature by 2-4 ^{0}C.
- Deep rooted shade trees help in conserving soil moisture during the period of stress
- Shade trees (Albizzia chinensis) add 2500-5000 kg/ha organic matter to the tea fields through droppings of leaves, twigs and pods.

- The rate of development and proliferation of feeder roots of tea is high under shade trees.
- Shade trees absorb/reflect over 70% of harmful infra-red radiation from the solar spectrum thus protect the tea bushes from sunscorch damage.
- Positive effects contributed by shade trees cannot be replaced by additional application of nutrients.
- Optimum shade density reduces tranpiration loss of water enhances photosynthetic efficiency and helps in the mobilisation of nutrients.
- Under shade trees a large proportion of the assimilates is diverted to pluckable shoots thereby increases the yield.
- Effective shade reduces the incidence of red spider and other mites.

Propagation of Shade Trees

In Northeast India, shade trees are propagated by seed. It is always advisable to obtain genuine shade tree seeds from own sources. For this purpose, mother trees that possess all the desirable characters of an ideal shade tree are selected in the tea estate from which seeds are collected. In leguminous trees, the seeds are enclosed inside pods. As soon as the pods turns brown, they are collected from the selected mother plants during January-February.

The pods are dried thoroughly under sun, and the seeds extracted by hand pounding. The seeds should be stored in wooden bins/gunny bags in well- ventilated cool dry room.

Pest, diseases and weeds in the Nursery

Sucking pests like Psyllids and leaf defoliators do considerable damage to the germinating seedlings in their early stage. Red rust is the most serious disease of shade nurseries and is the principal cause of failure in many instances. Therefore, from the beginning care must be taken to keep the nurseries free of pests by taking timely action. In addition, proper weed control and drainage should be given attention.

MANURING

It is important to ensure adequate replenishment of nutrients in the soil removed by harvest. A series of field experiments conducted under different agro climatic conditions of Northeast (N.E.) India have recorded response not only to N but also to annual application of P and K. The response curves obtained from these experiments have indicated that annual application of balanced dose of NPK is required to harvest maximum crop. Results have shown that for sustaining a crop of 2300 kg made tea per hectare (KMTH) in different regions, in general, a dose of nitrogen not exceeding 140 kg, phosphate 20-50 kg and potash not exceeding 140 kg would suffice. Survey data from a large number of tea estates also corroborated these results and in 80% of the surveyed gardens, this NPK dose was found to be optimum for producing 2500-3000 KMTH. The possibility of sustaining a productivity level of 3500 KMTH or more by marginal increase of N over 165 kg along with 50 kg phosphate and 165 kg potash was also indicated from the data. The current recommendation on manuring is based on a number of trials conducted at Tocklai and a commercial tea estates of Northeast India.

In the Darjeeling hills, the agro climatic conditions are different from those of plains. The cool temperature slows down the organic matter decomposition and the productivity levels are lower as compared to the plain districts. Therefore, the N levels should not exceed 110-120 kg/ha. The manuring schedule recommended for Darjeeling is shown in the table below:

Time of Manuring

Fertilisers should be applied only when the tea bushes are ready to utilise them. As a general rule, the best time for fertiliser application is after the first rain in spring has moistened the soil to a depth of 45 cm and there is some new growth in case of unpruned tea. In case of pruned and skiffed teas, the fertilisers should be applied after the bushes have produced two new leaves. A weed free clean ground is desirable at the time of manuring.

Method of Manuring

In the mature tea, after complete coverage of ground, fertilisers are applied uniformly on the ground as broadcast. In the hills of Darjeeling and other areas, fertilisers should be applied in a half circular band on the up slope keeping a distance from the collar.

It is recommended to apply N and K fertilisers in two splits if the dose exceeds 100 kg/ha. This will ensure adequate supply of nutrients throughout the growing period and higher utilisation efficiency. The 60% of the dose of N and K should be applied in March-April (1st split) and the second split (i.e. the remaining 40%) should be applied in August-September. The entire quantity of phosphatic fertiliser should be applied in the beginning of the season along with first split of N and K. In very light texture soil, the number of splits may be increased.

Sulphur

The tea soils of Northeast India also were depleted of sulphur to different extent due to replacement of SOA by urea for quite sometime. Studies showed that soil application of sulphur @ 20-45 kg/ha improves yield and quality of tea. The current recommendation is that sulphur should be applied at 20 kg S/ha only if the soil test indicates values lower than the critical limit (40 ppm available sulphur). It can be applied to well drain tea fields as broadcast along with the NPK fertilisers in spring. Studies have also shown that available sulphur is correlated with high organic matter and response in the Dooars has not been consistent.

Micronutrients

Among various micronutrients tried only zinc has shown consistent response and is recommended particularly in unpruned / skiffed teas. Foliar application of zinc sulphate at 1-2% concentrations (w/v) has been found beneficial and produced up to an average of 10% increase in yield. The present recommendation is not to exceed 12.5 kg zinc sulphate per hectare per year, which is to be applied in 4-6 sprays (1-2% concentrations) in early (March onwards; 2-3 sprays) and late season (September onwards; 2-3 sprays). Indiscriminate use of zinc sulphate sprays should be avoided as it may lead to high levels in made tea. Foliar application of manganese, boron and molybdenum has shown varied responses and are at present not recommended as a regular practice.

Foliar nutrition

Foliar nutrition is beneficial under stress condition or coinciding with physiological changes in the bush. NPK mixture 2-1-2 or 2-1-3 where potash status is low, @ 0.5-1 percent can be sprayed during this period. Foliar sprays of potash @ 1% MOP (w/v) during moisture stress period can enhance the water use efficiency of the bushes. MOP (1%) and $MgSO_4$ (1%) sprays from Jan to March help ameliorate the ill effect of moisture stress in UP teas.

IRRIGATION

Distribution of rainfall in North East India is highly uneven. During the period from October-February, rainfall is scanty and the quantity varies from 5 to 10% of the annual rainfall. In this period average evaporation exceeds average precipitation by 8 to 357 mm. Therefore, conservation and supplementation of soil moisture by irrigation becomes necessary during this period of moisture stress.

Field Capacity (FC), Permanent Wilting Point (PWP) & Available Moisture

The amount of water retained after drainage of saturated soil is called field capacity (FC) moisture. At field capacity a loam or clay soil retains moisture at about 1/3 rd atmospheric pressure, whereas in sandy soil it may be as low as 0.1 atmosphere. At permanent wilting point, the soil moisture tension reaches about 15 atmosphere. Field capacity moisture and wilting point moisture are influenced by soil texture. Finer the texture higher is FC and PWP moisture content in soil. The range of moisture between field capacity and wilting point is available to plant roots. Fifty percent of this available moisture is considered as readily available moisture for working of irrigation schedule.

Method of Determination of Irrigation Requirement

Plant water use may be measured or estimated. Direct measurement requires sophisticated apparatus. Therefore it is usual to rely on indirect measurements like monitoring of soil moisture status, or on estimates based on meteorological observations. Less sophisticated equipment like tensiometer can be used to monitor soil moisture status. However, the meteorological observations seem to be the only practical approach for large tracts under tea. The environmental factors determine the potential evapotranspiration, the plant factors determine the water requirement for growth, and the soil factors determine the water holding capacity and moisture release characteristics. There are several models available for indirect measurement of evapotranspiration (ET). Estimates of Penman's ET based on meteorological data have been worked out for different tea growing regions of North East India:

Regions	*Oct*	*Nov*	*Dec*	*Jan*	*Feb*	*Mar*	*Apr*
South Bank	127	89	68	64	86	135	158
North Bank	128	89	66	64	87	139	153
Cachar	144	111	83	82	108	157	174
Dooars	138	100	76	75	95	150	164

The irrigation requirement should also take into account the rainfall during the period of moisture stress. In North East India high intensity rainfall causing surface runoff occurs during monsoon. The intensity, duration and rainfall-runoff analysis during November to April shows that almost the entire rain received during this period can be taken as effective rainfall (ER) from irrigation point of view. Regional analysis of effective rainfall for different return periods during droughty months was carried out at Tocklai from long-term rainfall data. The difference of ET and ER for the period October to April can be considered as the total irrigation requirement.

Soil moisture measurement

Time of irrigation can be determined by judging the soil moisture content. Soil moisture can be determined by gravimetric method, resistance meters, neutron probe and tensiometers. Out of these equipments the less sophisticated equipment viz. tensiometers can be used for timing of irrigation.

For young tea areas the tensiometer should be installed at 20 cm depth in the year of planting, and subsequently lowered every year by 10 cm as the root system develops.

For mature tea, the tensiometer should be installed with its tip down to 90 cm depth below the ground surface. It will be desirable to install an additional tensiometer at ¼ the depth of root zone i.e. at 25 cm to provide information on soil moisture in the most active root zone. When shallow tensiometer indicates suction equal to 0.5 then irrigation should be started.

Tensiometer Reading	*B. Range of soil moisture*
0.0 - 0.1	Soil is saturated
0.1 - 0.2	Soil is in field capacity range
0.2 - 0.5	Indicates soil moisture availability to the plant
0.5 - 0.6	Usual range for starting irrigation
0.6 and above	Moisture stress range

Methods of Irrigation

Sprinkler irrigation is the most widely accepted method in tea gardens of North East India. Drip irrigation is confined to seed baris.

Sprinkler system

This system simulates the natural rainfall and is easy to operate. However, it is capital intensive and requires high amount of energy. The rotating-head sprinkler is the most commonly used system. It consists of a head with one or two nozzles, which is rotated slowly by the action of water passing through it. It irrigates roughly a circular patch of land around the sprinkler. Rotating sprinkler operates under wide range of pressure and discharge. Sprinklers are classified broadly into three main groups according to the operating pressure.

The ratio of effective water application to total application from the sprinklers is known as the application efficiency, and this is usually around 70-80%. It is necessary to seek professional advice in choosing the appropriate equipment and designing a sprinkler layout.

Irrigation in young tea

In planning irrigation first priority should be given to the newly planted young tea. Among tea areas those having coarse texture soil and in the hills especially the south facing slopes should also get priority.

Mature Tea

Maximum response of irrigation is generally obtained in the best sections of existing mature tea areas. For this it will be also essential to identify and remove other limiting factors. The best results are expected to come from irrigating unprune or early light skiffed teas. In general, depending upon rainfall received in October irrigation should commence from November and continue till March/April. The first application in November can be a little more than the estimated field irrigation requirement followed by five more applications, each at an interval of three weeks. In severely drought prone years, irrigation in April may be necessary but after the rainfall in April exceeds 75-125 mm, irrigation can be discontinued. As in the case of young tea, irrigation schedule should not be interrupted except after heavy rainfall exceeding 38 mm and then for not longer than 2-3 days. The quality of the water used for irrigation should be checked prior to use.

Drainage

The objective of drainage planning is to remove the excess moisture in the root zone within a specified time so that the soil returns to field capacity without any harm to the tea plants. Quick and safe disposal of water from the catchments is the basis of drainage design. Developing outlet will augment the rate of outflow and thereby reduce water logging. Without a proper outlet, drainage planning can not be successful and thus for an efficient drainage layout, development of outlet should be taken up first with the ultimate objective of lowering the water table below the root zone depth. Studies at Tocklai indicate that improved drainage can increase tea yields to the extent of 30-35% over a period of time. The water table midway between two drains should be maintained at 90 cm below the ground level and it should not rise above this level for more than two days after heavy rains.

Drainage investigation

A drainage index map of the area showing all topographical features and outlet conditions are essential for drainage planning. Fluctuation of water table during the rainy season should be recorded using observation wells and seepage pressure with piezometer. This information together with soil properties like texture and hydraulic conductivity as well as records of highest flood level are important in designing an effective drainage system.

Open ditch drainage system

Open drainage system consist of a net work of field drains, collector drains and the main drain which discharges into natural drainage channels such as hullahs, streams or rivers. They are advantageous for removing large volumes of water from tea fields with minimum cost.

Pipe drainage system

A pipe drainage system is a drainage network in which slotted PVC, corrugated plastic and cement-asbestos pipes, etc., are laid at 1.2 to 1.5 m depth to remove excess water from the root zone. Around the pipes, a layer of graded gravels is used as filter material to prevent clogging of the slots or joints. The advantages of the pipe drains are (1) there is no loss of planting area, (2) maintenance cost is almost nil if properly laid, (3) effective in unstable sandy soils where open drains can not be maintained and (4) does not interfere with farming operations. However, the initial cost of pipe drainage is very high.

PESTS AND DISEASES

The tea growing environment in the North East India is conducive to a large number of pests and diseases. Detail studies have been made at Tocklai on the biology and control of tea pests during the last decades. Adequate and timely measures are to be taken to reduce crop loss due to pests and diseases.

Mode of dispersal of some major tea pests

	Mode of dispersal
Red spider mite	Wind, Pluckers, Cattles, Weed host
Scarlet mite	Weed host
Bunch caterpillar	Bush to bush migration of caterpillar, Migration of moths
Looper caterpillar	Migration from shade trees, Migration of moths
Red slug caterpillar	Migration of caterpillar through ground, Migration of moths
Nettle grub	Bush to bush migration of caterpillar, Migration of moths
Flush worm	Shoot to shoot migration of caterpillar, Migration of moths
Thrips	Migration of adults
Helopeltis	Migration of adults

Important weed hosts of tea pests in North East India

Weed hosts	*Pests*
Ageratum conizoides	Root knot nematode
Borreria hispida	Root knot nematode and scarlet mite
Commelina bengalensis	Root knot nematode and scarlet mite
Clerodendron infortunatum	Scarlet mite
Malastoma malabethricum	Helopeltis, Scarlet mite and Red spider mite
Mikania micrantha	Helopeltis
Pouzolzia indica	Root knot nematode and scarlet mite
Urena lohata	Red spider mite
Oxalis corymbosa	Root knot nematode
O. acetocella	Helopeltis

Pesticides invariably leave residues and their indiscriminate use may render the teas unsuitable for consumption and trade. With ever growing concern over pesticide residues and the rising costs of the pesticides, the concept of pest control has undergone radical changes. Therefore, monitoring of pests for their early detection, integrated management of pests (IPM) and discretion on the choice of pesticides to be used on tea is of utmost importance.

Cultural Operations

1. Bush sanitation/Cold weather practices
2. Ground sanitation
3. Balanced nutrition
4. Soil rehabilitation prior to replanting.
5. Improved drainage
6. Proper shade management
7. Soil amelioration

Mechanical and Physical Methods

1. Hand collection
2. Light trapping

Regulatory Methods

Only healthy, pest and disease free planting materials should be procured/used.

Biological Methods

Biological methods of control involve the conservation, preservation and introduction of natural enemies of tea pests, like predators, parasitoids and pathogens for suppression of pests within tolerable levels. More than one hundred indigenous natural enemies (predators, parasitoids and pathogens) have been recorded against various tea pests.

Pesticide can have detrimental effects on natural enemies thereby disturbing the bio-diversity and the balance in natural ecosystem. It is therefore necessary that the indigenous predators, parasitoids and pathogens are adequately preserved. Less toxic, selective pesticides should be preferred as much as possible, to keep the natural enemies active in tea areas.

Chemical Control & Choice of Pesticides

Different groups of synthetic organic insecticides have been recommended against major and minor pests in tea after regular bioassay tests and multilocation field screenings for efficacy and residues. The maximum permissible residue limits for pesticides in tea fixed by different international regulatory bodies have restricted the choice of pesticides.

TRA has organised awareness campaigns among the tea planters of this region through seminars and conferences highlighting the need for safer plant protection schedules for minimising residues so that tea continues to be a health drink. TRA has also generated data on residues of commonly used pesticides in tea through extensive supervised field trials. A number of invoice tea samples have also been monitored for pesticide residues recently, which reveals that most of the samples have residues well below the MRL limits.

Steps to minimise pesticide residues in tea are:

- Monitoring and early detection of pests
- Integrated pest management
- Use of safer pesticides, botanicals and bioagents
- Avoid repeated spraying of the same pesticide
- Allow sufficient waiting period between spraying of pesticides and plucking
- Spot treatment

Blister Blight

Blister blight is a seasonal disease and can spread very rapidly by means of air borne spores. Mature tea in general and teas recovering from pruning in particular suffer from blister blight. The fungus attacks young succulent growth on all teas especially when the environment is foggy, moist and cool. A single spore can produce a blister when deposited on tender shoots in about 10-21 days from the time of infection.

Control measures

This disease can be controlled by adopting chemical as well as cultural control strategies simultaneously. Some of the predisposing factors are dull, humid, cloudy and cool weather prevailing for several days, heavy shade etc. In such conditions prevails, monitoring of blister prone sections should be thoroughly undertaken. The shades should be thinned.

Normally 4-6 rounds Copper are necessary in Darjeeling. In Assam and Dooars, the spraying can be discontinued once the damp weather condition is over and it becomes hot and sunny for a few consecutive days. The order of priority for chemical treatment is:

nurseries → pruned young tea → early medium pruned tea → light pruned tea → unprune tea.

Black Rot

The disease is caused by fungi Corticium theae and Corticium invisum. Both the fungi produce similar effect on the bush and some times occur together. They attack the maintenance leaves, causing gradual deterioration in the health of the bush and consequent loss of crop. The disease is first noticed in May/June/July although germination from the sclerotial stage starts in March/April. The fungus remains active right up to September and then over winters in the sclerotial form in the cracks and crevices of the stem. Transformation in to sclerotial stage is a means for survival during the unfavourable weather condition. When favourable condition prevails, they germinate and re-infect.

Cultural control

1. Pruning or skiffing of the severely affected sections.
2. Improvement of aeration by lopping side branches and 'matidals'.
3. Thinning out of the dense shade.
4. Improvement of drainage.
5. Alkaline wash after pruning.
6. Adoption of shorter pruning cycle in chronic and severely affected sections.

Chemical control

Thorough drenching of all the stems and decomposing pruning litters below the bush with copper oxychloride (COC) one week after the first spell of rainfall in February/ March. Spray two blanket rounds of COC or Copper hydroxide or Hexaconazole at 15 days interval during May-June. Spot spraying of these two fungicides should continue till September if the disease incidence is severe. Application of two monthly rounds of COC or Carboxin during winter will inhibit sclerotia formation.

Control by bioagents

Two fortnightly rounds of 10% Bacillus subtilis broth can be applied in lieu of chemical fungicide during both active and dormant phase of the disease. Spraying should be directed towards small stems and under surfaces of the leaves, which bear fungal pathogens.

Branch canker and Thorny stem blight

Branch canker and Thorny stem blight are important stem diseases of tea prevailing in plains of Assam and Darjeeling hills, respectively. Branch canker is caused by Poria hypobrunnea and the causal organism of thorny stem blight is Tunstallia aculeata. These pathogens are typical wound parasites and enter the tea plants normally through heavy pruning cuts. Hence it is necessary to protect the fresh wounds with protective paints to stop the entry of these pathogens.

Control measures

For an effective control of these diseases careful removal of all dead wood at each pruning and subsequent protection of the pruning cuts with protective paints/sprays is necessary. These protective paints may be bituminous paint, COC or Trichoderma bioagents. During recent times Trichoderma treatment has been found to be the best method of protecting the pruning cuts from the damage of both Poria and Tunstallia. For painting of large pruning cuts soon after rejuvenation and medium pruning 80 litres of 20% Trichoderma spore suspension using 16L or kg of the bioagent formulation/ha is required (considering 8000 plants/ha). For spraying of entire bush frame after light pruning/cut across pruning, 400 litres of 5% Trichoderma spore suspension using 20L or kg of the bioagent formulation/ha is required using a NMD 60/450 nozzle.

Red Rust

It is a disease of young stems mostly, the ultimate symptoms of which normally manifest on the leaves. The causal organism of this disease is Cephaleuros parasiticus, an alga, causes severe damage especially to the young teas. Tissues of the stem are killed in patches and cause dieback. The leaves of the infected branches variegate with yellow patches. The alga produces brick red or orange coloured fructifications in patches on the infected stems of the tea bush.

Cultural control

1. The predisposing factors to be identified and corrected are i) poor drainage, ii) low soil fertility, particularly potash, iii) improper soil acidity, iv) inadequate shade and v) continuous use of green crops like Tephrosia candida, T. vogelli etc.
2. Pruning of severely affected sections.

Chemical control

In susceptible areas young teas should be sprayed with four rounds of COC/Copper hydroxide during mid April to mid July. First two rounds at 15 days interval and the subsequent rounds at monthly interval. Young shade trees/ green crops in nurseries should be painted with COC to avoid fresh infection.

2% MOP and Urea can be sprayed as a rehabilitary measure in the severely affected sections. Spraying of fungicides should be directed towards the young stems and laterals bearing rusty fructifications.

Dieback of Primaries and seed decay

During the last couple of years, dieback of primaries in certain succulent TV clones and seed decay in some biclonal seed stocks have been causing serious concern. The causal organism of these maladies is identified as Fusarium solani, a parasitic fungus. Generally this pathogenic fungus is predominant in the humid temperate climate of North East India. The susceptible clones and seed stocks to Fusarium solani are recorded as TV 19, TV 23, TV25, TV 26, TV 29, S3A1 parent of seed stock 491 and both the parents of seed stock 520. Symptoms of these maladies are as follows :

On primaries : Blackening of the petioles of leaves, which gradually extends to the nodes and inter-nodes, followed by wilting of the primaries. White cottony fungal growth is seen on the dying tissues, which turns brown at maturity accompanying with development of small pink perithecia of Necteria

On seeds : Blackening of fruit carp followed by immature cracking and dropping of seeds. Seed becomes light pinkish with powdery coverage due to the fungus sporulation in the later stage of infestation.

Control measures

Several chemicals have been evaluated for controlling Fusarium solani both in laboratory as well as in field. In these trials Carbendazim, Hexaconazole and Tridemorph and their combinations were found effective in controlling the pathogen.

Primary root diseases

The most common primary root diseases of tea plants in the plains of North East India are "Charcoal stump rot" and "Brown root rot". "Black root rot" is occasionally found in Darjeeling gardens while Red root rot is of rare occurrence. The root diseases can spread by means of direct contact or from left over diseased woody material. However, the spread of Charcoal stump rot and Black root rots may take place by air borne spores as well. The diseased materials may also be carried by rain to healthy teas. Unfortunately, there is no known method of detecting root infection by examining the above ground portion of the plant. Depending on the age and size of the bush it may require six months to four years to kill a infected plant. The tea roots remain in close proximity and hence by the time a dead plant is noticed, it is possible that its immediate neighbours have also become infected. Adoption of proper cultural practices is a vital point for control of primary root diseases as no effective chemicals are available at the moment to control these pathogens.

Cultural control

1. When the disease occurs on individual bush or in patch, the diseased plant together

with a ring of apparently healthy bushes surrounding it should be uprooted removing all bits of roots.

2. When a large area is infected the diseased patches should be isolated by 90 - 100 cm deep and 30 cm wide trench surrounding the area until uprooting can be undertaken. This trench should be connected to the nearest drain.
3. In case of old tea, due for uprooting, only the diseased plant is to be uprooted and the whole area should be kept isolated by an encircling trench as described above. The area should be uprooted two years before replanting.
4. All vacant areas thus created should be kept under a green crop for two years before infilling or replanting.

Biological control

Trichoderma bio agents have been found effective in controlling primary root diseases namely, Charcoal stump rot and Brown root rots. Planting pit treatment for infilling and replanting, 10% spore suspension should be used for treating the excavated soil of planting pits @ 100ml/pit. Alternatively, compost or well rotten cattle manure should be mixed thoroughly with the spore suspension and the treated manure should be mixing thoroughly with the excavated soil of the planting pit @ 200g/pit. The treatment will require 30L or kg of Trichoderma biocide/ha.

For top dressing around infills already planted in mature tea (1 year or below) should be sprayed around the collar region on the round @ 100 ml/plant using 5% spore suspension during Apr/May or Sep/ Oct before the soil gets dry. Forking of soil is a must before spraying.

Secondary Root Diseases

Violet root rot and Diplodia

Violet root rot caused by Spherostilbe repens and attacks of Diplodia sp. on weak plants are the common secondary root diseases of tea. Violet root rot disease is very common on stiff clayey waterlogged soils. It can easily be controlled by improvement of drainage. Diplodia fungus can live on dead stem and root tissues, which are deficient in reserved starch material. Healthy plants are not affected by this fungal pathogen. No chemical treatment is necessary for controlling both these diseases.

Young Tea Management

Various field management practices are followed in post-planting care to encourage early establishment and vigorous growth of tea plants as well as to increase their radial spread and longevity. These are shade, weed control, pest and disease control, drainage, irrigation, manuring and bush frame formation.

After planting, the area should be mulched adequately with green vegetative matter leaving a 10 cm distance from the collar of the plants uncovered. Sowing of green crops, e.g., *Crotalaria anagyroides, Tephrosia candida, Priotropis cytisoides* etc., in the young tea fields between alternate tea rows provides shade and generates mulching materials. This is particularly useful in droughty areas. The green crop has to be periodically lopped and the lopping left in situ to decompose.

The young plants should also be protected from pests and diseases using high volume sprays of suitable pesticides. If tea plants are affected by red rust, apart from spraying with copper fungicide as a control measure, it is also necessary to look for correction of drainage, shade, soil pH, and damage by drought etc.

Plants like *Indigofera teysmanii, Melia azadirach, Gliricidia sepium* etc., should be planted for providing temporary shade. They will be ready to provide shade within a year of planting and should be kept lopped periodically. Indigofera teysmanii is the most common species used for temporary shade and is propagated by means of seed, air-layering or stem cutting.

Weed growth is most vigorous in young tea fields during April to October warranting adequate control measures. Soon after planting, the ground is cheeled and mulched heavily or sprayed with a pre-emergent herbicide (simazine or oxyfluorfen). Subsequently hand weeding in rings of 15 cm radius around the collar of the plant is done. Herbicides spraying in young tea field is done with great care and using a protective shield.

Young tea fields should be adequately drained to avoid retention of excessive soil moisture and to promote root development. Ground should remain levelled, without any ridges and depressions, to prevent localised water logging. During dry periods, the first preference to irrigate should be given to young teas.

PRUNING

Pruning is one of the most important operations, next to plucking, which directly determines the productivity of tea bushes. It is a necessary evil in the sense that it has to be carried out periodically in spite of huge crop loss it results. If pruning is delayed, in other words as the age of wood from pruning increases, the size and weight of growing shoots on plucking surface decreases. There is preponderance of banji shoots on plucking table as more and more buds fail to grow with loss of vigour of growing apices. Therefore, to maintain the vegetative growth, pruning is necessary. The objectives are:

- To renew the wood.
- To provide stimulus for vegetative growth.

- To divert stored energy to production of growing shoots.
- To correct past defects in bush architecture.
- To maintain ideal frame height for economic plucking.
- To improve bush hygiene.
- To reduce the incidence of pests and diseases.
- To regulate the crop.
- To facilitate consolidation by infilling of vacancies.

Types of Pruning

Light prune

Tea bushes are usually pruned every 3 or 4 years at 4 -5 cm above the last pruning cut. This type of pruning is called light prune (LP). The time period from one light prune year to another is called one pruning cycle and LP is a thus, natural sequence given at the end of a pruning cycle. It helps to renew the wood, regulate crop distribution, reduce pests and diseases and maintain ideal frame height of the bushes.

Height reduction prune and medium prune

However, when the tea bushes grows tall and plucking becomes difficult, they are brought down to an optimum height by height reduction prune (HRP) at 60-70 cm, or medium prune (MP) at 45-60 cm above ground. Both HRP and MP help in rejuvenating the tea bushes that have become old and their yields have started declining. MP removes the knots and unproductive excess woods and facilitate consolidation by infilling of vacancies. Medium pruning thus provides opportunity for taking many corrective measures for improving the health and production capacity of old tea sections. Such sections are brought back to the normal 3 4 year pruning cycles in about 4-5 years.

Heavy prune

Heavy prune (HP) is given at 15-45 cm for complete renewal of frame. In collar prune, all above ground parts of the tea bushes are cut down and this operation is carried out only when the root system is considered strong enough to withstand the shock and initiate new growth. In practice however, very low pruning is generally avoided now a days as it results in heavy mortality, particularly in Assam jats growing in poorly shaded sandy soil. In between two successive prune (LP) years, tea bushes are given lighter forms of cuts which are termed as deep skiff (DS), medium skiff (MS), light skiff (LS), level of skiff (LOS) or untouched which is called unpruned (UP).

Deep Skiff

Deep skiffing (DS) of tea bushes is done normally between 12-15 cm above the last LP mark. The DS helps to regulate crop distribution and to reduce the ill effects of drought, excessive creep and the height of plucking table.

Medium Skiff

Medium skiff (MS) is normally given at 5 cm over last Deep skiff mark. The objective of MS is to regulate crop distribution, reduce the ill effects of drought, reduce the incidence of excessive banji formation and reduce the height of plucking table.

Level of Skiff and Light Skiff

Level of Skiff (LOS) is given 4-6 cm above the tipping mark mainly to level the plucking surface. Light Skiff (LS) is usually given up to 1 cm above the previous tipping height.

Time of pruning and skiffing

In general, pruning is carried out when the tea bush is dormant and there is a good reserve of root starch. In draughty areas of Northeast India, pruning is done in mid-December to mid-January avoiding shoot initiation during drought. Medium pruning can be carried out in mid-December to end of January. Skiffing is performed at anytime between mid-December and end January; deeper the cut earlier is the skiffing time. In Darjeeling, pruning can be started as early as November. The best time for pruning may differ with cultivars, which can be ascertained by root starch reserve test.

Resting before Pruning

For LP generally resting is not required. However, if there is inadequate foliage, the tea bushes are given rest by stopping plucking from 3 weeks before pruning. Very weak tea plants and those due for MP should be rested 5-8 weeks prior to pruning. Vigorously growing well-shaded tea bushes are pruned first. For MP/HP in the year of prune additional doses of potash and phosphate are also applied and in inadequate shaded areas, temporary shade of Indigofera teysmanii at 3-4 m apart should be planted a year before pruning.

Pruning and Bush Sanitation

After pruning at predetermined height, bushes are sprayed with an aqueous solution of copper fungicide on the same day. This reduces chances of damage by Poria, Aglaospora etc. Knife cleaning out (KCO) can be started any time after prune, and should be completed before bud-break starts. KCO involves removal of knots and of dead, diseased and unproductive wood. All large cuts should be indopasted within 24 hours. In MP some of the branches may have to be cut below the general pruning height to get clean wood. To prevent the chances of sun scorch, the bushes may be covered with the pruning litter immediately after the prune. At times they may also require demossing and lime-caustic-wash. Bushes, which are dead, badly damaged and affected by primary root rots should be uprooted and removed out of the section. In bushes attacked by termites the earth runs should be removed, and the bush and the surrounding area treated with recommended insecticide.

The caustic wash is prepared using 6 kg washing soda and 2kg quick lime in 100 lit of water and is used after keeping the preparation overnight.

When to go for medium or rejuvenation prune

Only those sections, which show potential for improving after the prune, should be considered for heavy prune. Some of the important criteria for judging the suitability to go for heavy, prune are as follows:

- The section should not be affected by primary root diseases.
- There should be scope for immediate improvement of drainage in the area.
- Vacancy should not be normally more than 30 per cent.
- Not more than 25 per cent of the remaining bushes should be affected by Poria and termites down to the collar region.
- Bushes show a steady decline in yield over the last 5 years or so.
- Recovery potential should be good. It is essential to build up enough starch reserves by giving adequate rest prior to pruning.

Pruning Cycle

In a pruning cycle, all forms of pruning/skiffings are introduced sequentially for distribution of crop and renewal of wood producing crop/quality. In comparison to annual prune, which causes a heavy rush only during the main season, well-balanced extended pruning cycles will provide an even distribution of crop resulting in an even flow to the factory throughout the season. The choice of pruning cycles depends on a number of factors, viz., crop, labour, quality and market requirements, pests and diseases, climate and soil, bush height, age, vigour and kind of tea, etc. In the plains of Northeast India under normal growing conditions 3 4 year pruning cycle and for Darjeeling depending upon elevation a 4-5 year pruning cycle is suitable.

PLUCKING

Plucking in tea is synonymous with harvesting in other crops. The tender apical portions of shoots consisting of 2-3 leaves and the terminal buds are nipped off in plucking. The plucked shoots are manufactured to produce tea.

Removal of the apical portion of a tea shoot stimulates growth of the dormant leaf and buds below the apex. The stimulated buds become active and start laying down initials of cataphylls, known as janams in North East India, of normal leaves and of another appendage intermediate between janam and leaf, which is known as fish leaf. While laying new initials, the bud swells up and after reaching a critical stage, starts unfolding janams, fish leaf and normal leaves in succession. All these appendages carry axil buds, which are capable of producing normal shoots of equal vigour. Advantage has been taken of this unique property in designing plucking system.

Plucking Systems

Three plucking systems are presently in vogue. They are:

- Janam plucking
- Fish leaf plucking
- Single leaf plucking

In theory, only janam should be left behind on the stubs of plucked shoots when following the janam plucking system, but in practice some fish leaves are also left on the bushes along with janams. This cannot be avoided for the sake of maintaining a flat plucking surface. Similarly, fish leaf plucking and single leaf plucking systems are generally a mixture of the two, with predominance of one or the other.

Whatever may be the plucking system, the harvested leaf should not be coarse as coarse leaf is harmful for quality of the product. Secondly, shoots below the plucking surface should be left alone.

Janam plucking is the system followed in North East India since under its soil environmental condition; the system has proved superior to the other systems. Other countries claim superiority of one or the other two remaining systems under their cultural conditions.

Due to janam plucking, the major bulk of tea produced in North East India is derived from the tiny janam axil buds.

Shoot Production

Light pruning stimulates the dormant buds on the short 2.5-3 cm pieces of stems known as sticks, to produce a new set of leafy branches, called primaries. When the primaries grow to a pre-set height above the bush frame, these are tipped (broken back) parallel to the ground surface. The average height of the fish leaf on the primaries is the usual height of tipping in N.E. India. All leaves left on the bushes below the tipping level are known as maintenance foliage. Tipping stimulates the buds on the axils of the leaves borne by the primaries, but the stimulus becomes weaker at increasing distance from the point of tipping. As a result, the top most point axil buds of all tipped primaries produce lateral shoots, which declines to 50% in the case of the second axil buds and 25% in the third.

The lateral shoots are plucked when they produce two or more leaves above the tipping level. Plucking of these lateral shoots of the first order originating from the axil buds on the primaries stimulates the growth of the second order laterals which when plucked produce the third order laterals and thus the production of successive orders of laterals continue until the end of the plucking season. In N.E. India up to 8th order laterals have been produced in a year.

Maintenance Foliage

Photosynthetic capacity develops gradually in a young, expanding tea leaf and it does not attain full capacity before attaining more than half its final size. It therefore, follows that the young shoots harvested for the manufacture of tea grow largely at the expense of the metabolites supplied by the mature maintenance foliage remaining below the plucking table. Hence adequacy of maintenance leaves must be assured to ensure sustained productivity of tea bushes. Under conditions of North East India, it has been observed that on an average five maintenance leaves per primary satisfy the requirements for health and productivity of tea.

Maintenance leaves on a tipped and plucked primary can remain for a maximum of 18 months, although its efficiency as a photosynthesising organ goes on diminishing after about 6 months. However, a very old maintenance leaf does not draw materials from other maintenance leaves. The leaf drops if it cannot generate enough materials to meet its own need.

SEED BARI

Tea seeds are produced in seed baries (orchards) planted specially for the purpose where the parents (selected generative clones) of approved stocks are established either planting directly or by grafting on any vigorous jat/clonal root stocks. Tea seeds are produced over the entire surface of a seed tree. Therefore, the planting distance between seed trees should be such that at maturity, the lower branches at the point of maximum spread of two neighbouring trees do not overlap. In general, spreading type of plants require about 5.5-6m space between them, while the erect type and Chinery plants may be planted at a closer spacing of 3.5-4.5m and 3-3.5m. The planting of seed bearers may be either square or triangular design. The number of plants varies from 330 to 1283/ha, depending on spacing and design of planting used. When the both parents are seed bearers, they are planted alternately in each row. If only one of the parents is a seed bearer, then the planting pattern is such that the male parent (pollen donor) is surrounded by a few seed bearers (female).

Seed baries can be raised on any soil suitable for tea cultivation. A well drained, fertile sandy loam to loam soil is most ideal.

Tea is a cross pollinated plant. Pollination is caused by pserphid, a small insect having a short flying range. It is therefore, necessary to isolate seed baries from other teas to avoid undesirable out crossing. A minimum distance of 10m should separate the seed baries from main tea fields.

Planting, after planting care and manuring are similar as that of young teas. Mature trees should be manured with NPK 2:1:2 mixture @ 100 kg N/ha/year in spring. Pruning is not recommended except to remove unwanted shoots to clear congestion which will facilitate aeration and light penetration. For higher seed production, 30-40% sunlight should reach the ground in a diffused manner. If there were large height difference between the clonal pairs (parents), balancing the height by lopping would facilitate efficient pollination. Old seed baries can be rejuvenated by heavy pruning (MP) as in tea under plucking. The major part of flower bud initiation (75%) takes place during March-June and about 97% of the flower bud formation is completed by end-September. Blossoming generally starts from early-October and ends by January with a peak period in November-December. Most of the seeds are produced from this peak flowering after 10 months. The seeds ripen by next October. The November-May period is very crucial for fruit development and moisture deficit in this period affects the seed yield adversely. The tea fruit is allowed to dehisce on the tree and the dropped seeds are collected from the ground. As these seeds lose viability rapidly, the normal practice is to collect them everyday. Tea seeds vary in size from 10 to 22 mm dia.

Grading and Sorting

After collection seeds are passed through a rotary type shifter to eliminate very small seeds. The remaining seeds are allowed to soak 2-3 hours in a trough of water. The seeds which sink (sinkers) are then spread on a plastic sheet, tarpaulin or on the concrete floor and examined for pest, disease and mechanical damage. The suspended or floating seeds (floaters) are discarded.

Storage of Seeds

Tea seed is stored for a minimum period in a cool place covered by a layer of semi-moist sand or sub-soil. Rapid drying during storage causes loss of viability.

Packing

After sorting, tea seeds are packed as soon as possible. For long distance transport, the seeds should be packed in wooden boxes in units of 20 kg. Seeds are spread in layers along with packing material and covering each layer with a thin sheet of tough paper. When the box is full, a sheet of paper is laid on top and the lid is nailed down. The different packing materials used are moist sand, sub-soil, powdered charcoal, ash and their mixtures. The moisture content of the packing material varies from 10-30%. If sandy loam sub-soil is used the optimum moisture content would be 10-12% while for powdered charcoal it may vary from 25-30%. Depending on the size of the grader used a kg of graded and sorted seed may contain 300-500 seeds.

PROCESSING AND CLASSIFICATION

A tea's type is determined by the processing through which it goes. Leaves of *Camellia sinensis* soon begin to wilt and oxidise if not dried quickly after picking. The leaves turn progressively darker as their chlorophyll breaks down and tannins are released. This process, *enzymatic oxidation,* is called *fermentation* in the tea industry, although it is not a true fermentation: it is not caused by micro-organisms, and is not an anaerobic process. The next step in processing is to stop the oxidation process at a predetermined stage by heating, which deactivates the enzymes responsible. With black tea this is done simultaneously with drying.

Without careful moisture and temperature control during manufacture and packaging, the tea will grow fungi. The fungus causes real fermentation that will contaminate the tea with toxic and sometimes carcinogenic substances, as well as off-flavors, rendering the tea unfit for consumption.

Tea leaf processing methods (Simplified)

Tea is traditionally classified based on the techniques with which it is produced and processed.

- White tea: Unwilted and unoxidised
- Yellow tea: Unwilted and unoxidised but allowed to yellow
- Green tea: Wilted and unoxidised
- Oolong: Wilted, bruised, and partially oxidised
- Black tea: Wilted, sometimes crushed, and fully oxidised
- Post-fermented tea: Green Tea that has been allowed to ferment/compost

Blending and Additives

Almost all teas in bags and most other teas sold in the West are blends. Blending may occur in the tea-planting area (as in the case of Assam), or teas from many areas may be blended. The aim is to obtain better taste, higher price, or both, as a more expensive, better-tasting tea may cover the inferior taste of cheaper varieties.Some teas are not pure varieties, but have been enhanced through additives or special processing. Tea is highly receptive to inclusion of various aromas; this may cause problems in processing, transportation and storage, but also allows for the design of an almost endless range of scented and flavored variants, such as vanilla, caramel, and many others.

Content

Tea contains catechins, a type of antioxidant. In a fresh tea leaf, catechins can compose up to 30% of the dry weight. Catechins are highest in concentration in white and green teas, while black tea has substantially fewer due to its oxidative preparation. Tea also contains theanine and the stimulant caffeine at about 3% of its dry weight, translating to between 30 mg and 90 mg per 8 oz (250 ml) cup depending on type, brand and brewing method. Tea also contains small amounts of theobromine and theophylline, as well as fluoride, with certain types of brick tea made from old leaves and stems having the highest levels.

Dry tea has more caffeine by weight than coffee; nevertheless, more dried coffee is used than dry tea in preparing the beverage, which mean that a cup of brewed tea contains significantly less caffeine than a cup of coffee of the same size.

Based on morphological differences between the Assamese and Chinese varieties, botanists have long asserted a dual botanical origin for tea; however, statistical cluster analysis, the same chromosome number (2n=30), easy hybridisation, and various types of intermediate hybrids and spontaneous polyploids all appear to demonstrate a single place of origin for *Camellia sinensis* — the area including the northern part of Burma, and Yunnan and Sichuan provinces of China. Yunnan Province has also been identified as "the birthplace of tea...the first area where humans figured out that eating tea leaves or brewing a cup could be pleasant". Fengqing County in the Lincang City Prefecture of Yunnan Province is said to be home to the world's oldest cultivated tea tree, some 3,200 years old.

REFERENCES

Claud Bald: Indian Tea. A Textbook on the Culture and Manufacture of Tea. Fifth Edition. Thoroughly Revised and Partly Rewritten by C.J. Harrison. Thacker, Spink & Co., Calcutta 1940 (first edition, 1933).

John C. Evans (1992): *Tea in China: The History of China's National Drink*,Greenwood Press.

Harler, C.R.: The Culture and Marketing of Tea. Second edition. Oxford University Press, New York and Bombay, Reprinted 1958 (First edition 1933, second edition 1956).

Yamamoto, T; Kim, M; Juneja, L R (1997). *Chemistry and Applications of Green Tea.* CRC Press. .

Lester Packer, Choon Nam Ong, Barry Halliwell (2004): *Herbal and Traditional Medicine: Molecular Aspects of Health,* CRC Press.

6

Coffee

Coffee is a brewed beverage prepared from roasted seeds, commonly called coffee beans, of the coffee plant. Due to its caffeine content, coffee has a stimulating effect in humans. Today, coffee is one of the most popular beverages worldwide. Coffee was first consumed in the ninth century, when it was discovered in the highlands of Ethiopia. From there, it spread to Egypt and Yemen, and by the 15th century, had reached Azerbaijan, Persia, Turkey, and northern Africa. From the Muslim world, coffee spread to Italy, then to the rest of Europe, to Indonesia, and to the Americas.

Coffee berries, which contain the coffee bean, are produced by several species of small evergreen bush of the genus *Coffea*. The two most commonly grown species are *Coffea canephora* (also known as *Coffea robusta*) and *Coffea arabica*; less popular species are Liberica, Excelsa, Stenophylla, Mauritiana, Racemosa. These are cultivated in Latin America, Southeast Asia, and Africa. Once ripe, coffee berries are picked, processed, and dried. The seeds are then roasted, undergoing several physical and chemical changes. They are roasted to varying degrees, depending on the desired flavour. They are then ground and brewed to create coffee. Coffee can be prepared and presented in a variety of ways.

Coffee has played an important role in many societies throughout history. In Africa and Yemen, it was used in religious ceremonies. As a result, the Ethiopian Church banned its secular consumption until the reign of Emperor Menelik II of Ethiopia. It was banned in Ottoman Turkey in the 17th century for political reasons, and was associated with rebellious political activities in Europe. Coffee is an important export commodity. In 2004, coffee was the top agricultural export for 12 countries, and in 2005, it was the world's seventh-largest legal agricultural export by value.

Indian coffee is the most extraordinary of beverages, offering intriguing subtlety and stimulating intensity. India is the only country that grows all of its coffee under shade. Typically mild and not too acidic, these coffees possess an exotic full-bodied taste and a fine aroma.

Indian coffee has a unique historic flavour too! It all began with a long, arduous journey around four hundred years ago, when the legendary saint Bababudan brought seven magical beans from distant Yemen and planted them in the Chandragiri hills of Karnataka. The sensations of aroma, flavour, body and acidity that you enjoy with each coffee experience is rooted in these mystical beginnings. It is often said, the Indian coffee grower pours his life into the crop. Is it any wonder then that India has consistently produced and exported a remarkable variety of high-quality coffees for over one hundred and fifty years!

India cultivates all of its coffee under a well-defined two-tier mixed shade canopy, comprising evergreen leguminous trees. Nearly 50 different types of shade trees are found in coffee plantations. Shade trees prevent soil erosion on a sloping terrain; they enrich the soil by recycling nutrients from deeper layers, protect the coffee plant from seasonal fluctuations in temperature, and play host to diverse flora and fauna.

Coffee plantations in India are essential spice worlds too: a wide variety of spices and fruit crops like pepper, cardamom, vanilla, orange and banana grow alongside coffee plants. India's coffee growing regions have diverse climatic conditions, which are well suited for cultivation of different varieties of coffee. Some regions with high elevations are ideally suited for growing Arabicas of mild quality while those with warm humid conditions are best suited for Robustas.

INDIAN COFFEE INDUSTRY

Significant increase both in production and exports notwithstanding, the international coffee market has been experiencing continuous fall in the commodity since the beginning of the current year. The composite indicator prices, according to the International Coffee Organisation's market report, has dropped to US cents 99.30 per LB in April, 2007 from US cents 105.81 per LB in January this year. Production of coffee during first six months of the coffee year 2006-07 stood at 9.29 million bags in March 2007 signifying an increase of 15 percent from 8.07 million bags recorded during comparable period of CY 2005-06. Exports in the first 6 months of coffee year 2006/07 (Oct-06 to Mar-07) were up over 18 percent from the same period last year - totalling 48.49 million bags compared to 40.98 million bags. Arabica exports in the 6-month period went up by over 13 percent from 27.39 million bags to 31.07 million bags, whereas Robusta exports increased by over 28% from 13.59 million bags to 17.41 million bags. Volume wise, the permits issued for coffee exports in fiscal 2006-07 is the highest ever (2,58,546 tonnes) surpassing the previous best of 2,46,908 tonne of actual exports in 2000-01. Export earnings in fiscal 2006-07 is the highest for the past 8 years i.e. since 1999-2000. The total export earnings will cross Rs.2000 crore for 2006-07.

According to latest ICO monthly report, world consumption in 2006 is estimated at 117.93 mn bags compared with 116.06 mn bags. While ICO member countries' domestic consumption in 2006 is estimated at little over 31 mn bags, the importing countries consumption is estimated at 86.84 mn bags. European Community's consumption is higher at 38.97 mn bags compared with 85.75 mn bags in 2005. India accounts for about 4.5 percent of world coffee production and the industry provides employment to 6 lakh people. Among the coffee growing states, Karnataka accounts for 70 percent of country's total coffee production followed by Kerala (22 percent) and Tamil Nadu (7 percent). Europe accounts for about 70 percent of India's total coffee exports. Of this again, 70 percent is shipped *via* Suez Canal. Major Indian coffee importing countries include Italy, Germany, Russian federation, Spain, Belgium, Slovenia, US, Japan, Greece, Netherlands and France.

BIOLOGY OF COFFEA PLANT

The *Coffea* plant is native to subtropical Africa and southern Asia. It belongs to a genus of ten species of flowering plants of the family *Rubiaceae*. It is an evergreen shrub or small tree that may grow 5 meters tall when unpruned. The leaves are dark green and glossy, usually 100–150 milimeters long and 60 milimeters wide. It produces clusters of fragrant white flowers that bloom simultaneously. The fruit berry is oval, about 15 milimeters long, and green when immature, but ripens to yellow, then crimson, becoming black on drying. Each berry usually contains two seeds, but 5–10% of the berries have only one; these are called peaberries. Berries ripen in seven to nine months.

Coffea arabica

Coffee is usually propagated by seeds. The traditional method of planting coffee is to put 20 seeds in each hole at the beginning of the rainy season; half are eliminated naturally. Coffee is often intercropped with food crops, such as corn, beans, or rice, during the first few years of cultivation. The two main cultivated species of the coffee plant are *Coffea canephora* and *Coffea arabica*. Arabica coffee (from C. *arabica*) is considered more suitable for drinking than robusta coffee (from C. *canephora*); robusta tends to be bitter and have less flavour than arabica. For this reason, about three-quarters of coffee cultivated worldwide is *C. arabica*. However, *C. canephora* is less susceptible to disease than *C. arabica* and can be cultivated in environments where *C. arabica* will not thrive. Robusta coffee also contains about 40–50% more caffeine than arabica. For this reason, it is used as an inexpensive substitute for arabica in many commercial coffee blends. Good quality robustas are used in some espresso blends to provide a better foam head and to lower the ingredient cost. Other cultivated species include *Coffea liberica* and *Coffea esliaca*, believed to be indigenous to Liberia and southern Sudan, respectively.

Most arabica coffee beans originate from either Latin America, eastern Africa, Arabia, or Asia. Robusta coffee beans are grown in western and central Africa, throughout southeast Asia, and to some extent in Brazil. Beans from different countries or regions usually have distinctive characteristics such as flavour, aroma, body, and acidity. These taste characteristics are dependent not only on the coffee's growing region, but also on genetic subspecies (varietals) and processing. Varietals are generally known by the region in which they are grown, such as Colombian, Java or Kona.

The *Coffea* plant is native to subtropical Africa and southern Asia. It belongs to a genus of ten species of flowering plants of the family *Rubiaceae*. It is an evergreen shrub or small tree that may grow 5 meters tall when unpruned. The leaves are dark green and glossy, usually 100–150 milimeters long and 60 milimeters wide. It produces clusters of fragrant white flowers that bloom simultaneously. The fruit berry is oval, about 15 milimeters long, and green when immature, but ripens to yellow, then crimson, becoming black on drying. Each berry usually contains two seeds, but 5–10% of the berries have only one; these are called peaberries. Berries ripen in seven to nine months.

Cultivation

Coffee is usually propagated by seeds. The traditional method of planting coffee is to put 20 seeds in each hole at the beginning of the rainy season; half are eliminated naturally. Coffee is often intercropped with food crops, such as corn, beans, or rice, during the first few years of cultivation.

The two main cultivated species of the coffee plant are *Coffea canephora* and *Coffea arabica*. Arabica coffee (from *C. arabica*) is considered more suitable for drinking than robusta coffee (from *C. canephora*); robusta tends to be bitter and have less flavour than arabica. For this reason, about three-quarters of coffee cultivated worldwide is *C. arabica*. However, *C. canephora* is less susceptible to disease than *C. arabica* and can be cultivated in environments where *C. arabica* will not thrive. Robusta coffee also contains about 40–50% more caffeine than arabica. For this reason, it is used as an inexpensive substitute for arabica in many commercial coffee blends. Good quality robustas are used in some espresso blends to provide a better foam head and to lower the ingredient cost. Other cultivated species include *Coffea liberica* and *Coffea esliaca*, believed to be indigenous to

Liberia and southern Sudan, respectively. Most arabica coffee beans originate from either Latin America, eastern Africa, Arabia, or Asia. Robusta coffee beans are grown in western and central Africa, throughout southeast Asia, and to some extent in Brazil. Beans from different countries or regions usually have distinctive characteristics such as flavour, aroma, body, and acidity. These taste characteristics are dependent not only on the coffee's growing region, but also on genetic subspecies (varietals) and processing. Varietals are generally known by the region in which they are grown, such as Colombian, Java or Kona. The coffee plant is a woody perennial evergreen dicotyledon that belongs to the Rubiaceae family. Because it grows to a relatively large height, it is more accurately described as a coffee tree. It has a main vertical trunk (orthotropic) and primary, secondary, and tertiary horizontal branches (plagiotropic).

Difference between Arabica and Robusta Coffee Beans

While there are several different coffee species, two main species of coffee are cultivated today. *Coffea arabica,* known as Arabica coffee, accounts for 75-80 percent of the world's production. *Coffea canephora,* known as Robusta coffee, accounts for about 20 percent and differs from the Arabica coffees in terms of taste. While Robusta coffee beans are more robust than the Arabica plants, but produces an inferior tasting beverage with a higher caffeine content. Both the Robusta and Arabica coffee plant can grow to heights of 10 meters if not pruned, but producing countries will maintain the coffee plant at a height reasonable for easy harvesting.

Plant Growth and Development

Three to four years after the coffee is planted, sweetly smelling flowers grow in clusters in the axils of the coffee leaves. Fruit is produced only in the new tissue. The Coffea Arabica coffee plant is self-pollinating, whereas the Robusta coffee plant depends on cross pollination. About 6-8 weeks after each coffee flower is fertilised, cell division occurs and the coffee fruit remains as a pin head for a period that is dependent upon the climate. The ovaries will then develop into drupes in a rapid growth period that takes about 15 weeks after flowering. During this time the integument takes on the shape of the final coffee bean. After the rapid growth period the integument and parchment are fully grown and will not increase in size. The endosperm remains small until about 12 weeks after flowering. At this time it will suppress, consume, and replace the integument. The remnants of the integument are what make up the silverskin. The endosperm will have completely filled the cavity made by the integument nineteen weeks after flowing. The endosperm is now white and moist, but will gain dry matter during the next several months. During this time the endosperm attracts more than seventy percent of the total photsynthesates produced by the tree. The mesocarps will expand to form the sweet pulp that surrounds the coffee bean. The coffee cherry will change colour from green to red about thirty to thirty-five weeks after flowing.

Coffee Plant Root System

The roots of the coffee tree can extend 20-25 km in total length and the absorbing surface of a tree ranges from 400 to 500 m^2 (Nutman). There are main vertical roots, tap roots, and lateral roots which grow parallel to the ground. The tap roots extend no further than 30-45 cm below the soil surface. Four to eight axial roots may be encountered which often originate horizontally but point downward. The lateral roots can extend 2 m from the trunk. About 80-90% of the feeder root is in the first 20 cm of soil and is 60-90 cm away from the trunk of the coffee tree. However, Nutman states that the greatest root concentration is in the 30 to 60 cm depth. The roots systems are heavily affected by the type of soil and the mineral content of the soil. To be thick and strong, the coffee roots need an extensive supply of nitrogen, calcium and magnesium. During planting the main vertical roots are often clipped to promote growth of the horizontal roots, which then have better access to water and added nutrients in the top soil.

Coffee Leaves

The elliptical leaves of the coffee tree are shiny, dark green, and waxy. The coffee bean leaf area index is between 7 and 8 for a high-yielding coffee. The coffee plant has become a major source of oxygen in much of the world. Each hectare of coffee produces 86 lbs of oxygen per day, which is about half the production of the same area in a rain forest.

OPTIMAL COFFEE ENVIRONMENT

For growing Arabica coffee beans, there are two optimal growing climates:

1. The subtropical regions, at high altitudes of 16-24°. Rainy and dry seasons must be well defined, and altitude must be between 1800-3600 feet. These conditions result in one coffee growing season and one maturation season, usually in the coldest part of autumn. Mexico, Jamaica, the S. Paulo and Minas Gerais regions in Brazil, and Zimbabwe are examples of areas with these climate conditions.
2. The equatorial regions at latitudes lower than 10° and altitudes of 3600-6300 feet. Frequent rainfall causes almost continuous flowering, which results in two coffee harvesting seasons. The period of highest rainfall determines the main harvesting period, while the period of least rainfall determines the second harvest season. Because rainfall is too frequent for patio drying to occur, artificial drying with mechanical dryers is performed in this type of coffee growing environment. Examples of countries that have this climate are Kenya, Colombia, and Ethiopia.

Robusta coffee is grown at much lower altitudes (sea level-3000 feet) in an area 10° North and South of the equator. It is much more tolerant to warm conditions than Arabica coffee.

Planting Coffee Beans

The following guide explains the fertiliser components recommended when planting coffee seedlings:

Before you plant coffee, Mavolta recommends that 80 g of P_2O_5, 12 g of K_2O, 200 g of dolomitic limestone (or 100 g of calcined limestone), 0.2 g of boron, 0.2 g of copper, and 1.0 g of zinc in a 0.4 x 0.4 x 0.4 meter hole (201) in each coffee planting hole. After the coffee plants are established, four applications of 5.0 g of nitrogen, 10-15 cm from the trunk, are recommended. In the first year 10 g of nitrogen and K_2O is applied 3-4 times. In the second year 12 g of nitrogen and K_2O is applied 3-4 times. After this time the amount of fertilizer recommended depends on productivity and is given in Table 1.

Table 1. Recommended amounts of N, P_2O_5, and K_2O relative to coffee plant productivity.

Productivity: 60kg bag/hectare²	*Nitrogen*	*Element P_2O_5*	*K_2O*
10	80-90	15-20	80-90
15	90-110	20-25	90-110
20	110-130	25-30	110-130
25	130-150	30-35	130-150
30	150-170	35-40	150-170

COFFEE PLANT PROPAGATION

For propagation of arabica coffee, ripe red cherries are collected, pulped, and the mucilage is removed by fermentation. The freshly picked coffee seeds (typically referred to as beans) can either be planted immediately or dried for later use. Coffee drying takes place on wire mesh trays in the shade. Correctly storing coffee beans is essential for a longer seed life. Dried coffee seeds can be used up to a year or more if properly stored.

Coffee Bean Germination Process

There are two basic methods for the germination of seeds. In one method, coffee seeds are pregerminated by spreading on a sand bed and covering with moist burlap bag sacks or straw. The seeds are watched closely and removed as soon as radicals emerge. An alternative method of germinating coffee beans is to mix the seeds with moist vermiculite or expanded polystyrene and keep in the polythene bag. Coffee seedlings are grown in nursery beds or polybags and are planted in the coffee fields when they reach 20-40 cm.

Growing Coffee Seeds in Nursery Beds

Once pregerminated, the coffee seedlings are planted in nursery beds containing soil consisting of well rotted cattle manure (10-20 liters per meter) and phosphate fertiliser (100 g per meter). Nursery beds should be built to be 1 meter wide and 50 cm deep and seedlings are spaced between 12-15 cm apart (for 20 cm tall plants) or 20 cm apart (for 30-40 cm tall plants). The nursery beds are shaded 50 % for the first couple months. Shading is reduced slowly and completely removed the last two months before planting coffee seedlings.

Growing Coffee Seeds in Polybags

Polybags, made of black diothene (200-gauge), are commonly used and filled with a mixture of topsoil, well rotted cattle manure, course sand, gravel, coffee pulp, and coffee husks. A ratio of three parts top soil to one part course sand and one part cattle manure is often used. A top dressing of nitrogen is applied by applying 20 g urea in 5.0 L of water per meter of bed.

Coffee Growing at Home

Growing a coffee plant at home is a rewarding experience that will help you learn and appreciate the work involved in producing a good cup of coffee. It is a very easy plant to take care of and is a great conversation piece, especially during flowering or cherry development.

Harvesting and preparing the coffee seeds

Ripe cherries should be harvested and picked from trees with good production and without any disease or other affliction. The cherry is the pulped by hand, washed with water, and fermented in a small container. The fermentation stage is complete when the mucilage no longer clings to the coffee. Wash away the fermented mucilage with fresh water. Any coffee beans that float at any stage of washing should be discarded. The beans must then be dried to about 20% moisture content on mesh screen in open and dry air, but not in direct sunlight. After pulping a coffee will have between 60-70% moisture content so you can determine the appropriate stopping point simply by weighing the beans. Otherwise you can simply bite open be bean and ensure that it is dry on the outside and slightly soft and moist on the inside. Alternatively, a pulped bean can be used immediately for planting and in some areas this is considered advantageous.

Germination

If coffee cherries are not readily available green coffee can be purchased from a local supplier, but it is essential that the bean is of a recent crop and recent shipment. I would recommend ordering green coffee from www.sweetmarias.com and asking for the most recent crop. The potential for germination will continue for almost four months, but after this time the germination rate is several fold less and germination time is significantly longer. Fresh seeds should germinate in 2.5 months, but old seeds can take as long as 6 months.It is advisable to pregerminate the seeds. First soak the seeds in water for 24 hours. Then sow the seeds in damp sand or wet vermiculite in which the excess water has been drained. Alternatively, you can place the seeds between moist coffee sacks, which should be watered twice a day and well drained.

Once the seed germinates very carefully remove it from the sand, vermiculite, or burlap bags. A hole about 1.25 cm deep should be made into a soil of friable loam soil with a high humus content. Rotted manure, bone meal, and dried blood can also be added. If this type of soil is not readily available try a light weight and porous soil. Place the seed flat side down (in pergamino if possible) in the hole and sprinkle soil to cover the hole. Do not press the soil down firmly. Placing a 1/2 inch of mulched grass on top will help preserve moisture, but should be removed when the seed has fully germinated.

The seeds should be watered daily. Too much water or too little water will kill the seed. The soil should remain well drained, but moist at all times.After germination the plant should either be left alone or carefully removed and planted in a soil with a low pH and high nitrogen content. The soil should be porous. Therefore, course sand or basalt gravel dust can be added. Manure can also be added. A fertiliser that is appropriate for orchids can be used sparingly for the coffee plant to maintain mineral levels and a low pH (acidic).

Care

The plant thrives under artificial plant lighting indoors. The outside temperature in countries outside the Tropic belt is too volatile and too cold to allow the tree to develop. I recommend watering the tree twice per week in what I call a full watering and a half watering. In a half watering, I simply add some water to the soil and allow it to drain. In a full watering I add water, allow it to drain, and then add water with fertiliser and allow it to drain. The key is to keep the soil most, but well drained.

After two or three years flowering and possibly cherries can be expected, but do not expect high-quality coffee unless you are at a high altitude and are monitoring the conditions of the artificial microclimate carefully. To spur flowering wait until the beginning of winter and significantly reduce watering for 2-3 months. When Spring begins water the plant well, which should shock it into producing flowers. From this point forward water well and regularly. Arabica coffee is self-fertilising so you will not need to worry about pollinating. Once the cherries mature you can harvest, pulp, ferment, dry, roast, and drink your own coffee production.

Coffee Fertiliser

Since the coffee hullls and pulp are rich in nutrients, many people often use coffee grounds as fertiliser. One 60 kg bag of coffee contains 1,026 g of nitrogen, 60 g of phosphorous, 918 g of potassium, 162 g of calcium, 90 g of magnesium, 72 g of sulphur, 0.96 g of boron, 0.80 g of copper, 3.6 g of iron, 1.2 g of manganese, 0.002 g of molybdenum, and 0.72 g of zinc. The pulp resulting from processing contains 1,068 g of nitrogen, 84 g of phosphorous, 2,250 g of potassium, 246 g of calcium, 78 g of magnesium, 90 g of sulphur, 2.04 g of boron, 1.08 g of copper, 9.0 g of iron, 1.80 g of Manganese, 0.004 g of Molybdenum, and 4.20 g of Zinc.

ARABICA COFFEE BEAN VARIETALS

Coffea arabica coffee beans

Although many varietals of Coffea Arabica exist, *C. arabica* varietal Arabica and *C. arabica* var. bourbon are considered to be the first coffee varietals. Other varietals are believed to be a product of these two cultivars. Production and resistance generally governs the types of coffee beans that a farm will choose to plant. Coffee quality is a secondary factor most of the time.

Coffee nean types

Typica - This is the base from which many coffee varietals have been developed. Like the other Coffea Arabica varietals that have been developed from it, Typica coffee plants have a conical shape with a main vertical trunk and secondary verticals that grow at a slight slant. Typica is a tall plant reaching 3.5-4 m in height. The lateral branches form 50-70° angles with the vertical stem. Typica coffee has a very low production, but has an excellent cup quality.

Bourbon - Bourbon coffee plants produce 20-30% more coffee than Typica, but have a smaller harvest than less most coffee varietals. Bourbon has less of a conical shape than Typica coffee plants, but has more secondary branches. The angles between the secondary branches and the main stem are smaller, and the branch points on the main stem are closely spaced. The leaves are broad and wavy on the edges. The fruit is relatively small and dense. The cherries mature quickly and are at a risk of falling off during high winds or rains. The best results for Bourbon coffee are realised between 3,500-6,500 feet. Cup quality is excellent and similar to Typica.

Caturra - Caturra is a mutation of Coffee Bourbon discovered in Brazil. It is a mutation with high production and good quality, but requires extensive care and fertilisation. It is short with a thick core and has many secondary branches. It has large leaves with wavy borders similar to Coffee Bourbon. It adapts well to almost any environment, but does best between 1,500-5,500 feet with annual precipitation between 2,500-3,500 mm. At higher altitudes quality increases, but production decreases.

Catuai - Catuai is a high yielding coffee plant resulting from a cross between Mundo Novo and Caturra. The plant is relatively short, and the lateral branches form close angles with the primary branches. The fruit does not fall off the branch easily, which is favourable with areas with strong winds or rain. Catuai also needs sufficient fertilisation and care.

Pache comum - Pache comum is a mutation of Typica coffee first observed on the farm El Brito, Santa Cruz Naranjo, Santa Rosa, Guatemala. Many consider the cup to be smooth or flat. This coffee varietal adapts well between 3,500-5,500 feet.

Pache colis - Pache colis was found in Mataquescuintla, Guatemala in a farm consisting of Caturra and Pache comum. The coffee fruits are very large and the leaves are roughly textured. Pache colis provides some resistance to phoma. It has secondary and tertiary branching, and typically grows to 0.8-1.25 m. It adapts well to altitudes of 3,000-6,000 feet with temperatures between 20-21°C.

Catimor - Catimor is a cross between Timor coffee (resistant to rust) and Caturra coffee. It was created in Portugal in 1959. Maturation is early and production is very high with yields equal to or greater than the yield of other commercial coffee varietals. For this reason the method of fertilisation and shade must be monitored very closely. The Catimor T-8667 descendants are relatively small in stature, but have large coffee fruits and seeds. The Catimor line T-5269 is strong and adapts well to lower regions between 2,000-3,000 feet with annual rainfall over 3,000 mm. T-5175 is very productive and robust, but can have problems at either very high or very low altitudes. At low altitudes there is almost no difference in cup quality between Catimor and the other commercial coffee varietals, but at elevations greater than 4,000 feet Bourbon, Caturra, and Catuai have a better cup quality.

Kent - Kent is used for its high yield and resistance to coffee rust.

Mundo Novo - Natural hybrid between Typica coffee and Bourbon coffee. The plant was first found in Brazil. The plant is strong and resistant to disease. Mundo Novo has a high production, but matures slightly later than other kinds of coffee. It does well between 3,500-5,500 feet with an annual rainfall of 1,200-1,800 mm.

Maragogype - This coffee varietal is a mutation of Typica coffee and was discovered in Brazil. The Maragogype coffee plant is large and is taller than either Bourbon or Typica. Production is low, but the seeds are very large. Maragogype adapts best between 2,000-2,500 feet. The cup characteristics are highly appreciated in certain coffee markets.

Amarello - This coffee varietal, as its name indicates, produces a yellow fruit. It is not widely planted.

Blue mountain - Blue mountain is a famous coffee varietal favoured for its resistance to the coffee berry disease and ability to thrive in high altitudes. It was first grown in Jamaica and is now grown in Kona, Hawaii. Blue mountain coffee, however, cannot adapt to all climates and maintain its high quality flavour profile.

HARVESTING COFFEE BEANS

Each year coffee is harvested during the dry season when the coffee cherries are brigh red, glossy, and firm.

Ripe cherries are either harvested by hand, stripped from the tree with both unrip and overripe beans, or all the coffee beans are collected using a harvesting machine These processes are called selective picking, stripping, and mechanical harvesting respectively. To maximise the amount of ripe coffee harvested, it is necessary t selectively pick the ripe coffee beans from the tree by hand and leave behind unripe green beans to be harvested at a later time.

In Brazil, harvesting the same coffee tree several times is less cost effective tha separating and discarding the unripe or overripe cherries. Therefore, Brazil typicall harvests using the stripping method when 75% of the coffee crop is perfectly ripe Stripping is feasible and cost effective in Brazil due to the uniform maturation of Brazilia coffees. In stripping, the coffee beans are pulled from the tree and fall to the groun where they are caught by sheets. The beans are removed from tree debris by tossin the coffee in the air allowing the wind to carry away sticks and leaves. The coffee i then put in 60 L green baskets, which is the tool are measurement used by coffe producers to determine wages. Some coffee estates, such as Fazenda Monte Alegre i Sul de Minas Brazil, have a computerised system to determine wages for picking coffe beans. This system accounts for the amount of coffee collected from each person, th difficulty of the coffee harvesting conditions, and the production of the region bein harvested.

COFFEE PROCESSING

Irrespective of the harvesting method, green coffee beans and overripe coffee cherrie inevitably end up mixed with the perfectly ripe cherries and must be separated durin coffee processing. Overripe coffee cherries, undeveloped coffee cherries, sticks an leaves float in water. Ripe coffee beans and green coffee cherries are dense and sinl Therefore, the first step in coffee production consists of separating the "floaters" fro the "sinkers." The coffee floaters are usually sent directly to the patio to be dried an are often slated for internal consumption. The ripe and green cherries can be sent t the patios to be dried using the natural process of preparing coffee or can be sent to th coffee pulping machines.

Coffee Processing Equipment

The first stage of coffee pulping is used to remove the green coffee cherries from th ripe cherries. In the coffee pulping machinery, the internal pressure is monitored to pus the coffee against a screen with holes only large enough for a coffee bean (not cherr to pass through. Since the ripe cherries are soft, they break and the coffee seed is release through the screen.

Pulping Coffee

The green cherries are hard and cannot be pulped. Instead of passing through the screen, the green coffee beans pass to the end of the barrel system and are separated from the ripe coffee beans. The pressure inside the barrel controls how many cherries will be pulped. A very high pressure will cause all of the cherries including the green beans to be pulped. It is necessary to continuously monitor the pressure so that about 3% of ripe cherries are not pulped and are removed with the green cherries. This margin of error ensures that no green cherries are mistakenly pulped. The pulp and coffee beans are then separated by centrifugal force and a barrel screen system.

Coffee Fermentation

The coffee beans covered in the slippery mucilage can be sent to the patios to dry as pulped natural coffees or can be sent to coffee fermentation tanks. The coffee fermentation tanks are used to remove the mucilage before drying. The pulped coffee beans are put into cement tanks with water and are allowed to ferment for 16-36 hours. On the way to the fermentation tanks, another density separation can occur. The highest quality coffees are the densest and should be separated and fermented in a different tank.

The coffee fermentation time depends on a number of factors including the amount of coffee fermenting, water temperature, and humidity. The mucilage is made up of pectin materials including protopectin (33%), reducing sugars including glucose and fructose (30%), non-reducing sugars such as sucrose (20%), and cellulose and ash (17%). Protopectin is not water soluble and will hydrolyse to pectinic acid in the fermentation tanks. Hydrolysis of the protopectin and degradation of the pectin by enzymes is the process that occurs to remove the mucilage during fermentation. Currently, the best way of determining the end of coffee fermentation is to feel the coffee beans to determine if they are still encased in mucilage. If the coffee beans are fermented for 36-72 hours, stinker beans develop. Lactic, acetic, and propionic acids are produced in this process and are believed to prevent the traditional fermentation taste by inhibiting mold growth that regularly occurs during drying on a patio in humid conditions.

From the coffee fermentation tanks, the beans are moved to drying patios and dried to 11-12% moisture content. A small portion of the lot is hulled and milled by a mini-huller. Three hundred grams of coffee is classified for defects (100 grams is often used), and the percentage of each screen size is determined. Then, 200-300 grams of coffee is roasted in a sample roaster and cupped to determine coffee quality. Ideally no lots will be mixed until the coffee has been classified and cupped. The coffee remains in pergamino until shipment time to help protect the flavour and aroma of the coffee. For a complete look at coffee production, view this video showing the steps of coffee processing: Medium (14.6 Mb) and Small (4.9 Mb).

FLAVOUR CHARACTERISTICS DUE TO COFFEE PROCESSING

Dry-Process: The dry-process (also known as the natural method) produces coffee that is heavy in body, sweet, smooth, and complex. The dry-process is often used in countries where rainfall is scarce and long periods of sunshine are available to dry the coffee properly. Most coffees from Indonesia, Ethiopia, Brazil, and Yemen are dry-processed.

Wet-Process: Wet-processing coffees is a relatively new method of removing the four layers surrounding the coffee bean. This process results in a coffee that is cleaner, brighter, and fruitier. Most countries with coffee valued for its perceived acidity, will process their coffee using the wet-process.

Pulped Natural: The pulped natural method consists of pulping a coffee, but emitting the fermentation stage to remove the silverskin. This results in a beverage that has characteristics of both a dry- and wet-processed coffee. It is often sweeter than wet-processed coffees, has some of the body of a dry-processed coffee, but also retains some of the acidity of a wet-processed coffee. This type of processing can only occur in countries where the humidity is low and the coffee covered in the sweet mucilage can be dried rapidly without fermenting. Brazil has made this method famous and produces some of the best pulped natural coffees in the world. All twenty winners of the Gourmet Cup competition in Brazil in 2000 processed their coffees using the pulped natural method.

Re-passed: There is another type of coffee that has emerged on the market called re-passed or raisins. These coffees are floaters and are usually discarded with the rest of the floaters. However, they have a flavour profile that some of the world's best experts find to be much sweeter than traditional pulped coffees. The cherries float because they have dried too long on the tree before being collected. This, however, allows the bean to interact with the mucilage for a longer amount of time before the start of fermentation.

The beans are removed from the rest of the floaters using a barrel system developed by Eduardo Sampio in Brazil. The coffees are then re-passed and pulped. They can then be washed or used as pulped naturals. The availability of the curiously sweet re-passed coffees is very limited since it is mainly experimental at this time. Ask your Brazilian supplier if they separate out this type of coffee and what flavour characteristics this coffee possesses. It may be another option for espresso blending and is likely to become the fourth category of coffee processing.

Comparison: The vast majority of coffee producers will claim the virtue of their processing method. In Guatemala, for example, dry processing is a bad word and rightly so. Due to their high humidity a dry processed coffee will almost definitely be fermented which is why only their lowest grade coffees are dried without pulping. However, in Brazil, dry processing results in a sweet, complex, and heavy-bodied coffee that is almost essential in any good espresso blend. The only conclusion that one can make is that every region has its own proper processing technique and that the processing technique should help attain the flavour profile that is desired by the producer and consumer.

Other flavour contributors: The processing method used on a coffee is usually the single largest contributor to the flavour profile of a coffee. The differences between a washed and dry-processed Brazilian coffee from Sul de Minas will generally be more distinct than the differences between two wet-processed coffees from two different regions. However, the microclimate and soil are the next major contributors to the flavour profile of a coffee and assuming processing is done correctly, they become the most important contributors to flavour profile.

COFFEE DRYING

Coffee Bean Moisture Measurement

Before shipment, coffee is dried and a coffee moisture meter is used to measure coffee bean moisture. Coffee must be dried from approximately 60% moisture content to 11-12% moisture content. Coffee is typically dried on large patios made of asphalt or cement and then transferred to mechanical dryers. The coffee on the drying patios is shifted every 30-40 minutes and is shaped into long rows of no more than 5 cm in height. Next to each row is open ground, which is warmed and dried by the sun. The coffee is then shifted onto the dry portion of the patio, and the section where it was previously is now allowed to dry in the sun. This helps accelerate the coffee drying process and prevents fermentation and moldy beans from developing. This method is widely used in Brazil, but less widely used in Guatemala or Costa Rica where the coffee is more often piled perpendicularly to the old piles. Drying coffee solely by patio takes 6-7 days for washed coffees, 8-9 days for pulped naturals (semi-washed), and 12-14 days for natural (dry-processed) coffees. This is why coffee beans are typically dried on a patio until they reach a moisture content of 15% and are then transferred to mechanical dryers. Once the coffee reaches a 25% moisture content or less, it can be piled at night and covered with cotton cloths to allow the coffee to breath. If it rains, these piles can also be covered with plastic. Coffee should not be covered with burlap sacks since this will impart a distinct burlap flavour and aroma to the coffee.

Coffee Drying Stages

In a study done in Kenya, Kamau reports that there are six stages to drying coffee.

1) Skin drying. Moisture 55-45%.
2) White Stage drying. Moisture 44-33%
3) Soft Black stage. Moisture 32-22%.
4) Medium Black Stage. Moisture 21-16%

5) Hard Black Stage. Moisture 15-12%

6) Fully dry coffee and conditioning. 11-10%.

In this study he found that sun drying coffee for stage three is mandatory for coffee quality. He also reports that as long as the temperature was between 40-50°C—which means a bean temp of 35°C—then the coffee quality will not be seriously compromised during the other coffee drying stages. The final two coffee drying stages (15-11% moisture) take just six hours at 40°C in a mechanical dryer.

Coffee Drying Equipment

There are several coffee dryer systems available. Many older dryers are converted grain dryers that are not as efficient as the new horizontal barrel dryers. The new coffee dryers are designed to mix the coffee evenly to ensure uniform drying. Drying coffee by using mechanical dryers accelerates the slowest part of the coffee drying process (15-11%) and helps prevent fermentation. In some environments that have a high humidity the entire drying process must take place in mechanical dryers. Mechanical dryers should never be set higher than 40-45°C and this question should be asked before buying any coffees that have been mechanically dried. At higher temperatures the germ is killed and the flavour potential of the coffee is ruined. At extreme temperatures the bean crystallises, and when smashed with a hammer, will break like glass.

Other Machinery for Drying Coffee

The best, but least utilised method of drying coffee is by using drying tables. In this method the pulped and fermented coffee is spread thinly on raised beds, which allows the air to pass on all sides of the coffee. The coffee is mixed by hand and the drying that takes place is more uniform and fermentation is less likely.

Most coffee from Africa is dried in this manner and select coffee farms around the world are following their lead. A relatively new method of drying coffee is solar drying. Solar coffee dryers are more economically sound than mechanical coffee dryers because they don't rely on fossil fuels or electricity. They are also more efficient than the patio drying technique because their hotter drying temperatures dry coffee beans at a faster rate. While solar coffee dryers have great potential for saving energy, they are not widely used today.

Other Machinery for Drying Coffee

The best, but least utilised method of drying coffee is by using drying tables. In this method the pulped and fermented coffee is spread thinly on raised beds, which allows the air to pass on all sides of the coffee. The coffee is mixed by hand and the drying that takes place is more uniform and fermentation is less likely. Most coffee from Africa is

7

Black Pepper

Black pepper (*Piper nigrum*) is a flowering vine in the family Piperaceae, cultivated for its fruit, which is usually dried and used as a spice and seasoning. The fruit, known as a peppercorn when dried, is a small drupe five millimetres in diameter, dark red when fully mature, containing a single seed. Peppercorns, and the powdered pepper derived from grinding them, may be described as black pepper, white pepper, red/pink pepper, green pepper, and very often simply pepper. The terms *pink peppercorns, red pepper,* and *green pepper* are also used to describe the fruits of other, unrelated plants.

Black pepper is native to South India and is extensively cultivated there and elsewhere in tropical regions. Black pepper is also cultivated in the Coorg area of Karnataka.Dried ground pepper is one of the most common spices in European cuisine and its descendants, having been known and prized since antiquity for both its flavour and its use as a medicine. The spiciness of black pepper is due to the chemical piperine. It may be found on nearly every dinner table in some parts of the world, often alongside table salt.

The word "pepper" is ultimately derived from the Sanskrit pippali, the word for long pepper via the Latin *piper* which was used by the Romans to refer both to pepper and long pepper, as the Romans erroneously believed that both of these spices were derived from the same plant.[citation needed] The English word for pepper is derived from the Old English *pipor*. The Latin word is also the source of German *Pfeffer*, French *poivre*, Dutch *peper*, and other similar forms. In the 16th century, *pepper* started referring to the unrelated New World chile peppers as well. "Pepper" was used in a figurative sense to mean "spirit" or "energy" at least as far back as the 1840s; in the early 20th century, this was shortened to *pep*.

Black pepper is produced from the still-green unripe berries of the pepper plant. The berries are cooked briefly in hot water, both to clean them and to prepare them for drying. The heat ruptures cell walls in the pepper, speeding the work of browning enzymes during drying. The berries are dried in the sun or by machine for several days, during and the volatiles have been exposed to the environment to undergo oxidation. An inert gas such as nitrogen can be used to help preserve the whole beans for an extended time. Some people claim that nitrogen could preserve coffee for up to two years, but after a few weeks the coffee no longer acts, tastes, nor smells like freshly roasted coffee. Whole bean coffee that has been opened and exposed to the environment should be kept for no longer than a week even if a vacuum seal is in place. Coffee that has been roasted very darkly is even more susceptible to oxidation and should be kept for an even shorter period of time. Finally, there is a significant loss in crema development for espresso coffee four days after roasting. Espresso coffee should be purchased immediately after roasting and consumed within a couple of days.

Colour Sorting Coffee Beans

Colour sorting is frequently used to remove the defective coffee beans that were not removed during coffee processing or hulling. Many countries sort coffee beans by hand due to inexpensive labour, but for most countries investment in a colour sorting machine, or colour separator, is necessary. There are several types of colour sorting instruments available on the market produced by two companies: Sortex and Xeltron.

The Colour Sorter

The methodology behind colour sorting machines is similar. The colour of each coffee bean is quickly measured using a type of CCD camera system and depending on this wavelength measurement the bean is either allowed to flow downward or is shot with a puff of air into the reject pile. Monochromatic machines remove white and black coffee beans since they have a lighter or darker hue than the average. The more specific bichromatic colour sorting machines detect and eliminate coffee beans that are white, unripe, broken, insect damaged, and black.

REFERENCESS

Dieter Fischer. "History of Indonesian coffee" Specialty Coffee Association of Indonesia. http://www.sca-indo.org/history-of-indonesia/. Retrieved on 2008-08-13.

Pendergrast, Mark (1999). *Uncommon Grounds: The History of Coffee and How It Transformed Our World*. Basic Books. ISBN 0-465-05467-6.

"Botanical Aspects". International Coffee Organization. http://www.ico.org/botanical.asp. Retrieved on 2007-07-19.

James A. Duke. "Coffea arabica L.". Purdue University. http://www.hort.purdue.edu/newcrop/duke_energy/Coffea_arabica.html#Ecology. Retrieved on 2007-07-20

Salvesen, David. "The Grind Over Sun Coffee". Smithsonian National Zoological Park. http://nationalzoo.si.edu/Publications/ZooGoer/1996/4/suncoffee.cfm. Retrieved on 2007-09-24.

dried in this manner and select coffee farms around the world are following their lead. A relatively new method of drying coffee is solar drying.

Solar coffee dryers are more economically sound than mechanical coffee dryers because they don't rely on fossil fuels or electricity. They are also more efficient than the patio drying technique because their hotter drying temperatures dry coffee beans at a faster rate. While solar coffee dryers have great potential for saving energy, they are not widely used today.

COFFEE MILLS

Molecular density sorting is one of the best ways to separate broken, small, undeveloped, and otherwise defective coffee beans. There are usually 2-3 stages of density sorting. In the first stage, the very dense rocks and stones are removed from the coffee. In the second stage, the coffee is hulled and the debris is removed. The third stage is most important and uses a densimetric table to separate the coffee into three or more densities. The settings on the machine must be continuously monitored to ensure that proper density sorting is occurring. The light and less dense coffee beans move to the right bottom side of the table, whereas the dense high-quality coffee beans move to the upper third of the table. This table will also remove any remaining debris that have made it through the previous density sorting stages. If the coffee mill is concerned with attaining the highest quality coffee possible, it may run the upper third coffee again to further separate any small deviations in density.

Storing Coffee Beans

Storing green coffee beans

Coffee must be stored in dry and cool conditions. Exposure to the sun or moisture will rapidly deteriorate the coffee. Storage in pergamino until right before the shipping time will help preserve the coffee. Burlap as are often used for coffee bean storage because they allow air flow. They also preserve the coffee longer than plastic or paper bags. Burlap bags should be aired on the patios before storing coffee to prevent a baggy flavour or burlap scent from being imparted to the coffee.

Coffee Seedlings Storage

Research in Kenya has shown that coffee seeds are useful for over two years if stored at 15°C at 41% moisture content in an airtight polythene bag.

Roasted Coffee Storage

Whole bean coffee maintains its freshness the longest. The freshness of ground coffee will be lost in a matter of minutes since its protective cellular structure has been broken which the pepper around the seed shrinks and darkens into a thin, wrinkled black layer. Once dried, the spice is called black peppercorn.

White pepper consists of the seed only, with the skin of the pepper removed. This is usually accomplished by process known as retting, where fully ripe peppers are soaked in water for about a week, during which the flesh of the pepper softens and decomposes. Rubbing then removes what remains of the fruit, and the naked seed is dried. Alternative processes are used for removing the outer pepper from the seed, including decortication, the removal of the outer layer from black pepper from small peppers through mechanical, chemical or biological methods.

In the U.S., white pepper is often used in dishes like light-coloured sauces or mashed potatoes, where ground black pepper would visibly stand out. There is disagreement regarding which is generally spicier. They have differing flavours due to the presence of certain compounds in the outer fruit layer of the berry that are not found in the seedGreen pepper, like black, is made from the unripe berries. Dried green peppercorns are treated in a manner that retains the green colour, such as treatment with sulfur dioxide or freeze-drying. Pickled peppercorns, also green, are unripe berries preserved in brine or vinegar. Fresh, unpreserved green pepper berries, largely unknown in the West, are used in some Asian cuisines, particularly Thai cuisine. Their flavour has been described as piquant and fresh, with a bright aroma. They decay quickly if not dried or preserved.

A product called orange pepper or red pepper consists of ripe red pepper berries preserved in brine and vinegar. Ripe red peppercorns can also be dried using the same colour-preserving techniques used to produce green pepper. Pink pepper from *Piper nigrum* is distinct from the more-common dried "pink peppercorns", which are the fruits of a plant from a different family, the Peruvian pepper tree, *Schinus molle*, and its relative the Brazilian pepper tree, Schinus terebinthifolius. In years past there was debate as to the health safety of pink peppercorns, which is mostly no longer an issue. Sichuan peppercorn is another "pepper" that is botanically unrelated to black pepper.

Peppercorns are often categorised under a label describing their region or port of origin. Two well-known types come from India's Malabar Coast: Malabar pepper and Tellicherry pepper. Tellicherry is a higher-grade pepper, made from the largest, ripest 10% of berries from Malabar plants grown on Mount Tellicherry. Sarawak pepper is produced in the Malaysian portion of Borneo, and Lampong pepper on Indonesia's island of Sumatra. White Muntok pepper is another Indonesian product, from Bangka Island.

BOTANICAL FEATURES

The pepper plant is a perennial woody vine growing to four metres in height on supporting trees, poles, or trellises. It is a spreading vine, rooting readily where trailing stems touch the ground. The leaves are alternate, entire, five to ten centimetres long and three to six centimetres broad. The flowers are small, produced on pendulous spikes four to eight centimetres long at the leaf nodes, the spikes lengthening to seven to 15 centimeters as the fruit matures. Black pepper is grown in soil that is neither too dry nor susceptible to flooding, moist, well-drained and rich in organic matter. The plants are propagated by cuttings about 40 to 50 centimetres long, tied up to neighbouring trees or climbing frames at distances of about two metres apart; trees with rough bark are favoured over those with smooth bark, as the pepper plants climb rough bark more readily. Competing plants are cleared away, leaving only sufficient trees to provide shade and permit free ventilation.

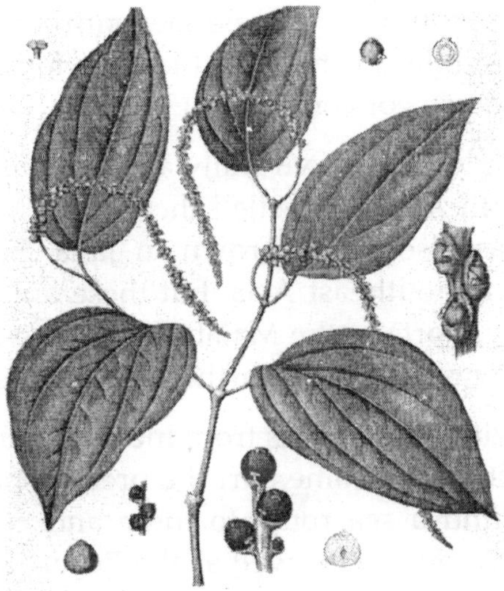

Paper Nigam

The roots are covered in leaf mulch and manure, and the shoots are trimmed twice a year. On dry soils the young plants require watering every other day during the dry season for the first three years. The plants bear fruit from the fourth or fifth year, and typically continue to bear fruit for seven years. The cuttings are usually cultivars, selected both for yield and quality of fruit. A single stem will bear 20 to 30 fruiting spikes. The harvest begins as soon as one or two berries at the base of the spikes begin to turn red, and before the fruit is mature, but when full grown and still hard; if allowed to ripen, the berries lose pungency, and ultimately fall off and are lost. The spikes are collected and spread out to dry in the sun, then the peppercorns are stripped off the spikes.

Black pepper is native to India. Within the genus Piper, it is most closely related to other Asian species such as Piper caninum.

USE OF PEPPER

Pepper has been used as a spice in India since prehistoric times. Pepper is native to India and has been known to Indian cooking since at least 2000 BC. J. Innes Miller notes that while pepper was grown in southern Thailand and in Malaysia, its most important source was India, particularly the Malabar Coast, in what is now the state of Kerala. Peppercorns were a much prized trade good, often referred to as "black gold" and used as a form of commodity money. The term "peppercorn rent" still exists today.

The ancient history of black pepper is often interlinked with that of long pepper, the dried fruit of closely related *Piper longum*. The Romans knew of both and often referred to either as just "piper". In fact, it was not until the discovery of the New World and of chile peppers that the popularity of long pepper entirely declined. Chile peppers, some of which when dried are similar in shape and taste to long pepper, were easier to grow in a variety of locations more convenient to Europe.

Until well after the Middle Ages, virtually all of the black pepper found in Europe, the Middle East, and North Africa travelled there from India's Malabar region. By the 16th century, pepper was also being grown in Java, Sunda, Sumatra, Madagascar, Malaysia, and elsewhere in Southeast Asia, but these areas traded mainly with China, or used the pepper locally. Ports in the Malabar area also served as a stop-off point for much of the trade in other spices from farther east in the Indian Ocean.

Black pepper, along with other spices from India and lands farther east, changed the course of world history. It was in some part the preciousness of these spices that led to the European efforts to find a sea route to India and consequently to the European colonial occupation of that country, as well as the European discovery and colonisation of the Americas.

Black peppercorns were found stuffed in the nostrils of Ramesses II, placed there as part of the mummification rituals shortly after his death in 1213 BC. Little else is known about the use of pepper in ancient Egypt, nor how it reached the Nile from India. Pepper was known in Greece at least as early as the 4th century BC, though it was probably an uncommon and expensive item that only the very rich could afford. Trade routes of the time were by land, or in ships which hugged the coastlines of the Arabian Sea. Long pepper, growing in the north-western part of India, was more accessible than the black pepper from further south; this trade advantage, plus long pepper's greater spiciness, probably made black pepper less popular at the time.

By the time of the early Roman Empire, especially after Rome's conquest of Egypt in 30 BC, open-ocean crossing of the Arabian Sea directly to southern India's Malabar Coast was near routine. Details of this trading across the Indian Ocean have been passed down in the Periplus of the Erythraean Sea. According to the Roman geographer Strabo, the early Empire sent a fleet of around 120 ships on an annual one-year trip to India and back. The fleet timed its travel across the Arabian Sea to take advantage of the predictable monsoon winds. Returning from India, the ships travelled up the Red Sea, from where the cargo was carried overland or via the Nile Canal to the Nile River, barged to Alexandria, and shipped from there to Italy and Rome. The rough geographical outlines of this same trade route would dominate the pepper trade into Europe for a millennium and a half to come.

With ships sailing directly to the Malabar coast, black pepper was now travelling a shorter trade route than long pepper, and the prices reflected it. Pliny the Elder's Natural History tells us the prices in Rome around 77 CE: "Long pepper ... is fifteen denarii per pound, while that of white pepper is seven, and of black, four." Pliny also complains "there is no year in which India does not drain the Roman Empire of fifty million sesterces," and further moralises on pepper:

It is quite surprising that the use of pepper has come so much into fashion, seeing that in other substances which we use, it is sometimes their sweetness, and sometimes their appearance that has attracted our notice; whereas, pepper has nothing in it that can plead as a recommendation to either fruit or berry, its only desirable quality being a certain pungency; and yet it is for this that we import it all the way from India! Who was the first to make trial of it as an article of food? and who, I wonder, was the man that was not content to prepare himself by hunger only for the satisfying of a greedy appetite?

Black pepper was a well-known and widespread, if expensive, seasoning in the Roman Empire. Apicius' De re coquinaria, a 3rd-century cookbook probably based at least partly on one from the 1st century CE, includes pepper in a majority of its recipes. Edward Gibbon wrote, in The History of the Decline and Fall of the Roman Empire, that pepper was "a favourite ingredient of the most expensive Roman cookery".

Pepper was so valuable that it was often used as collateral or even currency. The taste for pepper was passed on to those who would see Rome fall. It is said that Alaric the Visigoth and Attila the Hun each demanded from Rome a ransom of more than a ton of pepper when they besieged the city in 5th century. After the fall of Rome, others took over the middle legs of the spice trade, first the Persians and then the Arabs; Innes Miller cites the account of Cosmas Indicopleustes, who travelled east to India, as proof that "pepper was still being exported from India in the sixth century". By the end of the Dark Ages, the central portions of the spice trade were firmly under Islamic control. Once into the Mediterranean, the trade was largely monopolised by Italian powers, especially Venice and Genoa. The rise of these city-states was funded in large part by the spice trade.

It is commonly believed that during the Middle Ages, pepper was used to conceal the taste of partially rotten meat. There is no evidence to support this claim, and historians view it as highly unlikely: in the Middle Ages, pepper was a luxury item, affordable only to the wealthy, who certainly had unspoiled meat available as well. In addition, people of the time certainly knew that eating spoiled food would make them sick. Similarly, the belief that pepper was widely used as a preservative is questionable: it is true that piperine, the compound that gives pepper its spiciness, has some antimicrobial properties, but at the concentrations present when pepper is used as a spice, the effect is small. Salt is a much more effective preservative, and salt-cured meats were common fare, especially in winter. However, pepper and other spices probably did play a role in improving the taste of long-preserved meats.

Its exorbitant price during the Middle Ages—and the monopoly on the trade held by Italy—was one of the inducements which led the Portuguese to seek a sea route to India. In 1498, Vasco da Gama became the first European to reach India by sea; asked by Arabs in Calicut why they had come, his representative replied, "we seek Christians and spices." Though this first trip to India by way of the southern tip of Africa was only a modest success, the Portuguese quickly returned in greater numbers and used their superior naval firepower to eventually gain complete control of trade on the Arabian sea. It was given additional legitimacy by the 1494 Treaty of Tordesillas, which granted Portugal exclusive rights to the half of the world where black pepper originated.

The Portuguese proved unable to maintain their stranglehold on the spice trade for long. The old Arab and Venetian trade networks successfully smuggled enormous quantities of spices through the patchy Portuguese blockade, and pepper once again flowed through Alexandria and Italy, as well as around Africa. In the 17th century, the Portuguese lost almost all of their valuable Indian Ocean possessions to the Dutch and the English. The pepper ports of Malabar fell to the Dutch in the period 1661–1663.

As pepper supplies into Europe increased, the price of pepper declined (though the total value of the import trade generally did not). Pepper, which in the early Middle Ages had been an item exclusively for the rich, started to become more of an everyday seasoning among those of more average means. Today, pepper accounts for one-fifth of the world's spice trade.

It is possible that black pepper was known in China in the 2nd century BC, if poetic reports regarding an explorer named Tang Meng are correct. Sent by Emperor Wu to what is now south-west China, Tang Meng is said to have come across something called *jujiang* or "sauce-betel". He was told it came from the markets of Shu, an area in what is now the Sichuan province. The traditional view among historians is that "sauce-betel" is a sauce made from betel leaves, but arguments have been made that it actually refers to pepper, either long or black.

In the 3rd century CE, black pepper made its first definite appearance in Chinese texts, as *hujiao* or "foreign pepper". It does not appear to have been widely known at the time, failing to appear in a 4th-century work describing a wide variety of spices from beyond China's southern border, including long pepper. By the 12th century, however, black pepper had become a popular ingredient in the cuisine of the wealthy and powerful, sometimes taking the place of China's native Sichuan pepper.

Marco Polo testifies to pepper's popularity in 13th-century China when he relates what he is told of its consumption in the city of Kinsay (Zhejiang): "... Messer Marco heard it stated by one of the Great Kaan's officers of customs that the quantity of pepper introduced daily for consumption into the city of Kinsay amounted to 43 loads, each load being equal to 223 lbs." Marco Polo is not considered a very reliable source regarding China, and this second-hand data may be even more suspect, but if this estimated 10,000 pounds (4,500 kg) a day for one city is anywhere near the truth, China's pepper imports may have dwarfed Europe's.

Like many eastern spices, pepper was historically both a seasoning and a medicine. Long pepper, being stronger, was often the preferred medication, but both were used.Black peppercorns figure in remedies in Ayurveda, Siddha and Unani medicine in India. The 5th century *Syriac Book of Medicines* prescribes pepper (or perhaps long pepper) for such illnesses as constipation, diarrhea, earache, gangrene, heart disease, hernia, hoarseness, indigestion, insect bites, insomnia, joint pain, liver problems, lung disease, oral abscesses, sunburn, tooth decay, and toothaches. Various sources from the 5th century onward also recommend pepper to treat eye problems, often by applying salves or poultices made with pepper directly to the eye. There is no current medical evidence that any of these treatments has any benefit; pepper applied directly to the eye would be quite uncomfortable and possibly damaging.

Pepper has long been believed to cause sneezing; this is still believed true today. Some sources say that piperine, a substance present in black pepper, irritates the nostrils, causing the sneezing; some say that it is just the effect of the fine dust in ground pepper, and some say that pepper is not in fact a very effective sneeze-producer at all. Few, if any, controlled studies have been carried out to answer the question. It has been shown that piperine can dramatically increase absorption of selenium, vitamin B and beta-carotene as well as other nutrients.

Pepper contains small amounts of safrole, a mildly carcinogenic compound. Also, it is eliminated from the diet of patients having abdominal surgery and ulcers because of its irritating effect upon the intestines, being replaced by what is referred to as a bland diet.Pepper gets its spicy heat mostly from the piperine compound, which is found both in the outer fruit and in the seed. Refined piperine, milligram-for-milligram, is about one percent as hot as the capsaicin in chilli peppers. The outer fruit layer, left on black pepper, also contains important odour-contributing terpenes including pinene, sabinene, limonene, caryophyllene, and linalool, which give citrusy, woody, and floral notes. These scents are mostly missing in white pepper, which is stripped of the fruit layer. White pepper can gain some different odours (including musty notes) from its longer fermentation stage.

Pepper loses flavour and aroma through evaporation, so airtight storage helps preserve pepper's original spiciness longer. Pepper can also lose flavour when exposed to light, which can transform piperine into nearly tasteless isochavicine. Once ground, pepper's aromatics can evaporate quickly; most culinary sources recommend grinding whole peppercorns immediately before use for this reason. Handheld pepper mills, which mechanically grind or crush whole peppercorns, are used for this, sometimes instead of pepper shakers, dispensers of pre-ground pepper. Spice mills such as pepper mills were found in European kitchens as early as the 14th century, but the mortar and pestle used earlier for crushing pepper remained a popular method for centuries after as well.

CULTIVATION

Hot and pungent black pepper is one of the most popular spices in the world. Native to India, today it is cultivated all over the tropics as a commercial crop. India is the biggest producer. Pepper plants are climbers which grow to a height or length of 10 m or more. When its main stem is established, it grows lots of side shoots to create a bushy column.The plants form short roots, called adventitious roots, which connect to surrounding supports.

Although black pepper is cultivated in many tropical regions, it is native to Kerala State in India where it still occurs wild in the mountains.

- *Leaves* - arranged alternately on the stems. They are shaped like almonds and taper towards the tip. They are dark green and shiny above but paler green below.
- *Flowers* - grow in clusters along flowering stalks known as spikes. Between 50 to 150 whitish to yellow-green flowers are produced on a spike.
- *Fruits* - the flowers develop into round, berry-like fruits. There may be 50-60 fruits on each spike. They grow to a diameter of 4 to 6 mm, each containing a single seed. Fruits are green at first but they turn red as they ripen. These fruits are picked when either green or red to produce black and white pepper.

Other species exist that are also known as pepper or peppercorns and are used in a similar way to black pepper. Indian long pepper, *Piper longum*, is one example. It has a milder flavour than black pepper. It is native to Assam through to Myanmar, and is a cultivated crop in the drier regions of India.

Pink pepper is obtained from a plant called *Schinus terebinthifolius*, otherwise known as the Brazilian pepper tree. This plant grows as a tree and is in a different botanical family to black and long pepper. It is native to South and Central America. Its pinkish-red fruits often enter European markets where it is used as a black pepper-like flavouring.

Peppercorns should not be confused with chilli peppers from the *Capsicum* genus. Planting pepper is a lifetime livelihood if well developed. If the soil is good and care is good, the quality of the harvest is likewise good and so commands a good price in the market. When harvest is poor, even the stems of this plant, when dried, can also be used and sold including the chaff when powdered. Although these are not so strong in taste as the fruit itself, they are nevertheless sought because of the aroma they give to the food. So, even the powdered pulp gets sold.

Soil and Climate

Black pepper grows in almost all types of soil. However, it thrives well in loose, well-drained soil. It is best suited under humid climate with rainfall of 100 to 250 cm and in an elevation of 350 meters above sea level.

Seedbedding

Propagation of black pepper cuttings is prepared in seedbeds like vegetable crops. Till soil with garden hoe and reduce soil to very fine tilt. As soil is pulverised, add fine river sand to make bed more porous. Incorporate compost to make it more fertile. Drench seedbed with Captan spray solution at a concentration of 5 level tablespoonfuls Captan per gallon of water. Apply solution at the rate of one petroleum canful per 5 square meters of bed.

Propagation

Black pepper is propagated by sowing seed, marcotting and by stem cutting. Of the three methods, stem cutting is generally employed using both terminal stem and lateral branches. Select planting materials from high yielding hermaphrodite mother plant. Divide stem into cuttings each with 3-5 internodes and cuttings are rooted in sandy seedbed under shade. Insert cutting at 45degree angle 15 centimeters apart each way. When cutting has developed 4-7 new leaves, then it is ready for transplanting in the field. In marcotting method, secure black pepper branches on 1/2 inch wide bamboo pegs and cut each of them at fourth node from top of cacao or mabolo leaf filled with clean garden soil or moist sphagnum moss. If plastic sheet is available, cut into, convenient pieces and use for enclosing soil or moss and tie both ends with string or vegetable tying material. In 3-4 weeks, each marcot can be separated from mother plant and set in the field.

Land Preparation

For backyard planting select well-drained area and divide into equal areas of 2 x 2 square meter lots. To each corner of these lots, dig holes 60 cm sq about 40 cm deep, separate topsoil from subsoil. At center of hole, plant "kakawate" post 4 cm in diameter and 2 meters long for pepper vines to climb on as they grow. Fill up holes with 50-50 mixture of compost and topsoil. For large scale planting whether in cultivated or newly opened land, choose rolling area to have good drainage For newly opened area, other crops may be planted first for at least 3 years to free it from decaying tree stumps which may be sources of diseases. After 3 years, prepare land as backyard planting.

Transplanting

Transplant black pepper at the start of rainy season, 800 cm apart from the post with seedlings opposite each other. Around 3,200 seedlings are needed per hectare. It starts to bear fruits three years after planting. Open space can be utilised for planting short lived catch crops such as snapbeans, ginger, hot and sweet pepper at a distance of one meter away from black pepper rows.

Management

Black pepper needs weeding, mulching and pruning. When seedling reaches one or two feet high, nip off the top growing vine to induce more branching. Prune old and unproductive branches and crowding laterals. Farmers engaged in black pepper farming do not usually use extensive commercial fertilisers, but to ensure abundant harvest apply the following rates:

Plant Age	*Rate*
6 months old	100 grams of 14-14-14
1 year	220 grams of 14-14-14
2 years	500-700 grams of 14-14-14

Control of Pests and Diseases

Common pests are leafhoppers, root grubs, and African snails. Control leafhoppers by spraying with either Sevin or Malathion at the rate of one level tablespoonful per gallon of water. Drench hill with Aldrin spray at a concentration of on tablespoonful per gallon of water to control grubs. Collect snails and use them as fertiliser or poison with bait obtained at BPI. Black pepper, when properly planted, is not prone to any serious diseases outbreak.

HARVESTING

Peppercorn (berry-like fruit) matures in 5 to 6 months. The whole spike is ready for picking when:

- peppercorn in a spike turns cherry-red
- peppercorn turns from dark green to shiny yellowish green
- peppercorn has brownish cotyledon when pinched.

Harvest during sunny days. Open basket or sack used as container is tied to the waist of harvester. He twirls the spike with one hand while he holds the peduncle with the other hand.

DRYING

There are two ways of drying black pepper to yield two types of black pepper: the black peppercorn and the white pepper. Black peppercorn is black pepper dried under the sun or solar radiation to 12 % moisture content. Spread peppercorn on mat and remove spike, then peppercorn is winnowed, cleaned and stored in sacks. To produce white pepper, remove ripe berries from spikes, place in bags and soak in running water 1 to 2 weeks to soften skin. Remove skin by hands or by treading on berries with the feet. Wash and dry immediately smooth, white kernels.

Planting

1. Plant materials in pepper consist of the stem with 3 nodes. Separate these from the mother plant at the beginning of the rainy season, and plant in the nursery side by side for growing roots in a rooting bed.
2. The bed for growing these must be in the shade and elevated, surrounded by hollow blocks.
3. Make the bed three layers of soil so as to ensure good drainage such as:
   - river sand on top — 1 foot thick
   - small pebbles in the second layer
   - bigger stones at the bottom
4. Water a little — once in the morning and one in the afternoon. Cover with a plastic sheet.
5. In one month, roots will grow about an inch at the nodes.
6. This can now be transferred to plastic bags with holes at the bottom, half filled with sand and the other half with loose soil. When transplanting, thrust a stick into the plastic bag, and in its hole, put in the plant.
7. In 2-3 months when there are about 4-5 new growths (shoots) in the upper part of the plant, it can now be transferred in the fields.
8. Black pepper is a climbing plant, so it needs a trellis or tree to climb. The best is the madre cacao (kakawate). Its long branches are cut like poles and thrust into the ground, 2-3 m apart, in April and May. These branches will grow leaves and branches in the first rain in June. Change the branch that does not grow.
9. Plant two seedlings of pepper at the base of every pole. Remove them from the plastic bag and plant one feet away from the kakawate pole. Tie the stem to the pole to help it climb. In one week, the roots of the pepper will adhere to the pole and the plant will climb by itself.

Maintenance/Care of the Pepper Plant

1. Always remove weeds around the plant. Apply fertiliser about 2 ft. away from the plant; it is much better if this is mixed with the soil. Put the fertiliser at the beginning of the rains in June: 1 kilo of 14-14-14 or 16-16-16, and Â½ kilo in October. Without chemical fertiliser, the proper grains will not grow big and heavy.
2. Always prune the branches of the kakawate to allow sufficient sunshine on the pepper plant (but not so much pruning as to destroy the tree).

It is in the rainy season that pruning should be done after when the branches and leaves grow fast. Pruning is done about 5-6 times a year, or every 1-1/2 months during the rainy season. Control also the height of the kakawate (madre de cacao) because it is difficult to harvest pepper from a very high post, which should not be beyond 10 feet.

In Indonesia and Malaysia, they use posts made from strong wood that are treated against termites to make them last long. In this way, no pruning will be necessary.

Pests and Diseases of the Pepper - Two plant diseases afflict pepper, usually those whose soil has poor drainage. Pepper does not like soil that retains water, and may be the cause of its death. Other than these, pepper is resistant to pests and diseases, so pesticides are not necessary.

In 2 years, pepper bears fruit. Its fist fruits are not so plentiful, but in its 3rd year, harvest ranges from Â½ to one kilo per tree. On its 4th year, each plant can give one to one and a half kilo dried pepper.

1. The bunches of grains are harvested when these turn yellow. They do not ripen at the same time. It takes 3 weeks to one month to harvest them.
2. The harvested pepper is spread on cement to dry like palay or coffee. In good weather, this dries in 3 days. In rainy weather, this can be left alone even up to 5 days, but let it dry by itself when the rain stops. It will be worse to gather them and put them in a container where the grains will heat up, which hastens its rot.
3. Threshing - To separate the pulp and foreign matters from the grains, it is better to use a machine for the purpose because the machine does faster work with less cost for labor.
4. Winnowing is done normally in which the immature grain and light ones separate from the matured. In this way, high quality grains are segregated, which command a higher price. As mentioned earlier, even its chaff and stems, when powdered, are commerciable.

Black pepper (*Piper nigrum*) (Family: *Piperaceae*) is a perennial climbing vine grown for its berries extensively used as spice and in medicine. India is a leading producer, consumer and exporter of black pepper in the world. During 2005-06, 16,700 tonnes of black pepper products worth Rs. 14,050 lakhs were exported to various countries accounting for 6.0 % of export earnings among spices. Black pepper is cultivated to a large extent in Kerala and Karnataka and to a limited extent in Tamil Nadu and other states. The crop is grown in about 2,46,000 hectares with a production of 69,000 tonnes annually. Kerala and Karnataka account for a major portion (92 %) of production of black pepper in the country.

Climate and Soil

Black pepper is a plant of humid tropics requiring adequate rainfall and humidity. The hot and humid climate of sub mountainous tracts of Western Ghats is ideal for its cultivation. It grows successfully between 20° North and South latitude, and from sea level up to 1500 m above sea level. The crop tolerates temperatures between 10° and 40° C. A well distributed annual rainfall of 125-200 cm is considered ideal for black pepper. Black pepper can be grown in a wide range of soils with a pH of 4.5 to 6.5, though in its natural habitat it thrives well in red laterite soils. The black pepper growing areas in the West Coast of India include (i) coastal areas where black pepper is grown in homesteads (ii) midlands and where black pepper is extensively cultivated on a plantation scale and (iii) hills at an elevation of 800-1500 m above sea level, where the crop is mostly grown on shade trees in coffee, cardamom and tea plantations.

Varieties

A majority of the cultivated types are monoecious (male and female flowers found in the same spike) though variation in sex expression ranging from complete male to complete female is found. Over 75 cultivars of black pepper are being cultivated in India. Karimunda is the most popular of all cultivars in Kerala. The other important cultivars are Kottanadan (South Kerala), Narayakodi (Central Kerala), Aimpiriyan (Wynad), Neelamundi (Idukki), Kuthiravally (Kozhikode and Idukki), Balancotta, Kalluvally (North Kerala), Malligesara and Uddagare (Karnataka). Kuthiravally and Balancotta exhibit alternate bearing habit. In terms of quality, Kottanadan has the highest oleoresin (17.8%) content followed by Aimpiriyan (15.7%).

Sixteen improved varieties of black pepper have been released for cultivation. Panniyur-1 and Panniyur-3 are hybrids evolved at the Pepper Research Station, Panniyur (Kerala) and have Uthirankotta and Cheriyakaniakadan as their female and male parents, respectively. IISR Girimunda and IISR Malabar Excel are the two hybrids released from Indian Institute of Spices Research, Calicut.

Propagation

Black pepper vines develop three types of aerial shoots, namely (a) primary stem with long internodes, with adventitious roots which cling to the standards (b) runner shoots which originate from the base of the vine and have long internodes which strike roots at each node and © fruit bearing lateral branches. Cuttings are raised mainly from runner shoots, though terminal shoots can also be used. Cuttings from lateral branches are seldom used since they develop a bushy habit. However, rooted lateral branches are useful for raising bush pepper.

PRODUCTION OF ROOTED CUTTINGS

Traditional Method

Runner shoots from high yielding and healthy vines are kept coiled on wooden pegs fixed at the base of the vine to prevent the shoots from coming in contact with soil and striking roots. The runner shoots are separated from the vine during February-March, and after trimming the leaves, cuttings of 2-3 nodes each are planted either in nursery beds or in polythene bags filled with fertile soil. Adequate shade has to be provided and the polythene bags are to be irrigated frequently. The cuttings become ready for planting during May-June.

Rapid Multiplication Method

An efficient propagation technique developed at Sri Lanka has been modified for adoption in India for quick and easy multiplication of black pepper vines. In this method, a trench of 45 cm depth, 30 cm width and convenient length is made. The trench is filled with rooting medium comprising of forest soil, sand and farm yard manure in 1:1:1 ratio. Split halves of bamboo with septa or split halves of PVC pipes of 1.25-1.50 m length and 8-10 cm diameter provided with plastic septa at 30 cm intervals are fixed at 45° angle on a strong support.

Rooted cuttings are planted in the trench at the rate of one cutting for each bamboo split. The lower portions of the bamboo splits are filled with rooting medium (preferably weathered coir dust-farm yard manure mixture in 1:1 ratio) and the growing vine is tied to the bamboo split in such a way so as to keep the nodes pressed to the rooting medium. The tying can be done with dried banana leaf sheath fibers or coir rope. The cuttings are irrigated regularly. As the cuttings grow, the bamboo splits are filled with rooting medium and each node is pressed down to the rooting medium and tied. For rapid growth, a nutrient solution of urea (1 kg), super phosphate (0.75 kg), muriate of potash (0.5 kg) and magnesium sulphate (0.25 kg) in 250 litres of water is to be applied @ 0.25 litre per vine at monthly intervals.

When the vine reaches the top (3-4 months after planting of the cutting) the terminal bud is nipped off and the vine is crushed at about three nodes above the base, in order to activate the axillary buds. After about 10 days, the vine is cut at the crushed point and removed from the rooting medium and cut between each node. Each cutting with the bunch of roots intact is planted in polythene bags filled with fumigated potting mixture. Trichoderma @ 1g and VAM @ 100 cc/kg of soil can be added to the potting mixture. Care should be taken to keep the leaf axil above the soil. The polythene bags should be kept in a cool and humid place, or should be covered with thin polythene (200 gauge) sheet to retain humidity. The buds start developing in about 3 weeks and the polybags can then be removed and kept in shade. The advantages of this method of propagation are rapid multiplication (1:40), well developed root system, higher field establishment and vigorous growth as a result of better root system.

Trench Method

A simple, cheap and efficient technique for propagating black pepper from single nodes of runner shoots taken from field grown vines has been developed at the institute. A pit of 2.0 m x 1.0 m x 0.5 m size is dug under a cool and shaded area. Single nodes of 8-10 cm length and with their leaf intact, taken from runner shoots of field grown vines are planted in polythene bags (25 cm x 15 cm, 200 gauge) filled at the lower half with a mixture of sand, soil, coir dust and cow dung in equal proportion. The single nodes are to be planted in the bags in such a way that their leaf axil is above the potting mixture. The polythene bags with the planted single nodes are arranged in the pit.

After keeping the bags in the pit, the pit should be covered with a polythene sheet. This sheet may be secured in position by placing weights on the corners. The cuttings should be watered at least five times a day with a rose can and the pit should be covered with the polythene sheet immediately after watering. It is advisable to drench the cuttings 2-3 times with copper oxychloride (2g/litre).

After 2-3 weeks of planting, the cuttings will start producing roots which are visible through the polythene bags. After the initiation of roots the frequency of watering may be reduced to 3-4 times a day. After about 1 month, new shoots start emerging from the leaf axil. At this stage it is advisable to keep the pit open for about 1 hour per day so that the cuttings would harden and will not dry when they are taken out of the pit. The cuttings can be taken out of the pit after 2 months of planting and kept in a shaded place and watered twice a day. These cuttings will be ready for field planting after about 2½ months. By this method 80-85% success can be obtained. Foliar application of nutrient solution will also enhance the growth of the cuttings.

Serpentine Method

Cheaper propagation technique for production of rooted cuttings of black pepper is serpentine layering. In a nursery shed with roofing sheet or shade net, rooted black pepper cuttings are planted in polythene bags holding about 500 g potting mixture, which will serve as mother plants. As the plant grows and produces few nodes small polythene bags (20x10 cm) filled with potting mixture may be kept under each node. The node may be kept gently pressed in to the mixture assuring contact with the potting mixture with the help of a flexible twig such as mid rib of a coconut leaflet to enable rooting at that junction. Roots start growing from the nodes and the cuttings keep on growing further. The process of keeping potting mixture filled polythene bags at every node to induce rooting at each node is repeated. In 3 months the first 10 nodes (from the mother plants) would have rooted profusely and will be ready for harvest. Each node with the ploythene bag is cut just below the rooted node and the cut end is also buried into the mixture to induce more roots. Polythene bags filled with solarised potting mixture or soil, granite powder and farmyard manure in 2:1:1 proportion is recommended for producing disease free rooted cuttings. The rooted nodes will produce new sprouts in a week time and will be ready for field planting in 2-3 months time. Daily irrigation can be given with a rose can. On an average, 60 cuttings can be harvested per mother plant in a year by this method.

NURSERY DISEASES

Phytophthora Infections

Phytophthora infections are noticed on leaves, stems and roots of cuttings in the nursery. Dark spots with fimbriate margins appear on the leaves, which spread rapidly resulting in defoliation. The infections on the stem are seen as black lesions which result in blight. The symptoms on the roots appear as rotting of the entire root system.

Spraying Bordeaux mixture 1% and drenching with copper oxychloride 0.2% at monthly intervals prevents the disease. Alternatively, metalaxyl 0.01% (1.25 g/litre water) or potassium phosphonate 0.3% could also be used. The potting mixture may be sterilised through solarisation. To the sterilised mixture, biocontrol agents such as VAM @ 100 cc/kg of mixture and *Trichoderma* @ 1g/kg of soil (*Trichoderma* population @ 10^{10} cfu/g) may be added at the time of filling of nursery mixture in polythene bags and at regular intervals. *Pseudomonas fluorescens* (IISR-6) may be added to the potting mixture @ 1 g of product containing 10^{10} cfu/g to enhance growth and to suppress root pathogens. Application of *Trichoderma* and IISR-6 in the potting mixture at the time of planting and drenching IISR-6 at 1st and 2nd months after planting is recommended for producing disease free cuttings. Since the biocontrol agents mainly protect the root system, the aerial portion may be protected with chemicals. If Bordeaux mixture is used care must be taken to prevent dripping of fungicide to the soil. Alternatively, systemic fungicides such as metalaxyl (1.25 g/L) and potassium phosphonate (3 ml/L) which are compatible with *Trichoderma* may be used.

Anthracnose

The disease is caused by *Colletotrichum gloeosporioides*. The fungus infects the leaves causing yellowish brown to dark brown irregular leaf spots with a chlorotic halo. Spraying Bordeaux mixture 1% alternating with carbendazim 0.1% is effective against the disease.

Leaf rot and blight

The disease is caused by *Rhizoctonia solani* and is often serious in nurseries during April-May when warm humid conditions prevail. The fungus infects both leaves and stems. Grey sunken spots and mycelia threads appear on the leaves and the infected leaves are attached to one another with the mycelia threads. On stems, the infection occurs as dark brown lesions which spread both upwards and downwards. The new flushes subtending the points of infection gradually droop and dry up. Leaf spots caused by *Colletotrichum* sp. are characterised by yellow halo surrounding the necrotic spots. A prophylactic spray with Bordeaux mixture 1% prevents both the diseases.

Basal wilt

The disease is mainly noticed in nurseries during June-September and is caused by *Sclerotium rolfsii*. Grey lesions appear on stems and leaves. On the leaves white mycelium are seen at the advancing edges of the lesions. The mycelia threads later girdle the stem resulting in drooping of leaves beyond the point of infection and in advanced stages the rooted cuttings dry up. Small whitish to cream coloured grain like sclerotia bodies appear on the mature lesions. The disease can be controlled, if noticed early, by adopting phytosanitary measures. The affected cuttings along with defoliated leaves should be removed and destroyed. Later all the cuttings should be sprayed with carbendazim 0.2% or Bordeaux mixture 1%.

Viral infections

Vein clearing, mosaic, yellow specks, mottling and small leaf are the most obvious symptoms for identifying viral infections in the nursery. As viruses are systematic in nature, primary spread occurs through planting material since black pepper is vegetatively propagated. When infected plants are used as source of planting material, the cuttings will also be infected. Hence selection of virus free healthy mother plants is very important. Secondary spread of the disease occurs through insects such as aphids and mealybugs. Because of closed placing of seedlings in the nursery, chances of spread through these insects are more. Hence regular monitoring of the nursery for insects and spraying with insecticide like dimethoate 0.05% should be resorted to whenever they are seen. Besides, regular inspection and removal of infected plants should also be done.

Nematode infestation

Root-knot nematodes (*Meloidogyne* spp.) and the burrowing nematode, *Radopholus similis* are the two important nematode species infesting rooted cuttings in the nursery. The damage caused to roots by nematode infestations result in poor growth, foliar yellowing and some times interveinal chlorosis of leaves. The establishment of nematode infected cuttings will be poor when planted in the field and such cuttings develop slow decline symptoms at a later date. Nematode infestations tend to be more in rapid multiplication nurseries.

Soil solarisation can be done for sterilising the nursery mixture. The solarised nursery mixture may be fortified with biocontrol agents such as *Pochonia chlamydosporia* or *Trichoderma harzianum* @ 1-2 g/kg of soil, the product containing 10^6 cfu fungus/gm of substrate. Alternately, rhizobacteria like IISR 853 can be applied @ 1 g/bag (formulations containing $10^8 - 10^{10}$ cfu/g) at monthly intervals. A prophylactic application of nematicide is also necessary to check the nematode infestation. For this, make three equidistant holes of 2-3 cm depth in the bag around the cuttings and place phorate 10 G @ 1g/bag or carbofuran 3 G @ 3g/bag in these holes and cover with soil. A light irrigation may also be given to ensure adequate soil moisture.

Pollu Disease

This disease is caused by *Colletotrichum gloeosporioides*. It can be distinguished from the pollu (hollow berry) caused by the beetle by the presence of characteristic cracks on the infected berries. The disease appears towards the end of the monsoon. The affected berries show brown sunken patches during early stages and their further development is affected. In later stages, the discolouration gradually increases and the berries show the characteristic cross splitting. Finally, the berries turn black and dry. The fungus also causes angular to irregular brownish lesions with a chlorotic halo on the leaves. The disease can be controlled by spraying Bordeaux mixture 1%.

Spike Shedding

Spike shedding especially in varieties like Panniyur-1 at higher elevations like Kodagu and Idukki is one of the emerging diseases. It is seen in serious condition when the pre-monsoon showers are delayed and flowering and spiking occur during June-July. These spikes predominantly produce female flowers instead of bisexual flowers. Heavy spike shedding may occur due to lack of pollination. Irrigation of vines from second fortnight of March coupled with prophylactic spraying with Bordeaux mixture 1% or carbendazim 0.2% reduces the intensity of spike shedding.

Stunt Disease

This disease which is caused by viruses is noticed in parts of Kannur, Kasaragod, Kozhikode, Wayanad and Idukki Districts of Kerala and Kodagu, Hassan and Uthara

Kannada districts of Karnataka. The vines exhibit shortening of internodes to varying degrees. The leaves become small and narrow with varying degrees of deformation and appear leathery, puckered and crinkled. Chlorotic spots and streaks also appear on the leaves occasionally. The yield of the affected vines decreases gradually. Two viruses namely *Cucumber mosaic virus* and a *Badnavirus* are associated with the disease. The major means of spread of the virus is through the use of infected stem cuttings. The disease can also be transmitted through insects like aphids and mealy bugs.

Phyllody Disease

This disease which is caused by phytoplasma is noticed in parts of Waynad and Kozhikode districts of Kerala. The affected vines exhibit varying stages of malformation of spikes. Some of the floral buds are transformed into narrow leaf like structures. Such malformed spikes show leafy structures instead of floral buds, exhibiting phyllody symptoms. In advanced stages, the leaves become small and chlorotic, and the internodes are also shortened. The affected fruiting laterals give a witches broom appearance. Severely affected vines become unproductive. In severely affected vines the entire spike is converted into small branches which appear chlorotic and the vines decline rapidly. The infected vine becomes unproductive within 2 to 3 years. The infected vines are to be destroyed to prevent the further spread of the disease.

Slow Decline

Slow decline is a debilitating disease of black pepper. Foliar yellowing, defoliation and die-back are the aerial symptoms of this disease. The affected vines exhibit varying degrees of root degeneration due to infestation by plant parasitic nematodes. The diseased vines exhibit foliar yellowing from October onwards coinciding with depletion of soil moisture. With the onset of south west monsoon during May/June, some of the affected vines recover and put forth fresh foliage. However, the symptoms reappear in subsequent seasons after the cessation of the monsoon and the diseased vines gradually lose their vigour and productivity. The affected vines show varying degrees of feeder root loss and the expression of symptoms on the aerial parts occur after a considerable portion of the feeder roots are lost. The root system of diseased vines show varying degrees of necrosis and presence of root galls due to infestation by plant parasitic nematodes such as *Radopholus similis* and *Meloidogyne incognita* leading to rotting of feeder roots. The damage to feeder roots is caused by these nematodes and *P. capsici* either independently or together in combination. There is no spatial segregation of plant parasitic nematodes and *P. capsici* in the soil under field conditions. Hence, it is necessary to adopt a combination of fungicide and nematicide application for the management of the disease.

- Severely affected vines which are beyond recovery should be removed from the plantation and destroyed.

- The pits for planting should be treated with phorate 10 G @ 15 g or carbofuran 3 G @ 50 g at the time of planting.
- Nematode free rooted cuttings raised in fumigated or solarised nursery mixture should be used for planting in the field.
- Phorate 10 G @ 30 g or carbofuran 3 G @ 100 g/vine should be applied during May/June (with the onset of south west monsoon) and September/October. Along with nematicides the basins should be drenched with either copper oxychloride 0.2% or potassium phosphonate 0.3% or metalaxyl 0.125%.

In areas severely infested with root knot nematodes, cuttings of the resistant variety 'Pournami' may be planted. Biocontrol agents like *Pochonia chlamydosporia* or *Trichoderma harzianum* can be applied @ 50g/vine twice a year (during April-May and September-October). The fungus load in the substrate should be 108 cfu/g.While applying nematicides, the soil should be raked in the basin of the vine lightly without causing damage to the root system and the nematicide should be spread uniformly in the basin and covered with soil immediately. Sufficient soil moisture should be ensured at the time of nematicide application. The control measures should be taken up during early stages of the disease.

Insect pests

Pollu beetle

The pollu beetle (*Longitarsus nigripennis*) is the most destructive pest of black pepper and is more serious in plains and at altitudes below 300 m. The adult is a small black beetle measuring about 2.5 mm x 1.5 mm, the head and thorax being yellowish brown and the fore wings (elytra) black. Fully-grown grubs are creamywhite and measure about 5 mm in length. The adult beetles feed and damage tender leaves and spikes. The females lay eggs on tender spikes and berries. The grubs bore into and feed on the internal tissues and the infested spikes turn black and decay. The infested berries also turn black and crumble when pressed. The term pollu denotes the hollow nature of the infested berries in Malayalam. The pest infestation is more serious in shaded areas in the plantation. The pest population is higher during September-October in the field.

Regulation of shade in the plantation reduces the population of the pest in the field. Spraying quinalphos (0.05%) during June-July and September-October or quinalphos (0.05%) during July and Neemgold (0.6%) (neem-based insecticide) during August, September and October is effective for the management of the pest. The underside of leaves (where adults are generally seen) and spikes are to be sprayed thoroughly.

Top shoot borer

The top shoot borer (*Cydia hemidoxa*) is a serious pest in younger plantations in all black pepper areas. The adult is a tiny moth with a wing span of 10-15 mm with crimson and yellow fore wings and grey hind wings. The larvae bore into tender terminal shoots and feed on internal tissues resulting in blackening and decaying of affected shoots. Fully-grown larvae are grayish green and measure 12-15 mm in length. When successive new shoots are attacked, the growth of the vine is affected. The pest infestation is higher during July to October when numerous succulent shoots are available in the vines. Spray quinalphos (0.05%) on tender terminal shoots; repeat spraying at monthly intervals (during July-October) to protect emerging new shoots.

Leaf gall thrips

Infestation by leaf gall thrips (*Liothrips karnyi*) is more serious at higher altitudes especially in younger vines and also in nurseries in the plains. The adults are black and measure 2.5-3.0 mm in length. The larvae and pupae are creamy white. The feeding activity of thrips on leaves causes the leaf margins to curl downwards and inwards resulting in the formation of marginal leaf galls. Later the infested leaves become crinkled and malformed. In severe cases of infestation, the growth of younger vines and cuttings in the nursery is affected. Spray dimethoate (0.05%) during emergence of new flushes in young vines in the field and cuttings in the nursery.

Scale insects

Among the various scale insects recorded on black pepper, mussel scale (*Lepidosaphes piperis*) and coconut scale (*Aspidiotus destructor*) cause serious damage to black pepper vines at higher altitudes and also to older cuttings in nurseries in the plains. Females of mussel scales are elongated (about 1 mm length) and dark brown and that of coconut scales circular (about 1 mm in diameter) and yellowish brown. Scale insects are sedentary, remaining permanently fixed to plant parts and appear as encrustations on stems, leaves and berries. They feed on plant sap and cause yellowing and wilting of infested portions; in severe cases of infestation the affected portions of vines dry up. The pest infestation is more severe during the post monsoon and summer periods. Clip off and destroy severely infested branches. Spray dimethoate (0.1%) on affected vines; repeat spraying after 21 days to control the infestation completely. Initiate control measures during early stages of pest infestation. In nurseries spraying neem oil 0.3% or Neemgold 0.3% or fish oil rosin 3% is also effective in controlling the pest infestation.

Minor pests

Leaf feeding caterpillars, especially *Synegia* sp., damage leaves and spikes of younger vines and can be controlled by spraying quinalphos (0.05%). Mealybugs, gall midges and aphids infest tender shoots especially in nurseries. Spraying of dimethoate (0.05%) may be undertaken if infestations are severe. Mealybug infestation on roots can be controlled by drenching with chlorphyriphos (0.075%) and undertaking control measures against *Phytophthora* and nematode infections.

MANURING AND FERTILISER APPLICATION

Manuring and fertiliser application for pepper vines is to be done for proper establishment and growth of plants.

Recommended nutrient dosage for black pepper vines (3 years and above) are as follows.

NPK 50:50: 150 g/vine/year (General recommendation)

NPK 50:50: 200 g/vine/year (for Panniyur and similar areas)

NPK 140:55: 270 g/vine/year (for Kozhikode and similar areas)

Only one-third of this dosage should be applied during the first year which is increased to two-thirds in the second year. The full dose is given from the third year onwards. It is better to apply the fertilisers in two split doses, one in May-June and the other in August-September. The fertilisers are applied at a distance of about 30 cm all around the vine and covered with a thick layer of soil. Care should be taken to avoid direct contact of fertilisers with roots of black pepper. Organic manures in the form of cattle manure or compost can be given @ 10 kg/vine during May. Neem cake @ 1 kg/ vine can also be applied. Application of lime @ 500 g/vine in April-May during alternate years is also recommended. When biofertiliser like *Azospirillum* is applied @ 100 g/vine, the recommended nitrogen dose may be reduced by half to 70 g/ vine. In soils that are deficient in zinc or magnesium, foliar application of 0.25% zinc sulphate twice a year (May-June and September-October) and soil application of 150 g/vine magnesium sulphate, respectively is recommended.

ORGANIC PRODUCTION

For certified organic production of black pepper, at least 18 months the crop should be under organic management *ie* in the new plantations the first crop of pepper can be sold as organic, as the yielding starts from third year. To convert an existing plantation to organic, a conversion period of 36 months is set for the perennial crops. The conversion period may be relaxed if the organic farm is being established on a land where chemicals were not previously used, provided sufficient proof of history of the area is available. It is desirable that organic method of production is followed in the entire farm; but in the case of large extent of area, the transition can be done in a phased manner for which a conversion plan has to be prepared. The entire pepper holding can be converted to organic production when pepper is grown as sole crop. When grown in a mixed cultivation system, it is essential that all the crops in the field are also subjected to organic methods of production. Black pepper as a best component crop in agri-horti and silvi-horti systems, recycling of farm waste can be effectively done when grown with coconut, arecanut, coffee, rubber etc. As a mixed crop it can also be intercropped with green manure/ legumes crops enabling effective nutrient built up. In order to avoid contamination of organically cultivated plots from neighboring non-organic farms, a suitable buffer zone with definite border is to be maintained. In smallholder groups, where the pepper holdings are contiguous, the isolation belt is needed at the outer periphery of the entire group of holdings. Pepper grown on this isolation belt cannot be treated as organic. In sloppy lands adequate precaution should be taken to avoid the entry of run off water and chemical drift from the neighboring farms.

Management Practices

For organic production, traditional varieties adapted to the local soil and climatic conditions that are resistant or tolerant to diseases, pests and nematode infection should be used. All crop residues and farm wastes like green loppings, crop residues, grasses, cow dung slurry, poultry droppings etc. available on the farm can be recycled through composting, including vermicomposting so that soil fertility is maintained at high level. No synthetic chemical fertilisers, pesticides or fungicides are allowed under organic system. Farmyard manure may be applied @ 5-10 kg/vine along with vermi/ leaf compost @ 5-10 kg/vine based on the age of the vine. Based on soil test, application of lime/ dolomite, rock phosphate/ bone meal and wood ash may be done to get required quantity of phosphorus and potassium supplementation. When the deficient conditions of trace elements become yield limiting, restricted use of mineral/chemical sources of micronutrients and magnesium sulphate are allowed as per the limits of standard setting or certifying organisations. Further, supplementation of oil cakes like neem cake (1 kg/ vine), composted coir pith (2.5 kg/ vine) or composted coffee pulp rich in potassium and suitable microbial cultures of *Azospirillum* and phosphate solubilising bacteria will improve the fertility.

Use of biopesticides, biocontrol agents, cultural and phytosanitary measures for the management of insect pests and diseases forms the main strategy under organic system. Management of pollu beetle by Neemgold (0.6%) spray at 21 day intervals during July-October, shade regulation and that of scale insects by removing severely infected branches and spraying Neemgold (0.6%) or fish oil rosin (3%) are recommended. Application of biocontrol agents like *Trichoderma* or *Pseudomonas* multiplied in suitable carrier media such as coffee husk/ coir pith compost, well rotten cow dung or quality neem cake may be done regularly to keep the foot rot disease in check. To control fungal pollu and other foliar diseases spraying of Bordeaux mixture 1% may be done restricting the quantity to 8 kg copper per hectare per annum. Application of quality neem cake mentioned earlier along with the bioagents *Pochonia chlamydosporia* will be useful to check the nematode population and thereby slow decline disease.

Certification

Certification and labeling is usually done by an independent body to provide a guarantee that the production standards are met. Govt. of India has taken steps to have indigenous certification system to help small and marginal growers and to issue valid organic certificates through certifying agencies accredited by APEDA. The inspectors appointed by the certification agencies will carry out inspection of the farm operations through records maintained and by periodic site inspections. The grower has to document all the details with respect to Field map, Field history sheet, Activity register, Input record, Output record, Harvest record, Storage record, Pest control records, Movement record, Equipments cleaning record and Labelling records etc. Documentation of farm activities is must for acquiring certification especially when both conventional and organic crops are raised. Group certification programmes are also available for organised group of producers and processors with similar production systems located in geographical proximity.

REFERENCES

Dalby, Andrew (2002). *Dangerous Tastes*. Berkeley: University of California Press. http://books.google.com/books?id=7IHcZ21dyjwC.

Davidson, Alan (2002). *Wilder Shores of Gastronomy: Twenty Years of the Best Food Writing from the Journal Petits Propos Culinaires*. Berkeley: Ten Speed Press.

Jaffee, Steven (2004). "Delivering and Taking the Heat: Indian Spices and Evolving Process Standards". *An Agriculture and Rural Development Discussion Paper*. Washington: World Bank.

McGee, Harold (2004). "Black Pepper and Relatives". *On Food and Cooking (Revised Edition)*. Scribner. pp. 427-429.

Turner, Jack (2004). *Spice: The History of a Temptation*. London: Vintage Books. OCLC 61213802.

8

Cashew

Cashew as a marketable commodity, has a very important role to play in the liberalised Indian economy. With export earnings of Rs. 12,320 million in 1995-96, cashew ranked as one of the top agricultural export commodities. From the farmers' as well as from the exporters' point of view, the current emphasis that cashew is receiving as a horticultural crop from the research and development front, is a welcome sign. At present, India has a processing capacity of nearly seven hundred thousand metric tons and to meet the raw nut demand, the country depends partially on imports from several African, and in recent years, from south-east Asian countries. This has considerable drain on the country's foreign exchange reserves and there is an urgent need to increase local production to substitute imported raw material in order to derive the maximum benefits from a strong processing and marketing capability developed over the years by the Indian cashew industry.

Research work on cashew was initiated on a relatively small scale in early 1950's resulting in the development of several production techniques. These efforts were further strengthened when the national research mandate was delegated to the Central plantation Crops Research Institute (CPCRI), Kasaragod, in 1970 which spearheaded the All India Coordinated Spices and Cashew improvement Project from 1971. These research activities received further impetus with the implementation of a World Bank aided multi-State Cashew Project in the States of Andhra Pradesh, Kerala, Karnataka and Orissa from 1982-86. A National Research Center for Cashew was established at Puttur to increase the production and productivity of cashew with the mission-mode approach in 1986.

The cashew development component of the combined All India Coordinated Spices and Cashew Improvement Project was de-linked and an independent National Cashew Research project was initiated with the newly established National Research Center (NRC) for the crop at the same time. There are 8 research centers and one sub-center at present, located in 8 cashew growing States in the country. This can be considered as a milestone in cashew development with firmly established linkages with the

Directorate of Cashew nut Development Corporation and other extension agencies which assisted in the transfer of newly developed production technologies.

STATUS OF CASHEW PRODUCTION

Cashew is grown in the western and eastern coastal areas and further inland in some parts of Karnataka and Madhya Pradesh. Currently, the area under cashew is around 634,900 ha with a total production of 417,000 tons (Table 1). With 118,000 ha and a production of 140,000 tons, Kerala accounts for 18.6% of the area and 33.5 % of production respectively. The highest productivity is observed in Kerala and Maharashtra with over one ton per ha. The high yields in Maharashtra are primarily due to the fact that cashew production is of recent origin and the major part of the plantations have been established with high yielding clonal material. Even the orchards raised from seeds are from selected progenies. The current targets are set to produce 700,000 tons from 700,000 ha by the year 2000 AD.

Table 1. Area, Production and Productivity of Cashew in India (1965-96)

States	*Area (ha)*	*National % Area*	*Production (tons)*	*National % Prod.*	*Productivity (kg/ha)*
Kerala	118,600	18.6	140,000	33.5	1,180
Karnataka	83,900	13.2	37,600	8.9	448
Goa	49,600	7.8	17,800	4.3	359
Maharashtra	66,700	10.5	69,000	16.5	1,034
Tamil Nadu	77,360	12.2	30,930	7.4	400
Andhra Pradesh	118,080	18.6	71,700	17.2	607
Orissa	101,850	16.0	43,000	10.3	422
West Bengal	8,680	1.4	6,960	1.7	802
Others	10,200	1.6	840	0.2	82
Total	634,970	100.0	417,830	100.0	658

VARIETIES

Thirty-three cultivars have been released so far by the National Research Center as well as several Agricultural Universities. Most of the varieties have a mean yield of 8-10 kg per tree which gives over one ton per ha. In view of the export potential however, cultivation is recommended only for such cultivars that have a kernel grade of W-210 to W-240 (210-240 kernels per lb).

PROPAGATION

A number of propagation methods have been tested for the multiplication of cashew. Air-layering was found to be one of the popular methods among growers. This technique however, produced trees with poor anchorage as the root density was found to be low.

It also resulted in poor field establishment and high susceptibility to cyclones and drought conditions. Air layering was therefore found to be unsuitable for commercial exploitation. In the case of mound layering too, the absence of a tap root was found to be a disadvantage. Epicotyl grafting was another method that had limitations due to high mortality at transplanting and incidence of collar-rot at the nursery stage. Soft-wood grafting developed at the research centers was found to be the most viable method of propagation that was commercially acceptable; it gave a success rate of about 70 percent. The technique of soft-wood grafting described below is similar to epicotyl grafting except the difference in the age of the rootstock.

Cashew

Forty to sixty day old seedlings are used as rootstocks. Two pairs of leaves are retained and the seedlings are decapitated at the soft-wood apical region. Wedge-grafting is then carried out with a 4-5 cm cleft on the rootstock and with a small portion of the inner surface removed to facilitate a perfect union of the wedge-shaped scion, which has been prepared by shaving a portion of the bark and tissue on either side. The union is then secured by tying with a 15-30 cm polythene strip. The top of the scion is covered with a polythene cap to protect the apical portion of the scion from desiccation. The grafted plants are maintained in a lath or screen house for 8-10 days until sprouts emerge, and then the grafted plants are provided more sunlight and the caps removed. This wedge grafting technique is carried out by using only the soft-wood tissues of the stock and scion. The following management techniques are important in nursing young grafted plants.

- Grafts need to be watered frequently depending on the season.
- Excess water needs to be drained by providing drainage holes in polybags.

- Shoots on the rootstocks have to be nipped off frequently.
- Polythene wrapping at the union has to be removed about three months after grafting to prevent girdling.
- When the scion leaves turn from brown to green, rootstock leaves have to be removed (approximately 60 days after grafting).
- Flower shoots that sprout during the normal flowering season should be removed at the nursery stage.
- To prevent roots penetrating into the ground, grafted plants should be shifted frequently or placed on thick gauge black polythene sheets.
- Partial shade has to be provided to avoid sun-scorch by placing the grafted plants in a lath/screen house. Direct sunlight should be avoided as polythene bags tend to perish. Watering on alternate days should be done in summer.
- Regular insecticide sprays need to be given to control leaf sucking insects.
- When transporting grafted plants, terminal shoots and taproots should be protected.

ESTABLISHMENT OF CASHEW ORCHARDS

Large extents of land are available in Karnataka, Maharashtra and west Bengal for establishing new plantations under the cashew expansion program. There is also the possibility of expanding cashew cultivation into non-traditional areas in Madhya Pradesh, Bihar etc.

Selection of Site and Land Development

When selecting land for cashew, soils with salinity/alkalinity or waterlogging should be avoided. Soil depth, slope, course texture, soil fertility and water availability seem to impose very little limitations as cashew is a hardy crop. For establishing new plantations, land preparation should begin with the first pre-monsoon rains. Land should be cleared of shrub vegetation before digging pits for planting.

Planting Season

Planting of grafted plants is usually carried out during the monsoon season from July-August both in the west coast as well as in the east coast. Orchards should have pits dug to receive grafted plants well in advance of the main monsoon weather.

Spacing and Planting Systems

A spacing of 7.5m x 7.5m or 8m x 8m is recommended for cashew which gives a tree density of 175 and 156 trees per ha, respectively. High density planting at 4m x 4m giving a tree density of 625 trees per ha in the initial years and subsequently thinning in stages to reach a final spacing of 8m x 8m is also practiced in some areas. This enables higher returns during the initial years and as the canopies grow in volume, alternate trees are removed to achieve the desired final spacing. In level sites however, it would be advantageous to plant cashew at a spacing of 10m x 5m which will give a tree density of 200 trees per ha and at the same time providing sufficient space for growers to plant inter-crops during the initial years of establishment.

Planting of Cashew

Pits are usually dug at the onset of the pre-monsoon rains to a size of 60cm x 60cm x 60cm in light to medium soils. If a hard substrate like laterite is present pits may be 1m x 1m to compensate for the lesser depth of soil. It is preferable to dig pits 15-20 days before planting to expose planting holes to direct sunlight which can help remove termites and other harmful insects that can damage young plants, if present. When filling, top soil mixed with compost (5 kg) or poultry manure (2 kg) and 200g of rock phosphate are placed in the pits. Contour planting is usually followed in sloping areas. Standard conservation measures need to be followed on steep lands when establishing cashew plantations.

Young plants are planted in the months of July-August. Most nurseries supply 5-12 month old grafted plants in polybags. At planting, the polythene bag is removed without disturbing the ball of earth and the roots. Care is taken to place the grafted plant in the pits leaving the graft joint at least 5 cm above ground level. Normally the scion is staked to avoid damage from wind and the support should remain up to the third year from planting. Most orchard growers use a mulch around the planting hole to suppress weeds and conserve moisture.

Terracing and Bunding

In the western and eastern coastal areas cashew is grown mostly on sloping land. A considerable amount of nutrient leaching and soil erosion are common in such situations. Growers have been advised to construct terraces and contour pits to conserve runoff water. Studies on the extent of root distribution revealed that 90 percent of the root system was confined to a radius of 2m and a depth of 1 m. In order to achieve maximum utilisation of applied nutrients, fertiliser practices were confined to this part of the root zone. Before the advent of the south west monsoon in May-June, basins of about 1.5m radius are prepared for 2 year-old trees and subsequently widened to 2m in the third year. Terraces are made by removing the soil from the elevated portion above the tree trunks to create basins of 1.5-2m. Contour drains are also constructed to collect rain water above the tree-line and prevent soil wash from the slopes.

Application of Manures and Fertiliser

Application of 10-15 kg of farmyard manure or compost annually is generally recommended for cashew. In addition, the current fertiliser recommendation is 500g N (1.1 kg urea), 125g P_2O_5(625g rock phosphate) and 125g K_2O (208g muriate of potash) per tree per year. This has increased production in the All India cashew trials carried out at the research centers. These trials also showed that the cashew responds well to increased N applications up to 750g. Since local NPK fertiliser mixtures do not deliver the required nutrients, application of straight fertiliser is recommended.

Fertiliser is applied annually at the end of the rainy season into a shallow trench at the drip line of trees. It is also recommended that fertiliser be applied in split doses during pre-monsoon (May-June) and post-monsoon (September-October) periods to assure better uptake of nutrients. If a single application is done, the post-monsoon period is more suitable when ground moisture is adequate. One third the recommended dose is applied in the first year, two third the dose in the second year and the full dose thereafter (Table 3).

Table 3. Recommended Doses of NPK Fertiliser for Cashew (g/plant).

Year	Urea (gm)	Rock Phosphate (gm)	Muriate of Potash
1	330	200	70
2	660	400	140
3 onwards	1,100	625	208

Based on the results of research conducted by the National Coordinated trials, the following methods of fertiliser application are recommended to cashew growers. In the red loamy soils in low rainfall areas such as the east coast, fertilisers have to be applied and raked into the soil along the drip line of tree canopies. In laterite soils and steep lands of the west coast, fertiliser is applied in circular trenches of about 25 cm width along the drip line of trees. Trenches are filled and a mulch is applied to ensure soil moisture retention.

Weeding

Until tree canopies shade out the weeds, weeding is essential around the tree trunks up to a radius of about 2 m. The rest of the orchard requires slashing of under growth at least twice a year. The weeding cycles are generally confined to the pre-monsoon and post-monsoon periods to coincide with the fertiliser application.

Alternatively, weedicides may also be applied after slashing, well in advance of the rainy season if the under growth is too dense. The recommendation is to apply Agrodar-96 (2-4 D) at the rate of 4 ml/litre of water followed by Grammoxone at the rate of 5ml/ litre of water. Approximately 400 litres of spray is required to cover one ha. The spraying is repeated in the post monsoon season if the weed load is heavy.

Mulching

In low rainfall areas, mulching around the base of trees helps in the control of weeds, retention of moisture and modulation of soil temperature, especially in the hot summer months. This becomes an essential operation as cashew is usually planted in very dry areas where other crops are seldom grown. Most growers utilise the slashed weeds to mulch their orchards.

Training and Pruning

During the initial phase of orchard establishment, shoots arising on the rootstock have to be regularly removed to promote better scion growth, particularly in the first year after planting when scion rejection could occur if rootstock shoots are left unchecked. Training of young trees during the first three years is essential to develop uniform canopies. Training in the juvenile phase comprises of removing basal branches and water shoots. The plants are trained to a single stem and branches are allowed to grow about 0.75-1m from ground level. Deformed branches are also removed during the first few years. Since cashew trees tend to spread their canopies and lodge easily, proper staking is also essential. Trees are kept under check by topping off the main stem at a height of 4-5m from ground level. Orchard operations such as terracing, weeding, fertiliser application, nut collection and stem/root borer infestation control can be easily achieved if trees are properly trained. Pruning should be carried out in August-September at least once in three years when unwanted growth is removed to provide adequate sunlight into the canopy. Since fruiting is only encouraged from the third year, de-blossoming has to be carried out as flower clusters appear during the juvenile phase.

Plant Protection

Root and stem borer infestation is usually controlled with swabbing tree trunks with carbaryl (2%) or using a coal tar/kerosene suspension (1:2). After pruning of trees, a standard practice is to smear all cut surfaces with Bordeaux mixture paste (10%) to prevent fungal infections and die-back. A 1% Bordeaux spray is also administered if the cut surfaces are small.

More than 60 species of insect pests have been identified in cashew in India. The major pests are the tea mosquito, stem/root borer, leaf minor, leaf and blossom webber and flower thrips. No major diseases that cause economic losses have been reported so far in cashew.

For efficient management of the tea mosquito bug (*Helopeltis antonii*), it is important to check the build up of the pest population on the cashew crop as well as on the alternate hosts such as neem, drumstick, cocoa, guava etc. Tea mosquito bugs can be effectively controlled by three sprays at flushing, flowering and fruiting stages with endosulfan or monocrotophos (0.05%) for the first and second sprays and carbaryl (0.15%) for the third spray. In case of severe infestation, it may sometimes lead to die-back caused by a secondary infection of *Botrydiplodia theobromae*. In such instances it will require pruning of the diseased shoots and swabbing of the cut surfaces with 10% Bordeaux paste and spraying the trees with a 1% solution of Bordeaux mixture.

The stem and root borer (*Plocaecderus ferrugineus* L.), is capable of killing cashew trees. In severe cases of injury by this pest, gummosis of the stem and yellowing followed by drying of leaves can occur. The effective control measure is to remove immature stages of the pest and swabbing the trunk and exposed roots with carbaryl (0.2%) or neem oil (5 %) and application of Sevidol 8G (75g/tree) into the basin around the tree. Prophylactic treatment of swabbing the trunk up to one meter height with coal tar and kerosene in the ratio of 1:2 twice a year during March and November could also give effective control. The spray schedule indicated for tea mosquito bug will also be effective against the control of other foliage and inflorescence pests.

Cover-Cropping and Inter-Cropping

Popular cover crops for cashew plantations are *Peuraria javanica*, *Calapagonium muconoides* and *Centrosema pubescens* which improve the fertility and moisture balance and help conserve orchard soils. Cover crop seeds are generally sown with the advent of the monsoons at a seed rate of about 7 kg/ha. On degraded steep lands, cover crops are usually established on seed beds between tree rows. Inter-cropping has become popular with the systematic establishment of large-scale orchards. It is practiced in the first few years when there is sufficient space between crop rows with the main objective of deriving some income until the cashew starts giving economic returns. In Andhra Pradesh, popular inter-crops are horsegram, cowpea, groundnut etc. Casuarina is also a tree inter-crop planted at a spacing of 1.5m x 1.5m in cashew orchards in Andhra Pradesh and Orissa. In the west Godavari district of Andhra Pradesh, cashew is grown in combination with casuarina and coconut.

In Kerala and coastal Karnataka, pineapple is grown as a biennial crop in the initial 4-5 years and farmers find it far more profitable than crops such as redgram and cassava. Adopting a two-row system of planting in contour trenches, yields of 15-20 tons/ha have been achieved from this inter-crop. The pineapple inter-crop also indirectly benefits the main cashew crop as contour trenches help to conserve water and nutrients. When hedge-row planting at a spacing of 5m x 10m is used, inter-crops of Acacia, casuarina etc. are taken without any ill-effects on the main crop. These trees are planted about 3.5 m away from the cashew in two-row plots spaced 1m x 1m apart in the center of crop avenues.

Irrigation and Drainage

Cashew cultivation is generally carried out under rainfed conditions. In homesteads however, it is preferable to give some supplementary irrigation during the warm summer months from January to March. An application of about 200 liters of water per tree every fortnight was found to double cashew yields in trials conducted at the National Research Center at Puttur. In the sandy tracts of the East coast, although frequency and quantity of water applied varies, trees are watered during the summer months. Both in the homesteads and large-scale orchards, cashew is susceptible to waterlogging and proper drainage is essential in low lying areas.

HARVESTING OF NUTS AND CASHEW YIELDS

Bearing commences after the third year of planting and the trees will be in full production by the tenth year whilst the economic life of a tree is about 20 years. The main harvesting season is from February to May. Most farmers harvest their crop before they drop to prevent pilferage. This very often results in poor quality of the kernels. The optimum stage of harvest is when nuts drop to the ground. High quality nuts are obtained when freshly fallen nuts are separated from the cashew apples and sun dried for 2-3 days to bring down the moisture percentage from about 25 percent to below 9 percent. It is very essential to dry the nuts in order to prevent spoilage during storage. The drying process helps to retain flavor and quality of the kernels. When cashew apples are used for processing, harvesting has to be carried out before they drop. A simple test of maturity is to float nuts in water when mature nuts will sink while the immature and unfilled nuts will float. Nuts are usually gathered every week during the harvest season. Cashew apples for the fresh fruit market should be harvested daily.

Normally, about 92 % of the trees yield by the third year from planting. The average yield per tree increases from about 2 kg at 3-5 years to 4 kg at 6-10 years and 5-10 kg when trees are 11-15 years of age. Thereafter, trees yield in excess of 10 kg as the trees get older.

MARKETING

Raw cashew nuts are a seasonal commodity and the trading season is from March to May. Growers usually supply the primary or village markets where small traders collect and supply the urban markets. The cashew trade is seldom handled by exclusive traders. Usually, those traders who collect other plantation products also trade in cashew. Due to the highly competitive nature of the cashew trade growers have few marketing problems. When large quantities are collected by middlemen, the processors enter the marketing chain and make wholesale purchases. Grades and standards for cashew are yet to be introduced in India. Quality is generally determined by appearance and cutting tests that traders employ prior to purchase. The raw cashew nut market involves a large amount of capital where nearly 80 percent of the produce is transacted within a matter of 35 days. The current value of Indian production is estimated at around Rs. 10,000 million. This capital is made available by industry for procurement and processing operations. There are no growers' cooperatives or organisations for cashew marketing. In Kerala however, the government has been involved in the procurement process and supply to large-scale processors. This adversely affected the cashew trade and has now been replaced by a free market policy. In addition to the local production of nearly 430,000 tons, India also imports a considerable quantity of raw nuts from several African and South-east Asian countries to satisfy the national processing capacity of 700,000 tons established in the country.

NUTRITIONAL ASPECTS OF CASHEW

The cashew nut (*Anacardium occidentale*) produces nuts, the kernels of which have increased considerably in economic importance over the past few decades. Indigenous to Brazil, the cashew was taken to West Africa, East Africa and India by the Portuguese in the 15th and 16th centuries. It was noticed that the tree grew well on poor sandy soils along the coastal belt and was used by the Portuguese in Africa as an anti-soil erosion measure. The tree prospered and spread naturally, particularly in East Africa and India and the progeny of these wild cashew has formed the basis of the raw material for the cashew industry. Though the plant was primarily intended for checking soil erosion, it is now mainly grown for its commercially important kernel and shell oil.

After the war in 1945, world production and consumption of cashew nut increased sharply and it soon became the world's most important dessert nut after almonds. World consumption of cashew nuts has been increasing steadily from 125,000 tons in 1955 to 1,000,000 tons in 1995, and is estimated to be around 1,260,000 tons by the year 2005. The production of cashew nut has also kept pace with demand in Europe and Africa and the increase in production has been achieved mainly through extension of the area under the crop. The production of cashew nut in India, however, is far short of the capacity of the existing processing industries and consequently, the cashew industry in India has been stagnant for the past two decades. To increase yields, improved planting material and better crop management practices will have to be given some attention.

Other countries in South East Asia and the Pacific region including many islands in Indonesia where the cashew nut as well as the cashew apple are appreciated. In Malaysia, where its suitability for cultivation has been shown and the Philippines where its economic potential has yet to be exploited are somc of the new areas with future potential although labor requirements seem to be prohibitive, especially in Australia where conditions also appear suitable for its cultivation.

Trends in production of cashew are also related to consumption and these in turn will depend on the world economic situation. With the increase in the standard of living in developing countries, a large consumer market is developing, especially in cashew producing countries. There is also the need to explore the possibility of wider dietary utilisation of the oil obtained from the shell during processing of raw nuts, which is also a potential exportable product.

Cashew is a highly nutritious and concentrated form of food, providing a substantial amount of energy. The cashew nut kernel has a pleasant taste and flavor and can be eaten raw, fried and sometimes salted or sweetened with sugar. It also contributes as an important source of invisible fat in the diet, being widely used in a variety of ways. There has been a growing demand for cashew in many temperate countries where the demand is increasing. The nut contains an acrid compound which is a powerful vessicant that is abrasive to the skin. The cashew shell contains 25% of this reddish brown oil, industrially known as Cashew Nut Shell Liquid (CNSL) which is a by-product of the roasting process.

POTENTIAL FOR CASHEW NUT PRODUCTION DEVELOPMENT

In 1960-61, 228,000 tons of raw nuts were processed of which nearly 50 percent was imported. During the same year, 44,000 tons of processed kernels were exported which accounted for 77 percent of the total kernel output from the industry. The processing output has considerably increased in recent years and in 1995-96, about 640,000 tons were processed (Table 4) of which 65 percent raw nuts were obtained from local production. Domestic consumption has also increased considerably from 13,000 tons in 1960-61 to 92,000 tons in 1995-96, while the country also earned a foreign exchange equivalent of Rs. 12,320 million (US$ 352 million) through the export of cashew kernels and cashew nut shell liquid (CNSL). Export earnings reached a peak in 1994-95 when 77,000 tons of kernels were exported with a value of Rs. 12,440 million (US$ 355 million). With the establishment of new orchards using high yielding vegetatively propagated planting material, the future looks bright for the cashew industry in India.

Table 4. Raw Nut Availability, Processing and Export Statistics of India, 1960-96, (in '000 tons).

Year	*Domestic Production*	*Imports*	*Total Raw Nuts*	*Kernels*	*Exports*	*Domestic Consumption*
1960-61	111	118	228	57	44	13
1970-71	177	169	346	87	50	37
1980-81	185	16	201	50	32	18
1990-91	295	83	378	95	49	46
1995-96	418	222	640	160	68	92

A study of the industry prior to 1985 revealed that most of the plantations were of seedling origin and cashew cultivation was mainly carried out as an afforestation and conservation program for waste lands rather than an economic venture. Since productivity was not the basic objective of such a program, the cashew was maintained under highly neglected conditions. Poor soil fertility in cashew growing areas, seedling progenies of nondescript origin and neglect of the crop resulted in low productivity. This was prevalent in most of the cashew growing areas of Karnataka, Goa, Andhra Pradesh, Orissa and Tamil Nadu. Neglected trees established for conservation purposes hardly responded to the application of any inputs. Seedling progenies that were established in Maharashtra however, showed that these plantations responded to the application of modern inputs developed by research. These seedling progenies were from selected stock and orchards had received better attention from the beginning of their establishment. In order to ensure better productivity, all new plantings are encouraged to use vegetatively propagated material of recommended cultivars and the use of seedlings is completely discouraged.

One of the major thrusts being advocated at present is to rehabilitate existing unthrifty seedling plantations. It may however, not be profitable to attempt rehabilitation of senile and sparsely populated orchards. The working group responsible for the preparation of the 8th national plan revised yield estimates from 2 tons per ha to one ton per ha, mainly having these plantations in view. The current area of 635,000 ha is likely to reach 1 million ha by the turn of the century. Cashew is being considered as a candidate crop for rehabilitation of waste lands by many development planners. The issue is however, being debated whether more waste land be brought under cashew or rehabilitation of old orchards be undertaken using elite planting material. Waste lands that are now being considered are far more inferior to those areas which already have cashew plantations. The cashew crop has contributed in some ways to conserving the soil in the existing orchards. The logical alternative would therefore be, to utilise existing cashew lands for a development program using available technologies to reach the required production levels without expanding into any more degraded waste lands.

REFERENCES

Morton, Julia F. *Fruits of Warm Climates.* ISBN 978-0961018412

Pillai, Rajmohan and Santha, P. *The World Cashew Industry* (Rajan Pillai Foundation, Kollam, 2008).

Alexander H. Tullo (September 8, 2008). "A Nutty Chemical". *Chemical and Engineering News* 86 (36): 26–27.

9

Coconut

The coconut palm (*Cocos nucifera* linn.) is the most useful palm in the world. Every part of the tree is useful to human life for some purpose or the other. Hence, the coconut palm is endearingly called 'kalpavriksha' meaning the tree of heaven. The copra obtained by drying the kernel of coconut is the richest source of vegetable oil containing 65 to 70 per cent oil. Coconut is grown in numerous islands and also in the humid coastal tracks of tropical countries. India ranks second in the world production of coconuts 6088 million nuts from 108 million hectares.

The coconut palm, or 'kalpa vriksha' provides many necessities of life like food and shelter. of all the tropical palms it is the outstanding one. It is mainly cultivated for the nuts from which 2 important products 'copra' and fibre are obtained. Copra yields oil and oil-cake. The trunk of the mature palm is used as timber for houses and the leaves are used for thatching houses, fencing etc. The unopened spathe is tapped for toddy called 'neera'. Sweet toddy can be converted into jaggery and sugar, fermented toddy is a mild alcoholic drink and vinegar can be made from it. Water from tender coconut is a refreshing drink. In kerela the extraction of coir from the husk of nuts and the manufacture of coir products provides employment for thousands of people. The coconut shell is largely used for fuel and for the production of charcoal and making a variety of curios. The shell flour is used as a filler for plastics. Thus every part of the palm is useful in one way or the other.

CLIMATE

Coconut is essentially a crop of the humid tropics. It is mainly grown in the coastal plains. Rainfall is the most important factor affecting its growth. A raintall for 100 to 225 cm per annum, evenly distributed is most important. Coconut requires an equable climate. The optimum mean annual temperature for its best growth and maximum yield is 27 deg celsius. frost and low humidity adversely affect the yield of the palm. Persistent high humidity is also harmful and bud rot is more common under such conditions. The palm requires bright sunshine of 2000 hrs a year.

SOIL

The coconut palm adapts itself remarkably well and thrives in almost all types of well-drained tropical soils, such as coastal sand, red loam, laterite, alluvial and reclaimed soil of the marshy lowland. A water table that is too high and remains static for too long is harmful.

VARIETIES

There are only two distinct varieties of coconut, the tall and the dwarf. The tall cultivars that are extensively grown are the West Coast Tall and East Coast Tall. The dwarf variety is shorter in stature and its life span is short as compared to the tall. Tall x Dwarf (TxD), Dwarf x Tall (DxT) are the two important hybrids.

There are 10 different combination of hybrids, developed by Kerala Agriculture University and Tamil Nadu Agriculture University and released for commercial cultivation. They are high yielders under the good management conditions. Laccadive Ordinary, Andaman Ordinary, Philippines, Java, Cochin-China, Kappadam etc. are the other tall cultivars under cultivation.

COCONUT PLANTING

Planting Material

Since coconut is a cross pollinated perennial crop the selection of seeds is of vital importance. Selection has to be made at the mother-palm level and at the seedling stage. The mother palms should be healthy, high yielding and regular in bearing. Also factors such as proportion and distribution of female flowers, high setting percentage and a high copra content should also be taken into consideration. if prepotent palms can be identified by progeny testing the yielding ability of the planting material can be further assured. A careful selection of the seedlings is important. The seedlings should be healthy and have a minimum of six leaves when they are one year old. The early splitting of leaves is a desirable character. Nine month old seedlings having a minimum of four leaves can also be planted.

Preparation of Land

The clearing and levelling of land have to be done before preparing the pits for transplanting seedlings. The depth of the pit depends on the soil type. In the sandy loam of low water-table the planting pits of the size 1mx1mx1m is recommended. In laterite soils with a rocky substratum, deeper and wider pits 1.2x1.2x1.2m are necessary. At the time of planting the pits have to be filled up with loose soil leaving a final depth of about 60cm. However when the water table is high planting on the surface or on soil mounds may be necessary. In a littoral sandy soil, circular pits 1.5mx1.5mx1.5m are made and partially filled with alternate layers of red earth and coconut husks (150 husks).

The spacing of palms requires careful consideration. A spacing of 7.5 to 9 m accomodating 177 to 124 palms per hectare is adopted under the square system of planting. The triangular method accomodates 20 to 25 more palms per hectare. In the hedge system of planting, a spacing of 5 to 5.5m along the row is given. Larger number of palms can be planted in a unit area ensuring sufficient light on each palm.

Spacing depends upon the planting system, soil type etc. In general the following spacing are recommended under different planting system in sandy and laterite soils.

	Planting system	*Spacing*
1	Triangular	7.6m
2	Square	7.6x7.6m, 8x8m, 9x9 m
3	Single	6.5m in rows - 9m between rows
4	Double Hedge	6.5 to 6.5m in rows - 9m between pairs of rows

Time of Planting

Seedlings can be transplanted in the beginning of the south west monsoon. If irrigation facilities are available, it is advisable to take up planting at least a month before the onset of the monsoon so that the seedlings get well established before heavy rains. Planting can also be taken up before the onset of the north-east monsoon. In low-lying areas subject to inundation during monsoon period, transplanting may be done after the cessation of the monsoon.

Before planting the pits are filled up with top soil and powdered cow dung / compost up to a depth of 50 to 60 cm. Then take a small pit inside this, so as to accommodate the nut attached to the seedling. Plant the seedling inside this pit and fill up with soil. Press the soil well so as to avoid water stagnation. If there is chance for white-ant attack apply Sevidol 8G(5gm.) inside the small pit before planting. In laterite areas apply 2 kg common salt per pit for improving the physical condition of the soil. Burying 25 to 30 coconut husks per pit in layers will be useful for moisture conservation.

Transplanting and After Care

In well drained soils seedlings can be transplanted at the beginning of the south-westerly monsoon. If facilities for irrigation are available, it is advisable to take up planting at least a month before the onset of the monsoon. in the low lying areas, seedlings are better transplanted after the rainy season. The transplanted seedlings should be shaded and irrigated properly during the summer. Irrigation with 45 litres of water once in four days is the optimum in sandy soils. provision for drainage is also important. Apply the complete NPK fertiliser from the early stage. Large quantities of green leaf or compost to the soil can be applied where it has a poor organic matter content. The pits should be cleared of all the weeds and the soil washed down by rain. The palms should be frequently examined for insect attack or fungus diseases and the necessary remedial action adopted promptly.

Care of Young Palms

The transplanted seedlings should be shaded and irrigated adequately during the summer months. Also provide staking so that winds may not uproot the young seedlings. For the first two years after planting, irrigate the seedling twice a week during the dry summer months. Shading is a must to the transplanted seedlings.

Coconut tree

Manuring

Regular manuring from the first year of planting is essential to achieve higher productivity. For coconut 20 - 50kg. organic manure should be applied per palm per year with the onset of south west monsoon, when soil moisture content is high. Different forms of organic manures like compost, farm yard manure, bone meal, fish meal, blood meal, neem cake, groundnut cake etc. could be made use for this purpose. In addition to this the following Fertiliser Schedule is recommended.

Irrigation

Soil moisture very often limits coconut production in those areas where long spell of dry weather prevail or where the rainfall is scanty and ill-distributed. So irrigate the palms during summer months in basins around the palm. The irrigation requirement varies according to the soil type and climatic condition. Generally, an adult palm requires 600 to 800 litres of water once in four to seven days. Irrigate in basins of 1.8m radius and 10-20 cm depth. In coastal sandy soils, sea water can be used for irrigating adult palms. Do not irrigate seedlings and very young palms upto 2 year with sea water. In irrigated gardens interruption of irrigation would lead to serious set-back in yield and general condition of palms. Hence, when once started irrigation should be continued regularly and systematically. Drip irrigation is the best suited method of irrigation for coconut. It saves water, labour and energy.

Inter-cultivation

Only minimum tillage is required for coconut. Inter-cultural operations are mainly intended to control weeds and to provide aeration to the soil. If these objectives are met, any tillage system (ploughing / digging, making mounds) is as good as another and can be followed depending upon the local conditions.

Husk Burial

Burying fresh or dried coconut husks around the palm is a very beneficial practice particularly for moisture retention especially in drought prone areas. The husk can be buried either in linear trenches taken 3 m away from the trunk between rows of palms or in circular trenches taken around the palm at a distance of 2 m from the trunk. The trenches may be dug at 0.5 m wide and at the same depth. The husks are to be placed in layers with concave surface facing upwards and covered with soil. The beneficial effects of husk burial will last for about 5-7 years.

Green Manure and Cover Crops

This will help to increase the organic matter content of the soil and also will prevent soil erosion in coconut gardens. The following Green manure / cover crops are recommended for cultivation in coconut gardens.

- Crotalaria juncea (Sunnhemp)
- Tephrosia purpurea
- Gliricidia maculata
- Calapagonium muconoides
- Mimosa invisa

Sow the green manure / cover crops during April-May with the onset of pre-monsoon showers. The green manure crops should be ploughed in and incorporated in the soil during August-September.

Mixed / Inter / Multi-species Cropping System in Coconut Garden

Schedule for inter-mixed cropping may be drawn up based on the canopy size an orientation of palms. A variety of intercrops like pineapple, banana, elephant-foot yam groundnut, chillies, sweet potato, tapioca and different vegetables can be raised i coconut garden. In older plantation cocoa, cinnamon, pepper, clove, nutmeg etc. ca be grown as mixed crops. However, these inter/mixed crops are to be adequately an separately manured in addition to the manures applied to the coconut palm

Mixed Farming

Mixed farming by raising fodder grasses such as Hybrid Napier or Guinea grass alon with leguminous fodder crops such as Stylosanthes has been found to be profitabl Raising the above crops in one ha. of coconut garden can support three to four diar animals. The animals supply large quantities of cattle manure which when applied t the soil will improve its fertility status. This sort of mixed farming will improve the yiel of the palm.

Care of adult palms

The palms generally start bearing at the age of five or seven years after planting an the stabilised yield is obtained from about the tenth year onwards for 50 to 60 year regular intercultivation and manuring of the palm are essential for stepping up an maintaining the productivity of palms at a high level. Tillage including digging ploughing, piling of mounds, levelling of mounds, etc. and the making of shallow basin each with a radius of 2 m, at the beginning of the monsoon and filling them up at th close of the monsoon are beneficial. In sandy soils, which are of low fertility and d not have a luxuriant growth of weeds, regular intercultivation may not be quit necessary but in other types of soils, intercultivation is essential for keepir.g down th weeds under check and for creating a soil mulch.

The method of intercultivation mainly depends on the local conditions, soil type topography, size of holding, intencity of rainfall, etc. The recommended dose o fertilisers is 500g of N, 300g of P_2O_5 and 1,200g of K_2O per palm per year for the ordinar tall variety. For soil which are poor in organic matter, the application of green manur or compost at 50kg per palm is recommended. Green-manuring crops usually grow are cowpea, sunnhemp. Fertilisers may be applied in two spilt doses. After the receip of summer showers one third of ht edose of fertilisers is spread around the palm withi a radius of 1.8m and raked in. Circular basins 1.8m in radius and 25cm in depth ma be dug in July-August and the green leaf manure or compost applied and partiall covered. The remaining two-third dose of the fertilisers may be spread in the first wee of september and the basins closed. Under irrigated conditions and in areas with assure supply of moisture throughout the year the annual dose of fertilisers may be applie every three months in equal split doses.

In regions with a long spell of dry weather, there is a definite set back in the growth nd yield of the coconut palm. An increase of 50% to 100% in the yield of nuts can e obtained by giving summer irrigation. The quantity and interval between irrigations lepend upon the soil and the climatic conditions. In loamy soils copious irrigation , once week is sufficient. In sandy soils along the sea coast during summer without any dverse effects

Mango, jackfruit, arecanut and other trr crops and a number of annual crops are ommonly grown along the coconut palms in household gardens. recent studies have hown that under rainfed conditions tuber crops, banana, pineapple, upland rice and ulses can be grown successfully as intercrops in adult coconut plantations without any letrimental effects on the fertility of the soil and the productivity of the palms. Under rrigated conditions cacao forms a very beneficial crop combination with coconut. Other rops suitable for mixed cropping include pepper and tree spices, such as cinnamon, love, nutmeg. Mixed farming including the raising of milch animals on fodder grass nd legumes raised as intercrops are manured with cowdung and urine. This practice mproves the soil fertility by recycling organic wastes , enhances the yield of palms and enerates additional income and employment potential for the farmer's family.

Iarvest and yield

'he coconut palm is unique in that once it attains the normal bearing stage , it continues) bear a bunch of nuts in every leaf axil almost at monthly intervals all year round nd throughout its life extended over 50 to 60 years. The nuts mature nearly one year fter fertilisation. Generally, harvesting is done once in 45 to 60 days. Tender nuts in lemand as a delicious drink in West Bengal and Maharashtra are best harvested at the ge of 6-7months. Large scale harvesting of the tender nuts is not being done in other arts of the country. Nuts for culinary purposes, for making cup copra and good quality ibre are harvested at the age of about 11 months whereas those for ball copra and coarse oir are harveste only when they are fully ripe.

The average yield per hectare varies from 10,000 to 14,000 nuts per annum. From well maintained garden an annual yield of 25,000 nuts per hectare can be obtained. Jearly one-third of the annual yield is harvested during the three months— march, april md may—in conditions obtaining on the west coast.

LANT PROTECTION

ests

'he major insect pests of the coconut palm are the rhinoceros beetle, the red palm veevil, the black headed caterpillar, the cockchafer beetle and the coreid bug. Adopt uitable remedial measures as and when required.

Rhinoceros beetle

Symptoms

- The adult beetle bores into the unopened fronds and spathes.
- Attacked fronds when fully opened show characteristic geometric cuts

Control measures

- Hook out the beetle from the attacked palms using beetle hook.
- As a prophylactic measure, fill up the top most three leaf axils with Sevidol 8G(25g + fine sand (200g) thrice in April, September and December.
- Place 10.5g naphthalene balls in the leaf axils and cover it with fine sand.
- To be practiced once in 45 days.
- Spraying 0.01% Carbaryl (50WP) in the breeding sites of the beetle help destroy the larva.
- Biological control using the virus Baculovirus oryctus and
- Green muscardine fungus, Metarrizhium anisopliae.
- Practice clean cultivation.

Red [alm weevil

Symptoms

- Presence of holes on the stem, oozing out of viscous brown fluid and extrusion o chewed up fibres through the hole.
- Some times the gnawing sound produced by the feeding grubs will be audible
- In the advanced stage of infestation yellowing of the inner whorl of leaves occur
- The crown falls down or dry up later when palm is dead.

Control measures

- Practice clean cultivation by cutting and removing palms already damaged and the decaying stumps in the garden. Such palms should be split open and the differen stages of pest inside burned off.
- Avoid injury to the trunk as the pest lay eggs in these wounds. Wounds if any should be pasted with a mixture of carbaryl / Thiodan and soil. While cutting leaves retain at least 1m of petiole.
- Use pheromone trap for attracting weevils and kill the collected ones.

— If rhinoceros beetle attack is prevalent, follow the recommended measures.

— Use fungicides if leafrot / bud rot is noticed as the weevil lays eggs in such palms.

— Inject attacked palms with 0.1% Endosulfan (3ml / litre water) or 1% Carbaryl (20gm/litre). Plug the holes in damaged region and pour the insecticide suspension into a slanting hole made above the damaged portion using a funnel.

Black headed caterpillar

Symptoms

— Severity of attack is seen during January to May. The caterpillar feeds on green matter from lower leaf surface, remaining within silk and frass gallaries. In severe attack all the green matter of the leaves will be eaten up.

Control measures

— Bioligical control is very effective against this pest through release of parasitoids like Gorriozus nephantidis, Elasmus nephantidis and Brachimeria nosatoi.

— In case of severe attack, remove the affected leaves and destroy by burning. Then spray the under surface of leaves with 0.02% Dichlorvos (Dichlorvos 100EC).

Coreid Bug

Symptoms

— The attacked buttons do not develop resulting in immature nut fall. The nuts if developed may become barren.

Control measures

— Pesticide application is necessary if infection is severe. Spraying has to be done 3 times a year.

— Apply 0.1% Carbaryl or Endosulfan on the inflorescence after the receptive phase of the female flowers.

— Tying perforated polybags (2 bags/palm) containing 2.5g phorate on to the stalk of inflorescence is also effective.

Rats

Symptoms

— Attack tender nuts resulting in immature nut fall.

Control measures

- Entry of rats on to the trunk can be prevented by fixing mechanical barriers upto 2m height from ground level using 40cm sized G.I. sheets.
- Use rat traps.
- Poison baiting with zinc phosphide or warfarin.
- Fumigate the hiding places using Aluminium phosphide tablets.
- Place wax blocks containing the poison Bromodioline @ 30 blocks per ha (each weighing 10g) on 5 palms. Repeat after 12 days. Practice this two times for reducing rat population.

Lace Bug

Symptoms

- Lace bug sucks sap from coconut foliage; it acts as a vector in transmission of Phytoplasma from root(wilt) affected palms to healthy palm.

Control measures

- This can be controlled by spraying Endosulphan or Monocrotophos 0.01%.

Mealy Bugs and Scale Insects

Symptoms

- During summer months mealy bugs cause damage to spindle leaves, spathes and bunches and the scale insects make encrustations on the foliage.

Control measures

- Two rounds of spray with 0.1% Fenthion or 0.05% Monocrotophos.
- In the case of scale insects spraying with dimethoate or monocrotophos 0.05% is efficacious.

Termites

Symptoms

- It is estimated that nearly 20% of the coconut seedlings are damaged by termites particularly in laterite soil.

Control measures

- Adoption of field sanitation by disposal of organic matter in nursery soil and covering germinating nuts with a layer of river sand.
- Swab the affected trunk with the same chemical.

White Grub (Leucopholis coneophora Burm.) (Melolonthinae: Coleoptera)

Symptoms

- The white grubs are mostly found in sandy loam tracts of Kerala and Karnataka. It damage the roots. In seedling, it tunnels in to the bole and collar region.
- It has an annual life cycle with a grub period of 8 months. Peak grub population is observed from Sept. to Oct. Adult beetles emerge out of the soil after pre-monsoon showers in May-June during sunset hours.

Control measures

- Plouging and digging of soil during pre and post monsoon period will expose the insect for predation.
- Collection and destruction of the adult beetles during May-June.
- Setting up light traps to attract adult beetles.
- Application of phorate 10G @ 100g per palm should be mixed and raked in the top 15cm soil in May-June and Sept.-Oct. is recommended in coastal and Malanad areas. Irrigation is necessary after the pesticide application.

Eriophid Mite of Coconut

Symptoms

- The Eriophid Mite (Aceria guerreronis (Keifer) (Acarina: Eriophyidae) which was a minor pest of coconut in India has become a major pest on coconut recently.
- The attack of this mite was first recorded in 1965 in guerrero, Mexico. Later it was reported from Ivory cost, Brazil, Costa Rica etc. The first Indian report of this pest is from Ernakulam District of Kerala during 1998. The sporadic occurrence of this pest is reported from almost all the 14 districts of Kerala, Tamil Nadu, Karnataka, Andhra Pradesh and Kalpeni Island of Lakshadweep. As per the recent survey conducted in Kerala (13th -18th December 1999) about 589 lakh bearing palms are affected. This accounts for about 42% of the total yielding palm population and the estimated crop loss is in the tune of 21.8%. A roving team to study the intensity

of infestation has identified its occurrence from all the South Indian States and Union Territory of Lakshadweep.

— The mites are very minute in size and are not visible by naked eye. The size of this measures to 200-250 microns in length and 20-30 microns in width. The life cycle of this mite is completed in 10-12 days. It remains underneath the periyanth (cap) and cause injury by feeding on the soft paranchymatic tissues. Though the mites are microscopic their damage is enormous and hundreds of mites could be seen in each infested button and tender nut. Visible symptoms are brown discoloration noticed in patches of the husk. In case of severe attack the button sheds, resulting in very poor setting percentage. In other cases the nuts are deformed and undersized with poor development of kernel and husk.

— The mites spread through wind and its multiplication is at a high rate. Though this pest was noticed only in a limited area during 1998 it has become a major pest of coconut in India.

Control measures

Being a micro pest inhabiting under the periyanth, the control measurers at field level are not easy. However by adopting the following integrated plant protection measurers the mite population could be reduced considerably.

1. Adoption of phytosanitary measures in coconut gardens such as cleaning the crown of the palm, keeping the plantation clean and burning of all immature nuts fallen due to mite infestation.
2. Spraying biopesticides on the bunches
   a) 2% neem oil - garlic emulsion (20ml neem oil + 20g garlic + 5g bar soap in 1 litre water). Emulsion has to be prepared on the same day of application itself.
   b) Other neem based pesticides at 0.004% (Azadirachtin). If the pesticide formulation contains 1% Azadirachtin, 4ml has to be used in 1 litre water.

   Wherever spraying is difficult root feeding may be resorted to with Azadirachtin 5% formulation (7.5ml+7.5ml water) or Azadirachtin 1% formulation (10ml + 10ml water).
3. Spraying has to be done 3 times a year - December-February, April-June and September-October. While spraying, ensure that the spray fluid falls over the perianth region especially on button and tender nuts. On an average 1-1.5 litre spray fluid is required per palm. Care should be taken to harvest mature bunches before spraying.
4. Following palm health care practices also may be adopted.

a. Recycling of biomass generated within the coconut system by vermi compost method or by using lignin degrading fungus.

b. Raising of green manure crops in the coconut basins (like sunhemp, cowpea, calapagonium etc.) and incorporation into the soil in the basin itself, these act as mulch during summer and slowly decompose and provide nutrients when incorporated to soil.

c. Application of recommended dosages of fertilisers, in two split doses, as per the package of practices prevailing in the respective states.

d. Recommended level of irrigation during summer months i.e. 250-500 litres of water per tree per week (based on evapotranspiration in the given area).

e. Soil moisture conservation by following methods.

   i. Burial of coconut husk in the basin.

   ii. Mulching the basins (2m radius) with coconut leaves/green manure / green leaf manure.

   iii. Mulching with coir pith wherever available (2m radius).

Slug Caterpillar (Leucopholis coneophora Burm.) (Melolonthinae: Coleoptera)

Symptoms

- The white grubs are mostly found in sandy loam tracts of Kerala and Karnataka. It damage the roots. In seedling, it tunnels in to the bole and collar region.
- It has an annual life cycle with a grub period of 8 months. Peak grub population is observed from Sept. to Oct. Adult beetles emerge out of the soil after pre-monsoon showers in May-June during sunset hours.

Control measures

- Plouging and digging of soil during pre and post monsoon period will expose the insect for predation.
- Collection and destruction of the adult beetles during May-June.
- Setting up light traps to attract adult beetles.
- Application of phorate 10G @ 100g per palm should be mixed and raked in the top 15cm soil in May-June and Sept.-Oct. is recommended in coastal and Malanad areas. Irrigation is necessary after the pesticide application.

Diseases

The coconut palm is affected by a number of diseases, some of which are lethal while others gradually reduce the vigour of the palm causing severe loss in yield. A brief account of the important coconut diseases is given.

Bud Rot

Symptoms

1. The earlier symptom is the yellowing of one or two younger leaves. Black spots appear on spindle leaves.
2. In the later stages the spindle withers and drops down.
3. The tender leaf base and soft tissues of the crown rot into a slimy mass of decayed material emitting foul smell.
4. Ultimately the entire crown falls down and the palm dies.

Control measures

1. Remove all affected tissue of the crown and apply Bordeaux paste on cut end and provide a protective covering till normal shoot emerges.
2. Spray 1% Bordeaux mixture on the crown of the neighbouring palms as a prophylatic measure.
3. Spray with 1% Bordeaux mixture during May and September if the disease occur frequently.
4. Cut and burn severely affected palms which cannot be saved.

Leaf Rot

Symptoms

1. Blackening and shriveling up of distal ends of the leaflets in the central spindle and younger leaves which later break off in bits.
2. Gradual weakening of the tree resulting in decline yield.

Control measures

1. Improve general condition of palms through proper manuring and management.
2. Remove the decayed portions of spindle leaf and upmost two leaves only.
3. Pour fungicide solution of Hexaconazol (Contaf 5E) - 2ml or Mancozeb (Dithane M45/Indifil M45) - 3g in 300ml water per palm to the base of spindle leaf.

4. Apply 20g Phorate 10G mixed with 200g fine sand around the base of spindle leaf.
5. 2-3 rounds of spraying is sufficient in case of mild infection.

Stem Bleeding

Symptoms

1. Exudation of reddish brown liquid through cracks developing on the stem.
2. Decaying of tissues at bleeding point
3. Vigour and yield declining.
4. Development of big holes inside the trunk

Control measures

1. Chisel affected tissue and dress the wound with 5% Calixin (5ml in 100ml water). Apply coal tar after 2 days. Burn off chiselled pieces.
2. Avoid any mechanical injury to the stem.
3. To avoid spread of disease on to upper portion of trunk, root feeding with 5% calixin may be adopted 3 times a year - April-May, Spetember-October and January-February.
4. Along with 50kg organic manure, apply 5kg neem cake containing the antagonistic fungi, Trichoderma culture to the basin during September.
5. Provide adequate irrigation during summer and drainage during rainy season.
6. Apply recommended doses of organic manures and chemical fertilisers.
7. Coconut stem boring insects like Xyleborus, Diocalandra should be controlled by applying Carbaryl 50% WP on the trunk @ 3g per litre water.

Root (wilt) Disease

Symptoms

1. Abnormal bending or ribbing of the leaflets, flaccidity of the leaves, general yellowing followed by marginal necrosis of the leaflets.
2. Abnormal shedding of buttons, reduced leaves and crown, gradual reduction in yield.

Control Measures

Being a non-lethal, debilitating disease an integrated approach is to be followed for management of root wilt.

1. Recommended control measures for leafrot disease as this disease gets superimposed on most of the root wilt affected palms.
2. Follow integrated nutrient management
   - apply organic manure @ 50kg / palm / year.
   - apply balanced dose of chemical fertilisers i.e. 500g Nitrogen (1.1kg urea), 300g Phosphorus (1.7kg Mussouriphos), 1000g Potassium (1.7kg Muriate of potash) in two splits - 1/3rd during April-May and 2/3rd during September-October under rainfed condition and in four splits during January, April, July and October under irrigated condition. In addition to this 500g MgO (1kg Magnesium sulphate) also has to be applied along with second dose of fertiliser application.
3. Cut and remove disease advanced, uneconomical palms yielding less than 10 nuts per palm per year.
4. Grow green manure crops - cowpea, sunhemp (Crotalaria juncea), Mimosa invisa, Calapagonium mucanoides, Pueraria phaseoloides etc. may be sown in coconut basins during April-May and incorporated during September-October.
5. Irrigate coconut palms with at least 250 litre water in a week.
6. Adopt suitable inter/mixed cropping in coconut gardens.
7. Provide adequate drainage facilities.

Tanjavur wilt

Symptoms

1. Decay of finer nuts, withering, yellowing, drooping and drying of outer whorl of leaves and exudation of reddish brown fluid at the base of the trunk.

Control Measures

1. Practice growing banana as intercrop in coconut
2. Root feeding with Calixin (2ml in 100ml water) once in 3 months.
3. Drench the basin with 25 litre of 0.1% Calixin.
4. Apply neem cake containing Trichoderma @ 5kg /palm/year.
5. Apply recommended dose of organic manures.
6. Avoid flood irrigation and follow drip irrigation.
7. Practice clean cultivation and burn off diseased plant pests.

8. Isolate infected palms by taking trenches of 1m depth and 0.5m width around the palm at 1.5m away from it.

Mahali

Symptoms

1. Shedding of female flowers and immature nuts.
2. Lesions appear on the young fruits or buttons near stalk which later result in decay of the underlying tissues.

Control measures

1. Spray 1% Bordeaux mixture on the crown of the palm once before the monsoon and later after 40 days interval.
2. Remove and destroy fallen nuts.
3. In dwarf palms, apply Dithane M-45 in place of Bordeaux mixture.

Crown Chocking

Symptoms

1. Characterised by emergence of shorter leaves with fascinated and crinkled leaves.
2. The leaflets show severe tip necrosis and fail to unfurl. In many cases, it gives a choked appearance to the frond.
3. Ultimately the affected palm dies.

Control measures

1. Application of 50 g Borax at half-yearly intervals (Feb-Mar and Sept Oct) along with recommended fertiliser in the basins will control the disease when it is in the early stage. In root wilt affected areas a dosage of 200gm - 300gm per palm per year is recommended.

Leaf blight or Grey Leaf Spot

Symptoms

1. Minute yellow spots encircled by greyish bands appear on the surface of mature leaves of the outer whorl.
2. Later they become greyish white. These spots coalesce into irregular necrotic patches.

3. Complete drying and shrivelling of the leaf blade are common when the infection is severe.

Control measures

1. Removal of the older 2-3 disease affected leaves and spraying the foliage with 1% Bordeaux Mixture will check the spread of the disease.

COCONUT PRODUCTS

The main product obtained from the coconut is copra or the dried kernel of the nut. The cup copra is prepared by splitting the husked nut into two halves and drying them in the sun. After drying for a day or two the kernel gets detached from the shell. The drying is continued for another four to five days. Ball copra is prepared by storing fully ripe nuts for 8-12 months on a raised platform in a shed. They are sometime smoked or heated with a slow fire to accelerate drying. During storage water inside the nuts gets dried up and the kernel gets detached. Nuts are then husked and the shells are carefully broken. The ball copra thus obtained is clean, sweet and highly valued.

The coconut fibre is known as coir. For retting, husks are buried in shallow pits in low-lying areas subject to the tidal flow of backwaters for about 6-12 months. The coir obtained from dry husks would be of inferior quality. it is easier to extract coir from green husks after sufficient retting and the quality of coir is good. The coir is then dried in the sun and spun into coir yarn by hand or with a machine. Coir yarn is used as such or made into ropes, mats, nets, bags, etc., large number of coconut products are manufactured in the country which have both domestic and export market.

Vinegar and soft drink are manufactured in the country from coconut water. Tender coconut water concentrate is another product which is manufactured and marketed successfully. Know-how for the preservation and packing of tender coconut water has been transferred to six firms in the country. Nata-de-coco is a gelatinous delicacy formed by the action of a micro-organism Acetobactor xylinium in a culture medium of coconut water. The know-how for its manufacture is available with the Board.

REFERENCES

Campbell-Falck D, Thomas T, Falck TM, Tutuo N, Clem K (2000). "The intravenous use of coconut water". *Am J Emerg Med* **18** (1): 108–11.

Fife, Bruce (2005). *Coconut Cures.* Piccadilly Books, Ltd., pp. 17. http://books.google.com/books?id=JW8RtCJqZ8MC. Retrieved on 2008-04-04.

Werth, E. 1933. Distribution, Origin and Cultivation of the Coconut Palm (in periodical: *Ber. Deutschen Bot. Ges.*, vol 51, pp. 301–304) (article translated into English by Dr. Child, R. (Director, Coconut Research Scheme, Lunuwila))

10

Rubber

Natural rubber is found in the latex of as many as 895 species of plants belonging to 311 genera of 79 families. Of these Hevea Brasiliensis (Euphorbiacaeae), the para rubber tree is the most important source of rubber. Others are glaziovii (Euphorbiacaeae), Ficus elastica (Moracaeae), Parthenium argentatum and taraxacum kok-saghyz (compositae). Many other species such as euphorbia intisy (Euphorbiacaeae), cryptostegia grandiflora, and landolphia sp. have been tried as possible minor sources of natural rubber. Many plant species produce natural rubber. Considerations of quality and economics, however, limit the source of natural rubber to one species, namely Hevea brasiliensis. It is a native of the Amazon basin and introduced from there to countries in the tropical belts of Asia and Africa during late 19th century. It can be termed as the most far reaching and successful of introductions in plant history resulting in plantations over 9.3 million hectares, 95 per cent of it across the globe in Asia.

Hevea brasiliensis, also known as the Para rubber tree after the Brazilian port of Para, is a quick growing, fairly sturdy, perennial tree of a height of 25 to 30 metres. It has a straight trunk and thick, somewhat soft, light brownish gray bark. The young plant shows characteristic growth pattern of alternating period of rapid elongation and consolidated development. The leaves are trifoliate with long stalks. The tree is deciduous in habit and winters from December to February in India. Refoliation is quick and copious flowering follows. Flowers are small but appearing in large clusters. Fruits are three lobed, each holding three seeds, quite like castor seeds in appearance but much larger in size. The seeds are oil bearing.

The rubber tree may live for a hundred years or even more. But its economic life period in plantations, on general considerations is, only around 32 years - 7 years of immature phase and 25 years of productive phase. The economic life period of rubber trees in plantations is around 32 years - up to 7 years of immature phase and about 25 years of productive phase. The soil requirement of the plant is generally well-drained weathered soil consisting of laterite, lateritic types, sedimentary types, nonlateritic red or alluvial soils.

The climatic conditions for optimum growth of Rubber tree consist of (a) Rainfall of around 250 cm evenly distributed without any marked dry season and with at least 100 rainy days per annum (b) Temperature range of about 20°C to 34°C with a monthly mean of 25°C to 28°C (c) High atmospheric humidity of around 80% (d) Bright sunshine amounting to about 2000 hours per annum at the rate of 6 hours per day throughout the year and (e) Absence of strong winds. Many high-yielding clones have been developed for commercial planting. These clones yield more than 2,000 kilograms of dry Rubber per hectare per annum, when grown under ideal conditions.

In places like Kerala, where coconuts are in abundance, the shell of half a coconut is used as the collection container for the latex but glazed pottery or aluminium cups are more common elsewhere. The cups are supported by a wire that encircles the tree.This wire incorporates a spring so that it can stretch as the tree grows. The latex is led into the cup by a galvanised "spout" that has been knocked into the bark. Tapping normally takes place early in the morning when the internal pressure of the tree is highest. A good tapper can tap a tree every 20 seconds on a standard half-spiral system and a common daily "task" size is between 450 and 650 trees. Trees are usually tapped alternate or third daily although there are many variations in timing, length and number of cuts. The latex, which contains 25 - 40% dry rubber, is in the bark so the tapper must avoid cutting right through to the wood or the growing cambial layer will be damaged and the renewing bark will be badly deformed making later tapping difficult. It is usual to tap a pannel at least twice, sometimes three times, during the trees' life. The economic life of the tree depends on how well the tapping is carried out as the critical factor is bark consumption. A standard in Malaysia for alternate daily tapping is 25 cm (vertical) bark consumption per annum. The latex tubes in the bark ascend in a spiral to the right. For this reason, tapping cuts usually ascend to the left to cut more tubes.

The trees will drip latex for about four hours, stopping as latex coagulates naturally on the tapping cut thus blocking the latex tubes in the bark. Tappers usually rest and have a meal after finishing their tapping work then start collecting the latex at about midday. Some trees will continue to drip after the collection and this leads to a small amount of cup lump which is collected at the next tapping. The latex that coagulates on the cut is also collected as tree lace. Tree lace and cup lump together account for 10 - 20% of the dry rubber produced.

The latex can be collected in its liquid state. It is sometimes necessary to add a few drops of ammonia solution to the cup, or to the transport tank, to prevent precoagulation of the latex before it reaches the factory. It can also be left in the cup to coagulate naturally into cup lump for collection before the next tapping, although this will produce a lower grade of product. Latex is generally processed into either latex concentrate for manufacture of dipped goods or it can be coagulated under controlled, clean conditions using formic acid. The coagulated latex can then be processed into the higher grade technically specified block rubbers such as TSR3L or TSRCV or used to produce Ribbed Smoke Sheet grades.Naturally coagulated rubber (cup lump) is used in the manufacture of TSR10 and TSR20 grade rubbers. The processing of the rubber for these grades is basically a size reduction and cleaning process in order to remove contamination and prepare the material for the final stage drying. The dried material is then baled and palletised for shipment.

BOTANICAL FEATURES

The Pará rubber tree (*Hevea brasiliensis*), often simply called rubber tree, is a tree belonging to the family Euphorbiaceae and the most economically important member of the genus *Hevea*. It is of major economic importance because its sap-like extract (known as latex) can be collected and is the primary source of natural rubber.

Hevea brasiliensis

Hevea brasiliensis is a native of Brazil and was introduced in tropical Asia in 1876 throught he Kew Gardens(England). The tree is now grown in the tropical regions of Asia, Africa, America. It is a hardy, quick growing tree with a straight trunk producing branches 3-5 m above the ground and forming a spreading canopy. The optimum ecological requirements consist of a fairly distributed rainfall not less than 200 cm, a warm humid climate(21 to 35 deg celsius) and a well-drained deep loamy soil. It flourishes from the sea level upto altitudes of 450-600 m.

Once the trees are 5-6 years old, harvesting can begin: incisions are made orthogonally to the latex vessels, just deep enough to tap the vessels without harming the tree's growth, and the sap is collected in small buckets. This process is known as rubber tapping. Older trees yield more latex.

The wood from this tree, referred to as parawood or rubberwood, is used in the manufacture of furniture. It is valued for its dense grain, minimal shrinkage, attractive color and acceptance of different finishes. It is also prized as an "environmentally friendly" wood, as it makes use of trees that have been cut at the end of their latex-producing cycle.

The Pará rubber tree initially grew only in the Amazon Rainforest. Increasing demand and the discovery of the vulcanisation procedure in 1839 led to the rubber boom in that region, enriching the cities of Belém and Manaus. The name of the tree derives from Pará, the second largest Brazilian state, whose capital is Belém.

These trees were used to obtain rubber by the Olmec people of Mesoamerica as early as 3600 years ago. The rubber was used, among other things, to make the balls used in the Mesoamerican ballgame. There had been an attempt made, in 1873, to grow rubber outside Brazil. After some effort, twelve seedlings were germinated at the Royal Botanic Gardens, Kew. These were sent to India for cultivation, but died. A second attempt was then made, some 70,000 seeds being sent to Kew in 1875. About 4% of these germinated, and in 1876 about 2000 seedlings were sent, in Wardian cases, to Ceylon, and 22 sent to the Botanic Gardens in Singapore. Once established outside its native country, rubber was extensively propagated in the British colonies. Rubber trees were brought to the botanical gardens at Buitenzorg, Java in 1883. By 1898, a rubber plantation had been established in Malaya, and today most rubber tree plantations are in South and Southeast Asia and some also in tropical West Africa. Efforts to cultivate the tree in South America were unsatisfactory, because of blight.

PROPAGATION

The earlier plantations were raised from unselected seeds. The yielding capacity of these trees being low the production was poor in many of the plantations. Hevea is propagated through seeds and only by adopting vegetative methods. The viability of the seeds being short, it is necessary to sow them as soon as possible. Planting is done in specially prepared nursery beds. The soil is dug up, mixed with fertilisers and beds of 30-120cm width and of convenient length are prepared. The germinated seeds are then planted about 30x30cm apart. The seedlings are kept in the nursery till they attain a diameter of 20 to 25 mm above the collar.

The seedlings are usually transplanted in the field in june-july. Healthy seedlings are selected and stumped at a height of 50cm. After pulling out the roots of the seedling are trimmed and then planted in position in prepared pits about 4.5m apart to accomodate an initial stand of about 500 plants per hectare. Sometimes the seeds are sown in situ in the fields, using 2 or 3 seeds per pit.

Bud grafting is done using buds from selected mother trees. Budwood nurseries to supply budwood stumps are established simultaneously with the rootstock nurseries. Nursery seedlings can be bud grafted when they attain the required girth above the ground. Budded stocks are ready to be stumped about 4 weeks after budding. They are cut 10 to 15cm above the bud patch and the stumps are pulled out and transplanted in the field. Green budding with the buds stripped from green shoots has also been practiced. The plant density recommended is 420 to 450 plants per hectare. After planting the field it takes 7 years for the trees to come into tapping. The fitness for tapping is guaged by the girth of about 55cm and a height of 125 cm from the bud union. When 50-70% of the trees have reached this size, tapping is commenced. In India, Hevea seeds normally ripen during July-September when the seeds are collected and seedlings raised. All earlier plantations were raised from unselected seeds. The yield potential of these having been low, the production of those plantations was poor. Selection work on Hevea with a view to improving the planting materials and the introduction of vegetative propagation by budding led, in course of time, to the establishment of numerous valuable clones.

Clonal Seeds

Seeds of clones are termed clonal seeds. There are different kinds of clonal seeds - monoclonal, polyclonal, legitimate and illegitimate. Monoclonal seeds of clone Tjir 1 which gave rise to seedlings superior to seedlings from ordinary, unselected seeds were once encouraged for planting extensively in our country. Evolution of newer, improved planting materials then progressively replaced Tjir 1 clonal seedlings. Hybrid polyclonal seeds collected from approved polyclonal seed gardens are the only seed material recommended for planting now. For production of good quality polyclonal seeds, gardens have been established in Kanyakumari district. Polyclonal garden seeds in the name Prang Besar Isolated Garden (PBIG) seeds from Malaysia used to find limited market in India until a few decades ago.

Germination of Seeds

Rubber seeds lose viability very rapidly if left in the field. The seeds are therefore picked up daily during the seed fall season and quickly transported to nurseries for germination and planting. For germination, raised level beds with a 5 cm thick layer of river sand, 90 cm wide and of convenient lengths are used. Partial shade in order to prevent strong sun is necessary. The seeds are sown in a single layer touching one another and germination beds are kept moist, but not wet, by evenly sprinkling water during morning and evening. The seeds are covered with loosely woven coir matting or gunnies.

Germination starts 6 to 7 days after sowing. Seeds sprouted each day should be picked and planted in nursery beds or in the field as the case may be. Germinated seeds will have young roots emerging first. That is the ideal stage for nursery planting. The sproutings are tender and require careful handling. Usually, germinated seeds are carried to nursery beds in buckets half filled with water. About 75 percent germination is considered good. Pickings are done for about 21 days after sowing.

Nurseries

Nurseries are required for raising seedlings, budded stumps and budwood. As far as possible open and level land should be selected for raising nursery. Water should be easily available for irrigation. The soil should be deep, well drained and fertile. The land should be dug to a depth of 75 cm and all stumps, roots, and stones should be removed. Nursery beds should be prepared with 60 to 120 cm width and convenient lengths and with pathways laid in between to facilitate manuring, watering, weeding etc. Planting distances should vary according to the type of planting materials to be raised in the nursery. The ideal spacing for seedlings is 30 cm × 30 cm. For budwood nurseries, the plants may be at distances of 60 × 90 cm or 60 × 120 cm or 90 × 90 cm. Nursery management aims at the most rapid production of standard healthy planting materials. More intensive care can be exercised in a nursery than in a field. Plants which are obviously unsuitable can be eliminated at an early stage. Mainframe operations for nursery include weeding, mulching, irrigation during dry months, manuring and disease and pest control.

Budding

The principle involved in budding is the replacement of the shoot system of a plant with that of another more desirable plant. In this process, a patch of bark of the seedling plant (stock) is replaced by a patch of bark with a dormant bud (bud patch) taken from the clone to be multiplied. The bud patch gets attached to the stock permanently and becomes a part of it. The stock is then cut off above the budded portion and the grafted bud develops into a shoot (scion) exhibiting the characters of the plant from which it was taken. The new tree thus formed is a two-part tree, comprising a root system belonging to the stock plant and a shoot system contributed by the donor of the bud.

Depending on the colour and age of the buds as well as the age of the stock plants used, three types of buddings are mainly recognised. These are brown (conventional) budding, green budding and young budding. In the first method, older buds having brown colour are used while in the other two, green tender buds are utilised. Depending on the part of the stock where budding is carried out, buddings are classified into four types: base budding, crown budding, over budding and high budding. Base budding is carried out at the base of the stock plant and includes brown budding, green budding and young budding.

Brown budding

Brown budding is generally carried out by grafting brown coloured buds taken from budwood of about one year's growth onto stock plants of 10 months or more growth. Vigorously growing healthy stocks having a girth of 7.5 cm at the collar region are ideal for budding. Stocks should be budded when the bark peels off very easily. Peeling is usually good when the top whorl of leaves is well developed, but before further extension growth commences. Test peeling of a small patch of bark above 15 cm from the base is the sure method to assess the peeling quality of the bark. Since all stock plants may not attain this stage at the same time, more than one round of budding may be necessary to cover all the stocks.

Brown buds are usually obtained from brown budwood produced by budded plants raised in budwood nurseries. Buds found in the axils of fallen leaves are generally utilised for budding. Budwood should be collected when the top whorl of leaves have fully expanded but not hardened to ensure proper peeling of the bark and high budding success. Test peeling may be carried out before harvesting the budwood. Collection of budwood should be done with sufficient care so as to avoid bruising. As far as possible, budwood should be collected in the morning or evening, and should preferably be utilised for budding as soon as collected. If budding is delayed, special measures should be adopted for preventing moisture loss. Budwood is harvested as per the requirement and cut into pieces of convenient length, usually 1 m. Good quality budwood will have around 20 healthy, well formed buds per metre length. Weak and poorly formed buds should not be used. While handling budwood of different clones, proper labelling has to be done for identification.

Budding is usually carried out with a specially designed knife having two blades, called budding knife. However, an ordinary pen-knife with a blade of 7 to 8 cm length could also be used. The first step in the preparation of the stock plant for budding is thorough cleaning of the basal 15 cm to remove dirt, soil, etc. Cleaning can be done with cotton waste or rags. Then two parallel vertical cuts starting from about 2.5 cm above the collar are made. The length of these cuts should be a little more than 5 cm and 1.5 cm apart. Then a horizontal cut joining the bottom ends of the vertical cuts is also made. All the three cuts should be made deep enough to reach the wood. After making these cuts the latex is allowed to completely ooze out for a few minutes through the cuts. During this time the budder can mark a few more stocks in a similar manner. When the latex flow ceases, it is wiped off from the surface. The flap of bark separated by the three cuts is then gently lifted with the aid of the knife and peeled upwards. Alternatively, the upper ends of the vertical cuts may be connected by the horizontal cut and the flap peeled downwards. The practice of removing the flap completely is also adopted. The exposed region is called the budding panel.

The bud patch used for brown budding should have a length of about 5 cm and a width of about 1.5 cm. For preparing the bud patch, two parallel vertical cuts having a length of 5 cm are made on the two sides of a bud, 1.5 cm apart. Then two horizontal cuts are made connecting the lower and upper ends of these cuts. A little time is allowed for the latex to ooze out. During this time, incisions are made around neighbouring buds of the same budwood. When the oozing of latex stops, it is wiped off and the bud patch marked out by the four cuts is stripped off by gently pushing to one side.

After removing the bud patch from the budwood, the inner side is examined carefully for the presence of the core of the bud, which appears as a slight projection. If it is not present, the bud patch should be discarded. The bud patch should be handled with utmost care so as to avoid any damage to the cambium. It should always be held at the edges without touching the cambium. Foreign matter like water, soil and sweat should not be allowed to fall on the cambium. Similarly exposing the cambium to strong sunlight or dry wind can result in its drying. All these can cause damage to the cambium. Damage to the cambium of the budding panel also should be avoided. The four edges of the bud patch are then slightly trimmed.

The bud patch is then gently placed in the budding panel after lifting the flap. It should be placed in such a way that the bud is above the leaf scar and its inner side is in contact with the budding panel. Exposure of too much area of the budding panel around the bud patch is unfavourable for budding success. At the same time the edges of the bud patch and budding panel should not touch each other. After placing the bud patch in the budding panel in the above manner, the flap, if retained, is placed back over it and is then bandaged using polythene strips of 45 cm length, 2.5 cm width and 250 gauge thickness. Bandaging should commence at the bottom and move upwards in a close spiral.

During the first few turnings of the bandage, the lower end of the flap should be kept gently pressed over the bud patch to prevent it from slipping. Bandaging should be tightened to keep the cambium tissues of the budding panel and the bud patch in intimate contact with each other. The end of the tape is finally kept intact with a knot. In the field and along borders of nurseries it may be necessary to shade the bud patch against strong sunlight. This can conveniently be done by tying a rubber leaf over the bud patch.

If the budding is successful, the cambium of the stock plant and that of the bud patch unite and the bud patch establishes as a part of the stock. The process requires 15 to 20 days and the plant is left undisturbed for 20 days after which the bandage is removed. The flap, if it had been retained, is cut a little above the upper end of the bud patch and removed. Freshness of the bud patch indicates initial success of the budding. The final success is ascertained in a similar manner after another 10 days.

Green budding

Both the stock plant and budwood used for green budding are very young. Seedlings which are five to seven months old are used as stock. Buds are collected from six to eight weeks old budwood, also known as bud shoots or bud sticks. Buds found above the scale leaves of the shoots alone are used for budding. These buds are green in colour and hence the name green budding.

Young, vigorous seedlings raised in nursery or in polythene bags are used as stock plants for green budding. Plants having a girth of about 2.5 cm at the base, with brown bark up to a height of about 15 cm, can be used for this purpose. The stock plants require about four to five months to attain this size. By proper care, this period could be further reduced.

Green budwood is obtained from budwood plants (source bushes) grown in nurseries for this purpose. They are collected when six to eight weeks old, when they have a length of 30 to 60 cm with a whorl of leaves at the top. The bud shoots are harvested by cutting at the base with a sharp knife. For proper peeling of the bud patch, harvesting should be done when the leaves are copper brown to dark green in colour. After harvesting, the leafy portion of the shoot is cut off. The non-leafy portion shows two to five scale leaves with axillary buds which are utilised for budding.

After cleaning the basal portion of the stock, two vertical incisions, a little more than 5 cm long and 1 cm apart are made starting from a point about 2.5 cm above the collar region. The lower ends of these cuts are joined by a horizontal cut and a few minutes allowed for the cessation of latex flow. The flap is then cut off leaving a short "tongue" of about 1.5 cm at the top. The stock is now ready to receive the bud patch.

The bud patch can be stripped from the bud shoot in the same way as in the case of brown budding. However, other methods can also be employed for this purpose. In one such method, a 6 cm long bud patch shaped like a pointed tower is marked out. Then the top 1 cm is separated from the bud shoot and holding on to this portion (by touching the inner and outer sides), the whole bud patch is gently stripped off. The top 1 cm of the bud patch with damaged cambium is pruned off. In a third method the bud patch along with a thin slice of wood is first taken from the bud shoot. This is the bud slip. The two sides of the bud slip are trimmed to the required width. The bud patch is then gently separated from the wood by pulling them apart. While doing so care should be taken to see that the bud patch does not bent. Only the slice of wood is bend. After separation, the lower and upper ends of the bud patch are also trimmed. When finally prepared, the bud patch should have a length of approximately 5 cm and a width of 1 cm, so that it fits snugly into the budding panel.

The upper end of the bud patch prepared in the above manner is gently inserted under the 'tongue' and placed in the budding panel. Then the bud patch is secured firmly by bandaging with a transparent polythene strip as in the case of brown budding.

This strip should be about 25 cm long and 2 cm wide. Transparent tape is insisted upon as it allows light to fall on the green bud patch which in turn enhances budding success. For the same reason, no shading is given. Buds are examined three weeks after the budding by observing through the bandage or after removing the bandage. Retention of the green colour is the indication of budding success. Final observation on budding success is done after 10 more days. If the observations are made through the bandage, after the second observation, the bandage is removed. The plant is now ready for cutting back.

Budding can be carried out at any time of the year. However, too dry or very wet weather is unsuitable. Generally more success is obtained during rainy season than in summer. Experiments have shown that in India the period from April to October is generally suited for brown budding. The success rate of green budding is more during summer months also. However, heavy rainfall is not suited for budding. For best results, budding should be carried out either in the early morning hours or in the evenings. Both brown budding and green budding have certain advantages and disadvantages.

Advantages of Green Budding over Brown Budding

i) It utilises the growth of the stock more efficiently, i.e. when green budded plants are cut back, only a small amount of stock growth is lost,

ii) Green budding is simpler and faster than brown budding and hence more number of plants can be budded reducing the labour cost per budding,

iii) It gives higher percentage of success during summer than brown budding,

iv) Opening of the budding and cutting back can be done simultaneously, thus saving labour cost,

v) Yield of green buds from a unit area of nursery is two to three times that of brown buds,

vi) Since the polythene strip used for green budding is smaller, cost incurred for this material is reduced,

vii) After cutting back green buds develop earlier than brown buds,

viii) Green budding is more suitable for crown budding.

Major Defects of Green Budding Technique

i) Green bud shoots cannot be retained in the nursery for long after they become mature enough for harvesting,

ii) After harvesting, green sticks cannot be kept for long periods, unlike in the case of brown budwood

iii) Scion of green budding is less vigorous than that of brown budding and hence it requires very careful attention during the early period of growth, especially in the field.

Young budding

This is a kind of green budding carried out on very young plants less than two months old. Stocks are raised in small bags of lay flat size 33 x 15 cm. The plants are given intensive nursing such as foliar application of fertilisers and fungicides twice weekly and soil application of NPKMg mixture weekly. When seven to eight weeks old, they are green budded. Four weeks after budding, plants are cut back leaving a snag of 20 to 25 cm length. Buds on the snag are nicked or the shoots coming from then pruned off promptly. When the scion develops two or three whorls of leaves, the plants are transplanted to the field.

This technique has got certain advantages over the normal green budding technique. By adopting this technique bag plants could be produced within seven months after the planting of germinated seeds in the bags, which is usually done in August/ September. In our country, the time required for this is around nine months in the case of normal green budding. Since small bags are used for the production of plants transportation is easier.

The cost of production is also slightly less compared to the normal method. Since the stock plants required for young budding are raised by sowing seeds directly in the bag, these plants have a better developed root system than the plants raised from green budded stumps. A well-developed root system prevents breaking of soil core and ensures faster and easier establishment of plants after transfer to the field. However, under the climatic conditions existing in our country this method does not have much practical application because by the time the bag plants produced from young buddings are ready (February - March) the climate becomes unfavourable for field planting. Hence planting has to be delayed up to the onset of monsoon (June). Bag plants produced from green buddings also becomes ready by this time.

Crown budding

Replacing the undesirable crown of a high yielding clone with a desirable crown is of practical significance. In many of the modern clones, though the trunk possesses a capacity for high yield, the crown shows many undesirable characters like susceptibility to wind and diseases. An undesirable crown can be replaced by a desirable one through crown budding. The tree produced by crown budding is a three-part-tree comprising the root system of the stock plant, trunk of one clone and the crown of another clone. Thus the desirable characters of the trunk of one clone and crown of another clone are combined.

Crown budding is ideally carried out when the scion of the budded plant has attained a height of 2.4 to 3 m. One to two years are usually required for the plants to attain such a growth. The height of the plant is more important than the age. Crown budding may be commenced when 50 to 60 per cent of the plants in the field are buddable. Budding is carried out at a height of 210 to 240 cm on the inter-whorl region below the top whorl of leaves. It should be done only when the top flush of leaves are fully expanded and hardened. Stem tissue should be green or dark green at the time of budding. This ensures maximum budding success. Too tender or too mature stem tissues adversely affect budding success. If the topmost flush of leaves are not mature enough, budding could be done below the second whorl of leaves provided all other conditions are satisfied.

Plants having height up to 4.5 m can also be used for crown budding. In the case of such overgrown plants having green tissue at a height higher than the prescribed, the height of budding has to be raised correspondingly so as to carry it out below the top whorl. For crown budding, the green budding technique is followed. Since the budding has to be done at a higher level, a self-supporting ladder should be used. On no account should the plant be bent for budding or any other operations. If the budding is a failure, rebudding is done on the opposite side of the stem, 5 cm above or below the first budding.

Successfully budded plants are cut back leaving a snag of about 5 cm. Treating the cut ends of the stem with some wound dressing compound is desirable. After cutting back, usually many trunk shoots arise. All of them should be pruned regularly at fortnightly intervals with a knife having a long handle. This should be continued until the crown bud sprouts and the crown shoot grows to a length of about 2.5 cm. After that, two or three trunk shoots arising about 15 to 45 cm below the crown shoot are allowed to grow. These shoots should be spaced apart and be on different sides of the stem as far as possible. They should never be allowed to become more dominant than the crown shoot, as this may suppress the growth of the latter. For this, if necessary, the top portion of the trunk shoots may be pruned. About nine months after cutting back, when the crown-trunk union is firmly established, the trunk shoots are pruned. Necessary precautionary measures have to be taken for the protection of the crown shoot especially from wind damage and perching of birds. The crown shoot later on fully establishes itself and in due course develops to be the crown of the three-part tree. If the crown shoot is lost for some reason before the pruning of the trunk shoots, the most vigorous among them is again crown budded, if possible, or allowed to develop as the crown.

Over budding

Budwood plants are sometimes budded at higher levels for converting an existing budwood nursery of a clone to another clone without replanting. This is termed as over budding. The method adopted is to carry out budding at the basal portion of the brown budwood before harvesting it. The budwood is harvested after the new bud is successfully attached to the plant, by cutting above the budded portion. Since the budding is carried out on a well-established plant, the scion emerging from the new bud grows vigorously producing more budwood compared to a newly established budwood plant. Thus it is a quick and economic method for converting budwood nursery of one clone to that of another.

Tissue Culture

Propagation of rubber is possible through tissue culture also. Tissue culture or micro propagation is the technique of producing plants from small (micro) pieces of plant tissues. Studies on tissue culture of rubber plants were started in 1966. Different parts of the plant such as embryo, anther, shoot tip and integument can be used for tissue culture.

Rubber Research Institute of India has developed a technique for the production of tissue culture plants from shoot tips as well as somatic embryogenesis of different tissues. Rubber plants were developed by the somatic embryogenesis of anther tissue, integumental tissue, immature inflorecence etc. Attempts are being made for the tissue culture of other plant parts like leaf, floral buds, ovules and micro spores.

Key steps in tissue culture of hevea include collection of the explant, sterilisation, inoculation of the explant in a nutrient media supplemented with growth hormones and sucrose. The cultures were kept under optimum light and temperature conditions for the required period.

Plants were formed in about eight months. They were then transferred to small polybags and kept in a green house for hardening. Even for the same clone the culture conditions vary with the physiological stage of the explant, seasons, part of the plant used etc. Because of these variations separate protocols have to be developed for each clone. This is a laborious and time-consuming process. However, procedures have been perfected for the propagation of several important clones by optimising these parameters. As in the case of most other tree crops, multiplication rate in tissue culture is very low for rubber. Further in the post-laboratory stages mortality is very high. However, after overcoming all these hurdles the RRII has successfully developed a large number of plants through various methods of tissue culture. These plants have been established in the field

Preparation and Packing of Propagation Materials

The propagation materials handled by rubber growers are ungerminated seeds, germinated seeds, seedling stumps, brown budwood, green bud shoot, brown budded stumps, green budded stumps, polybag plants and stumped buddings. Specific techniques are required for the preparation of these materials. If not properly prepared, their quality could be reduced, which in turn adversely affects the establishment after planting. After preparation, these materials may require storing and/or transportation. During storage and transit they are likely to get damaged by loss of moisture or by breaking, rubbing, bruising, crushing, etc. To avoid these and to give ample protection to these materials, certain specific methods are adopted for packing and transporting.

Ungerminated seeds

Fresh and healthy seeds collected from the field can be kept under shade without much loss of viability for about seven days. Storing fresh seeds in water at ambient temperature increases their water content, which in turn prolongs the viability. By packing seeds loosely in well-aerated containers with powdered charcoal having 20 per cent moisture, 70 per cent viability can be retained up to 30 days. Storage of seeds at 4oC in sealed polythene bags is also considered to be a reliable method for retaining viability up to four months.

Immediately after collection, seeds are generally packed in powdered charcoal of 20 per cent moisture for transportation. Containers usually used are wooden boxes, double gunny bags, bags lined with polythene or polythene bags. For transporting over very long distance, seeds may be packed tightly in layers in aerated cases with damp sawdust-charcoal powder mixture, at least 2 cm thick between two layers of seeds.

Germinated seeds

Germinated seeds are collected from germination beds when the radicle just comes out of the seed. If the root is allowed to elongate it may get damaged while packing and transporting. To prevent this, germination beds are inspected every day and germinated seeds are picked up

Brown budwood

After harvesting, the brown budwood is cut into pieces of one metre length with a pruning saw for the convenience of handling. Immature top portion, which may be green or partially brown, is discarded. For use on the same day and transporting over short distance brown budwood is kept wrapped in wet sacking. For longer storage and transporting, their cut ends are sealed with melted wax and each piece covered with banana sheath, wet sacking, coconut fibre, or grass leaves. They are then tied into bundles of convenient size. By this method, viability could be retained up to three days. For storing up to 14 days and carrying over very long distances each piece is first wrapped with perforated polythene and then packed in boxes with wet sawdust or coconut fibre.

Green bud shoots

After collection of the shoots, their leaf bearing top portion is cut off. The leafless lower part having the scale leaves are used for taking buds. Since green bud shoots are tender it is better to use them for budding immediately after harvesting. Usually shoots harvested in the morning are used on the same day, preferably by noon. In the seedling nursery they are carried in trays or buckets containing water or kept rolled in wet sack or cloth without touching each other and always kept in shade. If packed with wet, aged saw dust in alternating layers after sealing the cut ends with wax, storage is possible up to six days.

Seedling stumps

Seedling prepared to a convenient size by pruning the stem and roots are called seedling stumps. Healthy and vigorous one year old seedlings are generally used for this. The seedlings should have a minimum girth of about 7.5 cm at the base and brown colour up to a height of 45 cm or more. For stumping, at first, the seedlings are cut back at some point between 45 and 60 cm, where the brown colour ends. Pruning is always done with a slanting cut, preferably above a whorl of buds. While cutting back green or partially brown stem should not be retained on the plant. Transpiration can take place through such regions and the resulting loss of water may lead to the drying of the stumps after planting. The plants are left in the nursery for 7 to 10 days. During this period a few buds below the cut end become activated and swell. At this stage the decapitated plants are pulled out without causing much damage to the roots and bark of the stem. The taproot is pruned to the maximum possible length, but not more than 60 cm and not less than 45 cm. The minimum of 45 cm is insisted for the sake of better establishment and the maximum of 60 cm is fixed for the convenience of handling. Lateral roots are pruned to a length of 10 to 15 cm. Plants infected by diseases or having defective roots are discarded. If more than one taproot is present the most vigorous one alone should be retained and all others pruned off at the base. After preparing the seedling stumps by proper pruning of roots and stem, the cut end of the stem is immediately sealed by dipping in molten paraffin wax. For transporting over short distances they are tied into bundles and then covered with a layer of grass or leaves. In this manner they can be stored up to three days. If transporting over large distances and storing for days together are required, it is better to avoid stumping the seedlings days before pulling out. Pruning of shoot and roots may be carried out after pulling out. The stumps in such cases are packed in boxes along with wet saw dust in alternate layers.

Brown budded stumps

Brown budded plant prepared for planting by pruning the stem and roots is known as brown budded stump. The ideal way to prepare a brown budded stump is to cut the stem of the plant at a height of about 7.5 cm above the upper end of the bud patch. The cut should have a downward slant of around 45o from the side of the bud to the opposite side. The plants are then pulled out and the taproot pruned to a length of 45-60 cm and the laterals to a length of 10-15 cm. While gripping the stem for pulling out the plant, special care should be taken not to exert any pressure on the bud patch. Otherwise there are chances for the bud to get damaged. In case it is found difficult to pull out the plant after cutting back the plant can be pulled out before cutting back and then pruned. If the budded stumps are intended for planting in polybags the taproot should be pruned to a length about 15 cm less than the height of the soil core. Rejection of defective plants, pruning of multiple taproots etc. are similar to those of seedling stumps.

Brown budded stumps are hardy and hence a proper method of packing retains the viability for a longer time. The cut ends of the stem are first sealed with melted wax. The bud patch is protected by covering with a small piece of banana sheath or folded rubber leaf. For overnight storing, they are kept erect in water as in the case of seedling stumps. For retaining their viability up to three days and carrying over short distances they are tied into bundles and each bundle covered with banana sheath, grass or leaves. At the time of planting the protective covering of the bud patch is removed. For long distance transport of budded stumps involving preservation over a few weeks, advance cutting back of budded plants before pulling out should be avoided. The freshly cut back and pruned budded stumps should have their cut ends sealed by molten wax and the whole stump given a light fungicide treatment. They are then packed in boxes with wet sterilised sawdust.

Green budded stumps

Green budded stumps are prepared from green budded plants in almost the same way as that of brown budded stumps. The major difference is that if cutting back is done before pulling out it should be done seven days before pulling out as green buds require lesser time for activation. Green budded stumps being comparatively tender and smaller with less reserve food stored in them, it is always preferable to plant them immediately after preparation. Packing methods adopted for green budded stumps are basically the same as that adopted for brown budded stumps. Protection to the bud patch can also be provided by retaining the polythene strip used for bandaging the bud patch at the time of budding. Standing in water enables preservation overnight. Additional steps like sealing of the cut end of root, tying into bundles and covering the bundles with grass etc. gives protection for two days. If packed in boxes with wet saw dust these could be safely stored up to six days and transported over long distances. In the case of both types of budded stumps described above the budded plant should be pulled out without much delay after budding. If retained in the nursery for a long time the dormancy of the buds will increase and consequently their sprouting after planting may be delayed.

Polybag Plants

While transporting polybag plants utmost care should be given to prevent any damage to the soil core. If the soil core is damaged roots may break and consequently the plant will be deprived of all the advantages of bag planting. Transporting over a short distance is done by carrying them on the head or shoulder. Vehicles like lorry, truck, tractor, etc. are used for transporting over long distances. While transporting by vehicles, the bags are stacked on the platform tightly to reduce their swaying and shaking to the minimum. Providing shade for protection from hot sun is also desirable. The bag should always be kept in a vertical position while loading, unloading and transporting. Carrying them in inclined or horizontal position may cause breaking of soil core. Care should also be taken to avoid tearing of the bags as it also can increase the chances for breaking of the soil core.

Stumped buddings

Stumped buddings are mainly of two types, mini stumps and maxi stumps. To prepare ministumps the scion is cut back when it develops brown colour up to a height of 60 cm from the bud union. Pruning is done at the point where brown colour ends preferably below a whorl of buds. The cut end is treated with any wound dressing material to prevent dehydration. Stem is then white washed with hydrated lime to avoid sun-scorching. Ten days are given for the activation of buds. Then the plants are pulled out and the roots pruned as in the case of budded stumps. If pulling out is found to be difficult due to the deep taproot it can be made easy by removing the soil at one side of the taproot up to a depth of 45 to 60 cm. A crowbar is inserted through this opening and the taproot is severed at the desired depth. This is called tailing.

For preparing maxi stumps cutting back of the scion is carried out when brown colour is formed up to a height of 240 cm. First step in the extraction of the plant is the tailing which is done five weeks before pulling out. Early tailing reduces the transplanting shock and enhances the development of new roots from the cut end after transplanting. After tailing soil removed for this purpose is placed back. Pollarding of the stem is done 10 days before pulling out at a height of 240 cm. where brown colour ends. As in the case of mini stumps pruning of stem is done below a whorl of buds. Wound dressing, white washing, pulling out and pruning of lateral roots are also undertaken as done for ministumps. Packing and transporting of mini stumps are similar to that of seedling stumps. Maxi stumps are not usually made into bundles due to their large size. They are packed head to tail on lorries or trailers, using grass or leaves as packing medium to prevent bruising and drying.

RUBBER GROWING REGIONS

The rubber growing regions in India can be classified under two major zones, traditional and non-traditional on the basis of agro-climatic conditions.Rubber cultivation in India has been traditionally confined to the hinterlands of the southwest coast, mainly in Kanyakumari District of Tamil Nadu and Kerala. There are hinterlands of coastal Karnataka, Goa, Konkan Region of Maharashtra, hinterlands of coastal Andhra Pradesh and Orissa, the northeastern states, Andaman and Nicobar Islands etc, where rubber is now being grown.

Agro-climatic Requirements

Humid tropical climate prevails in the rubber-growing tract. Average annual rainfall in the tract varies from about 2000-4500 mm. The southern parts of the traditional tract enjoy southwest and northeast monsoons almost equally while the northern areas receive mostly the southwest monsoon. From south to north the drought period extends from two to five months in a year and the distribution of rainfall becomes more uneven. However, variation in temperature and humidity in the rubber tract is not so marked as that of the rainfall. The temperature remains very warm and humidity very high throughout the year.

Climatic conditions for optimum growth of rubber tree

1. Rainfall of 2000 to 3000 mm evenly distributed without any marked dry season and with 125 to 150 rainy days per annum
2. Maximum temperature of about 29°C to 34°C and minimum of about 20°C or more with a monthly mean of 25 to 28°C
3. High atmospheric humidity of the order of 80%
4. Bright sunshine amounting to about 2000 h per annum at the rate of 6 h per day through all the months
5. Absence of strong winds

Only a few regions in India meet all these requirements. Fortunately rubber can be grown successfully under moderately deviating

Rubber Growing Soils

Soil in the rubber tract is generally highly weathered and consists mostly of laterite, lateritic types. Sedimentary types and nonlateritic red and alluvial soils are also seen in some non-traditional areas. The laterite and lateritic soils are mostly very porous, well drained, moderately to highly acidic, deficient in available phosphorus and varying in potassium and magnesium content. Red soil found in some areas is characterised by reddish to brown colour and fine loamy texture. This soil is generally acidic and highly deficient in available phosphorus.

Soil for rubber cultivation should have a minimum depth of one metre without any ıtervening hardpan or impenetrable layer. Water table should also be well below one ıetre so that at least one metre of soil with good aeration, essential for root penetration available.Well-drained soil is essential for optimum growth and yield of rubber plants. ı marshy areas, owing to poor physical properties and waterlogged conditions growth rubber is always found to be very poor.

EED MANAGEMENT

ommon Weeds

ɔrreria sp. (Button weed), Chromolaena odorata (Siam weed), Lantana aculeate .antana), Mimosa pudica (Touch-me-not), Clerodendron sp., Mikania micrantha /Iile-a-minute), Sida sp., Imperata cylindrica, Pennisetum sp. (Napier grass), Axonopus ɔ (Carpet grass), Paspalum sp., Digitaria sp. (Tropical crab grass), Cynodon dactylon ermuda grass) etc.

ethods of Weed Control

'eeds can be controlled by either manual methods or with the use of chemicals known herbicides. However, an integrated method involving a combination of these two ethods is more economical and eco friendly.

anual

ıis involves slashing, scraping, uprooting etc. As manual clean weeding will expose ils to erosion, mulching should be done immediately after weeding.

ıemical

eeds can be controlled with the use of herbicides or weedicides. There are two main pes of herbicides, the pre-emergent and post-emergent herbicides.

e-emergent herbicides

e-emergent herbicides are soil-applied chemicals used for killing weeds before they ıerge from soil. In newly prepared planting strips, the application of pre-emergent rbicides immediately after planting can check the weed growth for 6-7 weeks.

st-emergent herbicides

ıese herbicides are applied on the weeds and are used to kill the weeds, which have ıerged and are actively growing.

Integrated Weed Management

This approach involves an integration of manual and chemical weed control methoc along with the establishment of cover crop during the first year itself which helps i smothering weeds in the inter spaces and thereby reducing the usage of herbicide t the minimum. Application of herbicides in the plant basin alone and slash weeding th remaining inter plant area along the planting strip was found to be cost effective an sustainable.

Weed Management in Different Stages

Rubber plantations are infested by various weed species throughout the year and prese major problems of access and competition. Weeds compete with rubber for light, sc moisture and nutrients and also serve as alternate hosts for insects and pathoger Therefore, management of weeds is an important cultural operation to be followed rubber plantations.

Maintenance of a luxuriant ground cover in the early phase of plantation ca minimise the weed infestation to a great extent. However, during the first year till th cover crops are fully established, the weeds that are sprouting around the cover cro patches should be controlled for their early establishment. Once the ground cover established, the weed management operations need be concentrated only on th planting strips.

Manuring / Fertiliser Application

General fertiliser recommendation

The general fertiliser recommendation for rubber is derived based on the results of t fertiliser experiments conducted by the Rubber Research Institute of India on rubb of different age groups. Majority of the rubber growing soils belong to the laterite an lateritic types and hence this general fertiliser recommendation can be followe However, practicing discriminatory fertiliser recommendation based on soil and le analysis will be more advantageous.

The nutrient requirements of rubber plant vary considerably during the thr important stages of growth; the nursery, immature and mature stages.

Fertiliser recommendation for North-Eastern region

Most of the soils in the North Eastern region are highly degenerated due to shifti cultivation and heavy nutrient removal by thatch grass. The organic carbon status the soil is in the medium range. In general, the available phosphorus and potassiu levels are low but the available magnesium is high. Based on these, a separate fertili ecommendation with slightly higher levels of N, P and K was formulated for the main eld in the North Eastern region. The time and method of fertiliser application is same s that of the traditional region.

iscriminatory fertiliser recommendation

ertility status of the individual fields can be assessed by soil and leaf analyses. Economic esponse to manuring, especially for mature trees depends on many factors such as the resent yield and the genetic yielding capacity of the clone, the age and condition of ne trees, the tapping history, the nutrient supplying capacity of the soil, the nutrient tatus of the trees, and the manuring and soil management history. Assessment of utrient status of individual fields through soil and leaf analysis and fertiliser ecommendation based on the individual case history of the field is termed as iscriminatory fertiliser recommendation.

Discriminatory fertiliser usage has many advantages compared to general fertiliser pplication. Very often adoption of this method results in considerable saving of ertilisers and ensures optimum growth and yield. Adequate supply of nutrients and naintenance of proper balance in the ratio of N, P and K is possible by adopting iscriminatory fertiliser usage. The method also minimises or corrects problems like vind damage, panel coagulation and late dripping. It improves the quality of latex by reventing pre-coagulation of field latex and reduces instability of preserved latex that ccur mainly due to high Mg content or improper ratio of magnesium and phosphorus

Jutritional disorders commonly observed in India

Rubber plants are found to exhibit typical symptoms of nutritional disorders caused y deficient and/or excessive supply of individual plant nutrients. Deficiency symptoms lue to lack of magnesium, potassium and in some isolated cases zinc and manganese ave been observed in rubber nurseries and field plantings. Magnesium deficiency is he commonly observed nutritional disorder. The characteristic symptom of magnesium leficiency is the development of chlorosis (yellowing) in the interveinal areas on exposed nature leaves giving rise a herringbone pattern. The deficiency incidences are seldom een in Kanyakumari district and in the northern parts of rubber growing tract, onsisting of Thrissur, Palakkad, Malappuram, Kozhikode, Kannur and Kasaragod listricts.

Potassium deficiency is commonly found in rubber grown in highly impoverished oils. The characteristic symptom of potassium deficiency is the development of narginal and tip chlorosis which is followed by marginal necrosis. Only older leaves xhibit the deficiency symptoms. Size reduction of the leaves and the absence of ierringbone pattern of yellowing allow potassium deficiency symptoms to be listinguished from those of magnesium deficiency.

Zinc deficiency causes interveinal chlorosis of leaves. The outstanding features of thi deficiency are that the laminae become much reduced in breadth in proportion to the length and the young leaflet becomes incurved towards one another and present hooked or claw appearance. Zinc deficiency incidences have been noted so far only i the case of young rubber plants either in the nursery or in the field. In most cases thes deficiencies are noticed to be only transient. The cause of deficiency appears to be heav application of phosphatic fertlisers resulting in poor availability of Zinc.

The typical manganese deficiency symptom is an overall paling and yellowing c the leaf with bands of green tissue outlining the midrib and main veins. Though thi deficiency is widespread in India it has been found to be only very mild in intensity

Apart from these disorders connected with nutrition, problems such as pre coagulation of latex on the tapping panel and excessive drainage of latex causing drynes of trees have also been reported from rubber plantations. The pre-coagulation of late on the tapping panel has been found to be due to excessive supply of magnesium t rubber. Also, there are indications to believe that unbalanced nutrition can caus excessive drainage of latex resulting in the dryness of trees.

The incidence of nutritional disorders mentioned above are known to affect th growth and productivity of rubber to a great extent. Therefore, the planters are advise to consult the Rubber Research Institute of India if any of these disorders is notice in their plantations and to take necessary preventive measures without delay.

Latex is obtained from the bark of the rubber tree by tapping. Tapping is a proces of controlled wounding during which thin shavings of bark are removed. The aim c tapping is to cut open the latex vessels in the case of trees tapped for the first time o to remove the coagulum which blocks the cut ends of the latex vessels in the case c trees under regular tapping.

REFERENCES

"Elastomer-The rubber tree", *Encyclopaedia Britannica,* 2008

Winchester, Simon (2003). *Krakatoa: The Day the World Exploded, August 27, 1883*. HarperCollins. pp. 223-224.

Zhang, J., Huss, V.A.R., Sun, X., Chang, K. and Pan, D. 2008. Morphology and phylogenetic position c a trebouxiophycean green algae (Chlorophyta) growing on the rubber tree, *Hevea brasiliensis,* wit the description of a new genus and species. *Eur. J. Phycol.* 43(2): 185 - 193.

11

Tobacco

India is the largest producer and exporter of tobacco in the World. Total tobacco production in India is about 700 million Kilograms annually. Rich and varied Indian geographic and agro-climatic conditions foster consistent availability of wide range of tobaccos for export all through the year. Indian tobacco, by virtue of its qualities, sheer volumes and diversity, is progressing gracefully to occupy its rightful place in the world tobacco market.

Tobacco is a principal cash crop of National importance. It has been playing a prominent role in the development of Nation's Economy. Although the cultivation of Tobacco is restricted to 0.3% of the total cultivated area, it provides employment to large number of people on the one hand. On the other hand, it makes significant contribution to National Exchequer by way of excise revenue and foreign exchange earnings. Tobacco being a labour intensive crop provides employment to more than 60 lakhs people who are engaged in the farming curing, redrying, packaging, grading, manufacturing distribution, export and retailing activities. The bidi industry which provides employment to around 44.00 lakhs essentially unskilled rural folks mostly women is also arresting the influx of rural labour to urban centres.

Although there is nation wide anti-Tobacco campaign, the commercial importance of Tobacco can never be underestimated due to the revenue earning potentiality and employment generation capacity of the crop. Presently there is a call for substitution of Tobacco with other crops, but the research findings show that there is no economically viable alternative crop which is as remunerative as Tobacco to the farmer.

Botanically, the Tobacco plant belongs to the family Solanacea and genus Nicotiana. The genus embraces over 60species of which two alone are cultivated. India grows both the cultivated species, viz. Nicotiana tabacum and Nicotiana rustica The largest area is under Nicotiana tabacum which is grown allover the country where as Nicotina rustica is confined to North and North Eastern States: i.e. Uttar Pradesh, Bihar, West Bengal and Assam.

About 5 to 6 per cent of the total area under Tobacco is aounted for Nicotiana rustica varieties. The cultivation of Nicotiana tabacum has country-wise spread and this type also aounts for more than 80% of the exchange earnings.

Specific types and varieties of Tobacco have been developed for use in cigarette, bidi, cigar, cheroot, hookah, chewing, snuff and hookah paste. Rustica types are used inchewing and snuff whereas tabacum types are used for all purposes.

Table showing net return from Tobacco and next best crop in some of the Tobacco growing state is detailed below:

Table 1. Differences in Net Returns from Tobacco and Next Best Crop (Rupees per hectare)

State	*Crop Giving Next Best Returns (I)*	*Differences in Net Returns Between Tobacco and (I)*
Andhra Pradesh (Northern Black Soils)	*Bengalgram*	4405
Andhra Pradesh	Chillies	875
Karnataka	Cotton	4500
Gujarat	Pearl millet	2306
Bihar Maize+	Potato	2729
West Bengal	Potato	5134
Tamil Nadu	Groundnut	2825

Eventhough the cultivation of Tobacco is spread all over the country, the commercial cultivation of Tobacco is concentrated in States like Andhra Pradesh, Karnataka, Gujarat, Maharastra, Bihar, Tamil Nadu and West Bengal etc. Cigarette Tobacco is mostly cultivated in Andhra and Karnataka, whereas bidi Tobacco is grown in Gujarat, Karnataka and Maharastra. Cigar and Cheroot Tobacco are also grown in Tamil Nadu, Andhra Pradesh and West Bengal and Chewing Tobacco is grown in Tamil Nadu, Gujarat, Bihar, West Bengal and U.P. Hookah Tobacco is grown in UP, and West Bengal.

The total area and production of Tobacco in India for the year 1997-98 were 463.5 thousnad ha. and 646 million kgs. respectively.

India ranks 4th in the total Tobacco consumption in the world. But India's cigarette consumption ranks 11th in the world. Out of the total production, only 19% of the total consumption of Tobacco is in the form of cigarette whereas 81% is in other forms like, chewing, bidi, snuff, Gutkas paste, Jarda, hookah paste etc. The per capita consumption of cigarette in India is one of the lowest in the world in comparision to major Tobacco consuming countries like Zimbabwe, UK, Brazil, U.S.A and Pakistan. The annual level for demand of cigarette in India remains the same at 96 billion sticks as it was 15 years ago, despite the cumulative growth in population by nearly 35 percent during the same period. However the consumption of Tobacco has been a matter of national debate in view of the emerging anti Tobacco drive in the country.

Tobacco is traditional item of India's foreign trade. India is one of the leading Tobacco exporting countries in the world. India aounts for 5.8% of the international trade and ranks 5th after Brazil, U.S.A. Turkey and Zimbabwe. The principal market for India Tobacco are U.S.S.R, U.K, Japan and Middle East countries.

Market and marketing system play a dominant role inensuring remunerative price for commercial crop like Tobacco. There has been significant improvement in the marketing of VFC Tobacco with establishment of Tobacco Board. The production and marketing of VFC Tobacco have been statutorily regulated by the Tobacco Board. But the growers of other varieties of Tobacco are at the mercy of unscrupulous traders and middlemen. Excluding VFC Tobacco, the method of marketing of Tobacco in India differs from type to type and from State to State. In case of VFC Tobacco the Government of India and the Tobacco Board are announcing Minimum Support Price (MSP) from year to year with the objective of protecting the interest of the growers of VFC Tobacco. There is need for establishment of organised marketing system for all types of Tobacco so that the Indian Tobacco can achieve significant share in the International market.

Research on Tobacco has been playing an important role in the development of Tobacco varieties in India. India grows different variety of Tobacco under different agro-climatic conditions. As such the problem of improvement of different varieties of Tobacco in India are numerous and complicated. The main research work on Tobacco is being done at Central Tobacco Research Institute at Rajahmundry and its Research Stations spread throughout India. Apart from conducting research for development of different varieties of Tobacco for maximising production, the CTRI, Rajahmundry has been presently doing research for development of alternative crops to Tobacco. CTRI, Rajahmundry has also been entrusted with the research for development of alternative uses of Tobacco in view of anti-smoking campaign.

Earlier the Directorate of Tobacco Development was running two Non-Plan Schemes on bidi Tobacco viz. (i) Seed and Seedlings Scheme and (ii) Farmers Training Scheme at Gujarat Agricultural University, Anand, but both these scheme have been terminated by the end of March 2000.

As a part of all India Comprehensive Scheme to study the cost of cultivation of the principal crops, this Directorate is running the scheme for Assessing the Cost of Production of VFC Tobacco in Andhra Pradesh from 1974-75 onwards. Under this she me the data on cost of production are collected by cost aounting method by Field staff posted in selected clusters in Tobacco belt of Andhra Pradesh. After collection and compilation of such field data, this Directorate has been regularly sending such Information to Directorate of Economics and Statistics. Afterwards, the estimates of cost generated under this scheme are passed on to the Commission for Agricultural Cost and Prices (CACP) for its use in recommending the Minimum Support Price (MSP) on VFC Tobacco from year to year basis.

Tobacco Industry in India contributes in a unique manner to several important facets of the Indian Economy, covering revenue, export, employment, and GDP growth. The Tobacco industry in India mainly covers manufacturing of cigarette, bidi, cigar and cheroot, hookah, snuff and other chewing Tobacco likezarda, gutkha and other pan masala.

Cigarette industry in India is essentially capital intensive in nature. The growth of cigarette industry both in domestic and international market represents a big revenue opportunity for the economy. But the burden of Tobacco tax has increasingly shifted to cigarette with the removal of duty on raw Tobacco since 1979, resulting in discriminatory rates of duty compared to other Tobacco products.

India's share in world cigarette production has remained at around 1.7% whereas India's exports of around 2.8 billion sticks of cigarette per year aounts for less than 1% of the world export of cigarette. There is significant opportunity for cigarette industry to extent and cosolidate its position in intentional market due to some recent trend like withdrawal/reduction of agricultural subsidy and escalating cost in the traditional cigarette exporting countries.

Bidi industry is one of the foremost cottage industries in India. Total amount of bidi Tobacco production was 150million kg and that of bidi was around 700 billion pieces during1994-95. Around 37% of Tobacco production in India goes to bidi making as per Indian Market Research Bureau (IMRB) report 1996.The social significance of bidi industry derives from the fact that it generates more employment in the economy compared to cigarette industry. During 1993-94 the bidi industry generate daround 1310 million man days of employment. There is scope for development of bidi industry in view of excessive and biased tax treatment of cigarette.

Cigar and Cheroot industry is one of the oldest industries in India, and is mainly concentrated in Tamil Nadu and Andhra Pradesh. But presently this industry is in deep dwindling condition due to un organised marketing system. There is are cent spurt in the development of chewing Tobacco segment which uses modern packing technology to attract the consumer.

From the beginning of 17th century, Tobacco has been playing a important role in international trade. Recently there has been world-wide anti Tobacco campaign which resulted in modest fall in area and production of Tobacco. But consumption of Tobacco in the world has remain more or less stable. Similarly the world cigarette production grew steadily up to 1997 and started declining slowly during 1998 and its again shows upward trend during 1999.

As per the projection by F.A.O, the global Tobacco industry is poised to increase at the rate of 1.9% annually which will result in an deficit of 2% projected global demand. Assets such as experienced farming community confers a significant competitive advantage for India. With an increase in the world import requirements translating in to a rise in export potential, Indian Tobacco industry is presented with significant opportunities to consolidate and extent its position in the global market.

Even though the cultivation of Tobacco is spread all over the country, commercial cultivation of Tobacco is concentrated in States like Andhra Pradesh, Karnataka, Gujarat, Maharashtra, Bihar, Tamil Nadu and West Bengal. Cigarette Tobacco is mostly cultivated in Andhra Pradesh and Karnataka, Pradesh and Karnataka, Pradesh and Karnataka, whereas Bidi Tobacco is grown in Gujarat , Karnataka and Maharashtra. Cigar and Cheroot Tobacco are also grown in Tamil Nadu, Andhra Pradesh and West Bengal and Chewing Tobacco is grown in Tamil Nadu, Gujarat, Bihar , West Bengal and U.P. Hookah Tobacco is grown in UP, and West Bengal .

The total area and production of Tobacco in India for the year 1997-98 were 463.5 thousand ha. and 646 million kgs. respectively. India ranks 4th in the total Tobacco consumption in the world. But India's cigarette consumption ranks 11th in the world. Out of the total production, only 19% of the total consumption of Tobacco is in the form of cigarette whereas 81% is in other forms like, chewing, bidi, snuff, Gutka paste, Jarda, hookah paste etc. The per capita consumption of cigarette in India is one of the lowest in the world in comparison to major Tobacco consuming countries like Zimbabwe, UK, Brazil, U.S.A and Pakistan. The annual level for demand of cigarette in India remains the same at 96 billion sticks as it was 15 years ago, despite the cumulative growth in population by nearly 35 percent during the same period. However the consumption of Tobacco has been a matter of national debate in view of the emerging anti Tobacco drive in the country.

Tobacco is a traditional item of India's foreign trade. India is one of the leading Tobacco exporting countries in the world. India accounts for 5.8% of the international trade and ranks 5th after Brazil, U.S.A; Turkey and Zimbabwe. The principal markets for Indian Tobacco are U.S.S.R; U.K; Japan and the Middle East countries. Although there is nation wide anti-Tobacco campaign, the commercial importance of Tobacco can never be underestimated due to the revenue earning potential and employment generation capacity of the crop. Presently there is a call for substitution of Tobacco with other crops, but the research findings show that there is no economically viable alternative crop which is as remunerative as Tobacco to the farmer.

Tobacco Industry in India contributes in a unique manner to several important facets of the Indian Economy, covering revenue, export, employment, and GDP growth. The Tobacco industry in India mainly covers manufacturing of cigarette, bidi, cigar and cheroot, hookah, snuff and other chewing Tobacco like zarda, gutkha and other pan masala. Cigarette industry in India is essentially capital intensive in nature. The growth of cigarette industry both in domestic and international market represents a big revenue opportunity for the economy. But the burden of Tobacco tax has increasingly shifted to cigarette with the removal of duty on raw Tobacco since 1979, resulting in discriminatory rates of duty compared to other Tobacco products.

India's share in world cigarette production has remained at around 1.7% whereas India's exports of around 2.8 billion sticks of cigarette per year accounts for less than 1% of the world export of cigarette. There is significant opportunity for cigarette industry to extent and consolidate its position in intentional market due to some recent trend like withdrawal/reduction of agricultural subsidy and escalating cost in the traditional cigarette exporting countries.

Bidi industry is one of the foremost cottage industries in India. Total amount of bidi Tobacco production was 150 million kg and that of bidi was around 700 billion pieces during 1994-95. Around 37% of Tobacco production in India goes to bidi making as per Indian Market Research Bureau (IMRB) report 1996. The social significance of bidi industry derives from the fact that it generates more employment in the economy compared to cigarette industry. During 1993-94 the bidi industry generated around 1310 million man days of employment. There is scope for development of bidi industry in view of excessive and biased tax treatment of cigarette.

Cigar and Cheroot industry is one of the oldest industries in India, and is mainly concentrated in Tamil Nadu and Andhra Pradesh. But presently this industry is in deep dwindling condition due to unorganised marketing system. There is a recent spurt in the development of chewing Tobacco segment which uses modern packing technology to attract the consumer.

From the beginning of 17th century, Tobacco has been playing a important role in international trade. Recently there has been world-wide anti Tobacco campaign which resulted in modest fall in area and production of Tobacco. But consumption of Tobacco in the world has remain more or less stable. Similarly the world cigarette production grew steadily up to 1997 and started declining slowly during 1998 and its again shows upward trend during 1999.

As per the projection by F.A.O; the global Tobacco industry is poised to increase at the rate of 1.9% annually which will result in an deficit of 2% projected global demand. Assets such as experienced farming community confers a significant competitive advantage for India. With an increase in the world import requirements translating in to a rise in export potential, Indian Tobacco industry is presented with significant opportunities to consolidate and extent its position in the global market.

BOTANICAL FEATURES

Botanically, the Tobacco plant belongs to the family *Solanacea* and genus *Nicotiana*. The genus embraces over 60 species of which two alone are cultivated. India grows both the cultivated species, viz. *Nicotiana tabacum* and *Nicotiana rustica*. The largest area is under

Nicotiana tabacum which is grown all over the country whereas *Nicotina rustica* is confined to North and North Eastern States. i.e. Uttar Pradesh, Bihar, West Bengal and Assam. About 5 to 6 per cent of the total area under Tobacco is accounted for by *Nicotiana rustica* varieties. The cultivation of *Nicotiana tabacum* has spread country-wide and this type also accounts for more than 80% of the exchange earnings. Specific types and varieties of Tobacco have been developed for use in cigarette, bidi, cigar, cheroot, hookah, chewing, snuff and hookah paste. While rustica types are used in chewing and snuff the the the tabacum types are used for all purposes.

Nicotiana tabacum

The genus Nicoliana has more than 60 species, of which two are commercially cultivated for the production of tobacco. They are N. tabum and Nrustica. N. is widely cultivated in most countries of the world, and as far north as Sweden, and as far south as New Zealand. The cultivation of N.rustica is restricted to India, Russia and some other Asiatic countries. Though the tobacco plant is tropical in origin, its production at present is concentrated mostly outside the tropics, except in India. In tobacco production, India ranks third after the U.S.A. and China, with an annual production of 380 million kg of leaf from an area of about 445 thousand hectares. Almost all.states in India grow tobacco, but the important ones are Andhra Pradesh, Gujarat, Tamil Nadu, Karnataka, Bihar, West Bengal and Uttar Pradesh in that order, and account for about 90 per cent, of the annual production.

The species N. tabacum is grown in almost all the states, whereas the cultivation of N. rustic is confined to the northern and north-eastern states, where the temperatures are considerably lower during the growing season. India ranks fourth in the world in tobacco' exports, after the USA, Turkey and Greece, Pride exports about 80 million kg of tobacco to about 50 countries and earns about 635 million rupees of foreign exchange annually. About 85 per cent of the export consists of flue-cured Virginia tobacco, and India is the world's second largest exporter of this variety after the U.S.A.

For the improvement and development of different types of tobacco grown in the country, the Indian Central Tobacco Committee was established in 1945. In 1947, this Committee established the Central Tobacco Research Institute, Pajahmundry (Andhra Pradesh) for fundamental reference on all tobaccos and applied work on tobaccos grown locally. Subsequently, the committee had taken over the research station at Guatur (Andhra Pradesh) from the then lindustrial Agricultural research Institute, New Delhi, and started the Regional Research Station at Yedasandur (Tamil Nadu), Hunsur (Karnataka), Pusa (Bihar) and Dinhata (West Bengal) for the improvement of the types of tobacco grown in these states. Besides the above stations, a number of schemes run by the state governments for the improvement of the crop were financed by the Committee.

With the abolition of the Commodity Committees in 1965, the Central Institute and the regional research stations attached to it came under the control of the Indian Council of Agricultural Research and the development activities were taken over by the Ministry of Agriculture, Government of India.

Soil and Climate

India produces a wide range of commercial types of tobacco and each state specialise's in the production of specific types. However, all these types are mostly used for internal consumption except the flue-cured Virginia tobacco which is exported. The flue. cured Virginia tobacco is normally grown in other countries during the rainy season in light soil with a sandy surface, a friable sandy clay subsoil, low in organic matter content and acidic in nature. In India, on the other hand, it is mostly grown in heavy black—Boils of Guntur, Prakasam, Krishna, Bust and West Godavari, Khammam and Nellore districts of Andhra Pradesh in winter

Soil types and crop season

Northern Black Soils (NBS): The black soils located in the districts of Khammam, Karimnagar, Warangal, East and West Godavari constitutes the Northern Black soils. The crop is grown under rain fed conditions. The crop is planted in October and harvested in Dec-Feb. The per hectare yields are around 1200 Kgs. The suitable varieties are Jayashree, Hema and VT1158. The tobacco produced is good bodied, lengthy leaf, lemon orange flashy in colour with medium nicotines. The tobacco is mostly preferred by CIS and Eastern European Markets.The domestic cigarette industry uses very little quantity of tobacco from his region.

Central Black Soils (CBS): The black soils existing in Krishna and Guntur of Andhra Pradesh and Gadchiroli district of Maharashtra state constitute the central black soils. The main varieties grown are Godavari Spl., Hema, and VT1158. The crop is planted in Oct-Nov and harvested in Dec-Feb. The crop is grown under rain fed conditions. The average yields are about 1100 Kgs/Ha. The tobacco produced in this tract is almost similar to that of Northern black soils. Preferred for exports mostly to CIS, Eastern European and West Asia & North African markets. Japan also used to import sizable quantities of tobacco from select areas of this region in the past. The domestic cigarette industry also uses sizeable quantity.

Southern Black Soils (SBS). The black soils present in Prakasam and Nellore districts nearer to the coast constitute the Southern black soils. Presently the crop is cultivated in about 30,000 ha., agaist the potential of 50,000 ha. The crop is planted in Oct-Nov and harvested in Dec-Feb. The popular varieties grown are Jayashree, Jayashree(MR), Hema, VT1158. The average yields are around 1200 Kgs/ha. The tobacco produced is of lemon to lemon orange in color, medium bodied. This tobacco is rated better in quality than the NBS/CBS tobaccos and excellently amenable for blending with others styles of tobacco in Cigarette manufacturing. Preferred for export to West Asia & North African countries (Egypt and Libya), Nepal and to CIS countries. Sizeable quantities are also exported to West Europe. Half of the production goes for domestic cigarette manufacturing.

Southern Light Soils (SLS): The light soils existing in Prakasam and Nellore districts constitute the southern light soils (SLS). Presently this crop is cultivated in about 58,000 ha against potential of 80,000 ha. The soils are suitable for growing good quality tobacco as a semi monsoon crop. The suitable varieties are Jayashree, Hema and VT-1158. The crop is planted in September-October, harvested in Nov-Jan. The yields are around 1000 kgs/ha. The leaf is lemon to lemon orange in colour, thin to medium bodied, ripe and open grain with good aroma and low in nicotine. This crop is preferred for export to UK and other EEC countries and South East Asia. All the MNCs prefer this tobacco. The domestic manufacturers use substantial quantities of this region tobacco for home consumption.

Northern Light Soils (NLS): The light soils existing in East and west Godavari and parts of Khammam districts constitute the Northern light soils (NLS). Presently this crop is cultivated in about 18,000 ha against the potential of 30,000 ha. The soils are sandy to sandy loams. UK. The crop is raised with the help of 7-8 irrigations producing semi flavourful tobacco. The crop is planted in Sept-Oct, harvested in Dec-Feb. The yields are ranging between 1600-2000 kgs/ha depending on stage of topping. The suitable varities are K326, 16/103, MCNair-12, Gold Streak and NLS-4. The leaf is medium to heavy bodied, ripe open grained, orange to deep orange in color and fluffy in nature. Mostly exported to UK & other European countries & South East Asia. All the MNCs prefer this tobacco. Sizeable quantities are also used in domestic manufacturing.

Eastern Light Soils (ELS): The light soils existing in Vijayanagaram of A.P and Rayagada area of Orissa are termed as Eastern light soils. Presently the crop is grown in 700-800 ha and the production is at about 1 M.Kgs. Good potential exists to expand the area under tobacco cultivation in this region. The tobacco is similar to NLS in quality. Exported to UK & other European countries.

Karnataka Light Soils (KLS): The light soils existing in Mysore, Hassan, Shimoga, Davangere, Coorg, Chikkamagalore and Chitradurga districts constitute the light soils of Karnataka. The crop in this transitional zone is raised as a monsoon crop with sufficient rainfall during the crop growth period. Presently the crop is grown in about 7000 ha and much scope do not exists for further expansion. The crop is planted in the last week of Apr-May and harvested in July-Aug. The per ha yields are around 1250 kgs. The varieties under cultivation are spl. FCV, Bhavya, K326, Ratna, Swarna. The tobacco produced is light to medium bodied, lemon orange to orange in color, soft in nature, open grained with a nicotine of 1.5 to 2.2%. West European, some African, Middle East and South East Asian countires mostly prefer the tobacco.

Cultivation

Usually, 6 to 10 ploughings are given by way of preparatory cultivation. Experiments have shown that deep ploughing of the fields once in two years during summer, followed by 2 to 3 ploughings and 1 to 2 harrowings are beneficial for most of the tobaccos, particularly for the flue-cured virginia tobacco grown in Andhra Pradesh. Digging with a spade, followed by ploughing with a mould-board plough and a country plough and then a harroeing is recommended for Bihar. Farmyard manure is usually applied and the dose varies from 10 to 125 cartloads per hectare for different types of tobacco. In addition, oil-cakes or ammonnium sulphate was found beneficial for some tobaccos. Based on the research findings of the Central Tobacco Research Institute, Rajahmundry, and its regional stations, the following manurial schedule is recommended :

1. Seven-and-a-half tonnes of farmyard manure, six weeks before planting and 22 kg of N + 50 kg of P_2O_5 + 50 kg of K_2O per hectare, a week before planting for the flue-cured Virginia tobacco on black soils of Andhra Pradesh.
2. Ten to twelve tonnes of farmyard manure, six weeks before planting and 30 to 40 kg of N + 80 to 100 kg of P_2O_5 + 80 kg of K_2O + 12 to 15 kg of MgO, of which the entire quantity of Mg and P_2O_5 and half the quantity of N and K_2O are applied a week before planting and the remaining quantities of N and K_2O are applied 3 to

4 weeks after planting as top-dressing just before irrigation for the flue-cured Virginia tobacco on light soils of Andhra Pradesh.

3. Fifteen tonnes of farmyard manure or 300 kg of groundnut-cake 2 to 3 months before planting and 44 kg of N per hectare a week before planting for natu tobacco in Andhra Pradesh.
4. Forty-five to 90 kg of N per hectare, half as oil- cakes and, half as ammonium sulphate for the bidi tobacco in Karnataka.
5. One hundred and eighty kg of Wilha, half as castor-cake or groundnut cake and Half as ammonium sulphate, 2 weeks before planting for the bidi tobacco in Gujarat.
6. Twentyfive tonnes of farmyard manure and 75 kg of N/ha (213 as ammonium sulphate and 113 as groundnut. cake, half before planting and the remaining half 6-7 weeks after planting for cigar and cheroot tobaccos of Tamil Nadu.
7. Sixty tonnes of farmyard manure and 112 kg per ha, half as mustard-cake and half as ammonium sulphate for the chewing tobacco in Bihar.
8. Thirty tonnes of farmyard manure and 150 kg of N+150 kg of Pz0 g per hectare for hookah and chewing tobaccos; 17 tonnes of farmyard manure and 125 kg of N+112 kg of PzOs+224 kg of KaO per hectare for the wrapper tobacco in West Bengal.

Spacing

The tobacco seedlings are always planted in rows. The distance between the rows and between the plants within the two varies with the type of tobacco. Based on the research findings, a spacing of 80 cm × 80 cm for natu and the flue-cured Virginia tobacco in black soils and 100 cm × 60 cm for the flue-cured Virginia tobacco in light soils of Andhra Pradesh; 75 cm × 50 cm for cigar and cheroot tobaccos and 75 cm × 75 cm for the chewing tobacco in Tanil Nadu; 100 cm × 70 cm and 105 cm × 45 cm for the bidi tobacco in Gujarat and Karnataka, respectively , 90 cm × 60 cm for the chewing tobacco in Bihar, 60 cm × 45 cm for N. rustica and 90 cm × 90 cm for N. tobwum in West Bengal, are recommended.

Interculturing

When after the plants get established, i.e. about 20 days after planting, the tobacco fields are given frequent inter culturing and weeding to conserve; the soil moisture and cheek the growth of weeds. Tined harrows, blade harrows and the country plough are normally used for interculturing. Mulching the crop with paddy straw at 3,600 kg per hectare after the first interculture was found beneficial in the case of the flue-cured Virginia tobacco on black soils of Andhra Pradesh in increasing the yields and improving the quality by understanding soil moisture.

Irrigation

In Andhra Pradesh, flue-cured Virginia tobacco on black soils is not normally irrigated, but the crop on light soils is given up to six irrigations. The irrigation water should not contain more than 50 ppm of chlorides, as otherwise the leaves get burnt and other qualities suffer. In black soils also, in adverse conditions, one irrigation on 40-day-old plants is recommended. The cigar, cheroot and chewing tobaccos in Tamil Wadu and the chewing tobacco in Bihar are normally irrigated. The bidi tobacco in Karnataka, Maharashtra and Gujarat and the hookah and chewing tobaccos in West Bengal are not normally irrigated, but in years of deficient rainfall, one or two irrigations are desirable to maintain the yield.

Topping and Suckering

The removal of the flower head alone or along with some of the top leaves of the plant is known as topping. After topping, the axillary buds grow and their removal is known as suckering. Most of the types of tobacco grown in the country are topped and suckered for improving the size, body and quality of the leaves, except wrapper tobacco where the texture is one of the important criteria. The flue-cured Virginia tobacco grown in the black soils of Andhra Pradesh was also not topped till recently, but of late, topping is recommended to improve the body of the leaf. Even now, if the growth of the crop is rank and the colour of the leaf is dark green, topping is avoided. All other types of tobacco are topped . The level of topping varies from type to type and ranges from high topping in the case of the flue-cured Virginia tobacco, in which case only the flowerhead is removed, leaving all the 20-24 leaves intact. To the very low topping in the case of chewing tobacco of Tamil Nadu where only 7 to 8 leaves are retained on the plant. Based on the research findings, topping at 14-16 leaves for the cigar and cheroot tobaccos in Tamil Nadu, at 13-15 leaves for bidi, tobacco in Karnataka and Gujarat and the chewing tobacco in Tamil Nadu and at 8-9 leaves for N. tabacum and at 6-8 leaves for N. rustica in West Bengal, is recommended.

To get the full benefit of topping, the suckers are to be removed periodically. This is a laborious and time consuming job. Hence, a number of experiments for the suppression of suckers by the application of oils, chemicals, hormones, etc. were conducted and it was reported that the application of coconut-oil to the top six leaf axils soon after topping suppressed the suckers in the flue-cured Virginia tobacco. Two per cent naphthalene acetic acid triethenolemin applied to the topped portion of the-stem suppressed the majority of the suckers in cigar and cheroot tobaccos of Tamil Nadu. In Bihar, though vegetable oils and hormones suppressed the suckers by 34 and 47 per cent respectively, in the case of chewing tobacco, they did not prove to be of much practical importance to the cultivators. However, piercing the stem after topping improved the yield and quality and is widely practised. Naphthalene acetic acid and ESSO oil did not produce significant effect in controlling suckers in the case of the bidi tobacco at Anand.

Harvesting, Curing and Fermentation

The signs of maturity and the methods of harvesting differ with the type of tobacco. The flue-cured Virginia tobacco is harvested during December-March in Andhra Pradesh and during July-September in Karnataka. The leaves are considered ready for harvesting when the normal green colour changes to yellowish green or to light yellow. Harvesting starts from the bottom and each time 2 or 3 leaves are harvested. In another week, the next 2 or 3 leaves mature when they are harvested. Thus the 20-24 curable leaves, that become available per plant in the case of the flue-cured Virginia tobacco, are harvested in 6 to 8 primings at weekly intervals. Soon after harvesting, the leaves are strung on bamboo sticks at the rate oF about 100 leaves per stick and loaded in the barn for curing.

The bidi tobacco is harvested in January-February when the majority of the top leaves develop red rusty spots known as spangles. The cigar and cheroot tobaccos are harvested 90 to 100 days after plant when the leaves pucker and become brittle and yellowish-green. The chewing tobacco is harvested 110 to 120 days after plant when the leaves develop Pronounced puckering. The hookah tobacco (rustica) is harvested in May or June, and the tabacum harvested when broad flecks appear on the leaves. The whole plants are harvested in the case of the bidi, cigar and cheroot, chewing and hookah tobaccos. except in West Bengal where harvesting is done by priming.

Tobacco leaves are cured after harvesting in order to impart the required colour, texture and aroma to the final product. Different methods of curing are adopted for different types of tobacco, depending on its quality requirements and the use to which it is put to. Flue-curing, air-curing (or shade-curing), sun-curing (or rack-curing or ground-curing), smoke-curing and pitcuring are the different methods of curing in vogue in the country. The Virginia tobacco is cured in special chambers, known as barns with artificial heat passing through metal pipes, called flues. Hence this tobacco is known as the flue-cured Virginia tobacco and is exclusively used for manufacturing cigarettes. Curing takes 100-120 hours, after which the leaves are unloaded from the barn, bulked for a few days and then graded, depending on the colour. Since the quality of the leaf depends, not only on colour but also on its position on the plant, the plant-position grading is now recommended.

The wrapper tobacco of West Bengal and the Lanka tobacco of Andhra Pradesh are air-cured. The wrapper tobacco is cured under atmospheric temperature and at 70-80 per cent relative humidity in barns with sides and the roof closed. The lanka tobacco is cured by hanging the strung leaf in thatched sheds for 2 to 21 months after which the curing is completed in pits.

Most of the tobaccos grown in the country are suncured, as in this method the construction of costly strue tares can be avoided and also because the sun is quite-bright and hot during this part of the year. The cigar and chewing tobaccos of Tamil Nadu are cured in the sun for 15 to 20 days by stringing the whole plants on bamboo poles. The natu tobacum of Andhra Pradesh is cured by hanging the Strung leaves on sea golds in the open for 1 to 11 months. In the case of the bidi tobacco, the plants are cut and left upside-down on the same spot for 3 to 7 days. After the leaves are dry, the lamina is stripped off. The chewing tobacco in Bihar is cured in the sun, the cut plants in the field for 4 to 7 days. In Uttar Pradesh, the plants are cut and exposed to sun for 1 or 2 days, after which the leaves are separate and spread on the ground or made into heaps for a further drying. In West Bengal, the leaves are primed and kept on the ground upside-down for a few hours for drying, after which they are tied into bunches of 4 to 5 leaves and cured in sheds for 4 to 5 weeks.

Some of the chewing tobaccos grown in Tamil Nadu are smoke-cured in cylindrical buts with open tops. The cured leaves, except that of the bidi tobacco are then bulked into heaps of convenient size for fermentation which improves the quality and aroma of the leaf. The bulks are broken up and remade occasionally to avoid excess moisture and decay. The leaf is then graded and packed, as required by the trade.

Yield

The average yield of tobacco leaf per hectare is about 750 kg and 950 kg for the flue-cured Virginia and the natu tobacco respectively in Andhra Pradesh; 1,000 kg, 450 kg and 350 kg for the bidi tobacco in Gujarat, Maharashtra and Karnataka states respectively; 1,250 kg and 1,600 kg for the cigar and the root and chewing tobaccos, respectively in Tamil Nadu, and 950 kg, 850 kg and 800 kg for the hookah and chewing tobaccos respectively in Bihar, West Bengal and Uttar Pradesh.

Varieties

There are local or commercial varieties grown in the various tobacco-producing areas under both N. tabacum and N. rustica, Wo numerous W mention. It is with the help of the Central Tobacco Research Institute, Rajahmundry, and its regional stations and other organisations connected with the crop. From early times man has developed various types of tobacco for various uses. There are about sixty five valid species of tobacco but among these only two species viz; Nicotiana Tabacum and Nicotiana Rustica are cultivated extensively. India grows both these species, but, by far, the largest area is under Nicotiana tabacum which is grown all over the country. Nicotiana rustica requires cooler climate. Its cultivation is confined mainly to the northern and north-eastern region.

Cigarette, bidi, chewing, hookah, cigar, cheroot, snuff, natu, burley etc. are the various types of tobacco grown in different parts of the country. The plant characteristics, soil and climatic requirements, agronomic practices, harvesting and curing techniques and the chemistry of end-products are all different in different types of tobacco grown in different zones. The area of concentration of the various types of tobacco are indicated below and is also depicted in the Tobacco Map of India.

irginia Flue cured Tobacco

irginia Tobacco used in cigarettes is the most important commercial variety and is an xport-oriented crop this accounts for about / 21 per cent of Indias total production of obacco. About 40 per cent of the total Virginia tobacco production is used for domestic igarette industry while the rest is exported to markets all over the world. Virginia obacco is grown mostly in the States of Andhra Pradesh and Karnataka. In Andhra radesh the cultivation of Virginia tobacco is concentrated in the coastal districts of rakasam, Guntur, Krishna, East and West Godavari and Nellore and in Khammam, arimnagar and Warangal Districts of Telangana. Traditional Virginia tobacco used to e cultivated in black soil areas. In view of the changing tastes in the consumers reference for light bodied leaf with less nicotine/tar content in the international market, he cultivation of Virginia Flue cured Tobacco has been extended to the light soil areas lso.

latu Tobacco

ndhra Pradesh also produces about 30 million kg of sim cured country or natu tobacco vhich is used in the manufacture of cigarettes and tobacco mixtures for pipes. This type s mainly grown in the districts of Kurnool, Guntur, East and West Godavari and Krishna.

urely Tobacco

urely Tobacco which is also a cigarette type is grown to a small extent in East Godavari, Guntur, Visakhapatnam and Telangana areas of Andhra Pradesh.

idi Tobacco

idi Tobacco is the most important Non-Virginia tobacco produced in the country. More han 85 per cent of the worlds bidi production is accounted for by India. Bidi tobacco ccupies 30-35 per cent of the total area under tobacco in the country whereas in terms of production its share is around 35-40 per cent. Gujarat, Karnataka and Maharashtra re the main bidi tobacco producing States. In Gujarat, bidi tobacco is grown in Charotar rea comprising Anand, Borsad, Petlad and Nadiad Taluks of Kaira District and Baroda listrict It is mostly grown under irrigated conditions. In Karnataka bidi tobacco is mainly grown in Nipani area of Belgaum District on the banks of River Krishna and its ributaries. Kolhapur and Sangli are the two important districts for cultivation of bidi obacco in Maharashtra.

hewing Tobacco

hewing tobacco is mostly grown is the States of Tamilnadu, Uttar Pradesh, Bihar, West Bengal and Orissa. Coimbatore and Periyar in Tamilnadu, Farrukhbad in Uttar Pradesh,

Pradesh, Vaisali and Samastipur in Bihar, Cooch Behar in West Bengal, Koraput in Orissa are the important chewing tobacco growing districts.

Cigar Tobacco

There are three distinct types of tobacco used in the manufacture of cigar viz. filler, binder and wrapper. In India, good quality filler tobacco is also used as a binder. Wrapper tobacco of good quality is mainly grown in Cooch Behar district of West Bengal. Cigar filler tobacco is mainly grown in Dindigul and Trichy districts of Tamil Nadu.

Cheroot Tobacco

This type of tobacco is mainly grown in Andhra Pradesh and Tamil Nadu. Lanka tobacco is grown in the Godavari river belt is also used for the manufacture of cheroots. In Saled, Trichy and Dindigul districts of Tamil Nadu cheroot tobacco is grown. Natu tobacco grown in Guntur, Kurnool and Godavari areas of Andhra Pradesh is also used in the manufacture of cheroots.

Hookah Tobacco

The tobacco which is used in smoking through the hookah or hubble-bubble is called Hookah tobacco. Hookah tobacco is mostly N. rustica type and it is grown in cold climate. This type of tobacco is cultivated in Uttar Pradesh, West Bengal and Orissa and it is locally known as Calcutta, Vilayati and Motihari. The major districts cultivation hookah tobacco are Purnea in Bihar, Cooch Behar and Jalpaiguri in West Bengal, Kaira and Ahmedabad in Gujarat and Farrukhabad and Etah districts in Uttar Pradesh.

Snuff Tobacco

Tobacco used for snuff are grown in Andhra Pradesh, Kerala, West Bengal, Uttar Pradesh and Gujarat. There is no special variety grown exclusively for snuff tobacco in the country. Generally the tobacco varieties used for snuff manufacture are mainly consumed in a number of forms viz. hookah, chewing and bidi. The important snuff manufacturing States in India are Tamil Nadu, Gujarat and Uttar Pradesh. Chennai in Tamil Nadu State is an important centre for manufacture of snuff. Generally tobacco leaf which is medium to thick in texture and having a pungent aroma is selected for snuff manufacture.

PEST CONTROL

Mix endosulfan 4a% dust @ 40 gm per 10 sq. m bed in the top soil or drench with chlorpyrifos 20% EC. @ 2 ml/liter of water one day before sowing in the nursery to prevent damage to seeds and tiny seedlings from soil pests like ants, mole crickets, earthworms, etc. The pests commonly noticed in the nursery are (i) caterpillars (ii) Stem-borer (iii) Grasshoppers and (iv) Whitefly.

Caterpillars and Cutworms

Spodoptera Litura F. commonly known as tobacco caterpillar is the worst pest in tobacco nurseries. The well grown caterpillars are grey or dark brown with v shaped white mark on the front portion of the black head. they feed voraciously along the veins of leaves and also cut the stems of small and tender seedlings. Hence they are also known as cutworms. They can be controlled by (i) deep ploughing of nursery area during April-May which will expose pupae to high temperature and destroy them (ii) sowing castor around tobacco nursery 15 days before sowing of tobacco so that castor puts up sufficient foliage by the time Spodoptera infestation builds up. (iii) spraying of 1 percent Neem seed Kernel suspension (NSKS) in water on 3 weeks old seedlings and 2% NSKS on 4 weeks old seedlings. NSKS may be prepared by dipping a muslin bag containing 100 gm of neem seed kernel powder in a bucket containing 10 litres of waters and squeezing the bag for 15 minutes to get a light brownish liquid. The suspension can be directly sprayed on tobacco seedlings.

Stem borer

The stem-borer is the larva of a tiny brown moth called Scrobipalpa (Gnorimoschema) heliopa Low. The tiny caterpillars bore inside the stem and mid ribs and feed on internal tissues. As a result, swelling appears where the borer stays. It causes stunting and unusual branching of the seedlings. At the planted crop is also affected by this pest. Suffocated seedlings should be removed and destroyed. Spray Quinalphos 25% E.C. at 0.05% (20 ml in 10 litres of water) at 30 and 40 days after germination. Final spray should be given before pulling out seedling.In planted crop spray above insecticide at 20 and 30 days after planting. After completion of harvesting the stem borer affected plants should be uprooted and burnt to prevent carry over of the pest to the next season.

Grasshoppers

These are green or brown jumping insects. they feed on leaves making circular holes. Cultural practices like keeping the nursery area free of weeds and grass helps to keep away the grass-hoppers. Spray the surrounding vegetation with endosulfan 35% E.C. @ 15 ml. or chlorpyrifos 20% E.C. @ 25 ml. in 10 litres of water or dust the area with endosulfan 4% dust @ grams per 10 sq. m. area.

Whitefly

White flies are small fly like insects seen on the underside of leaves. They will fly immediately when the plant is disturbed. The adult fly sucks the sap from leaves and transmits the leaf curl virus diseases from infected to healthy seedlings. Normally this pest can be controlled by spraying any of the pesticides recommended for control of S. Litura F.

Control of diseases in Seedbeds

Damping off is the most common disease occurring in tobacco nurseries. It is a fungal disease caused by soil borne pathogen. Under wet weather conditions the disease spreads to the entire bed killing the seedlings. It can be controlled by application of Ridomil 25 WP at 0.2 or Bordeaux mixture 0.4% (40 gm of copper sulphate 40 gm of lime in 10 litres of water) or Fytolan or Blitox @ 0.2% (20 gm in 10 litres of water) to the seed beds 2 weeks after sowing. This may be repeated once in 4 days under normal weather conditions and once in 2 days under wet weather conditions.

Black Shank Leaf Blight

This is a soil borne disease. It frequently occurs during mid-nursery period causing leaf blight and blackening of roots and stems leading to death of seedlings.Black shank can be controlled by same schedule suggested for Damping off. Use Mc Nair-12 which is a black shank tolerant variety in irrigated light soils. Spray Fytolan or Foltaf @ 0.2% concentration when the seedlings are 50-60 days old.

Collarrot

Just like damping off sudden death of seedlings in patches is noticed in seed beds. Blackening of the collar region, wilting and rotting of leaves are the symptoms. Spray 0.4% Bordeaux mixture or 0.2% Ziram or 0.2% Biltox, 3 weeks after seed germination to control this disease. Repeat this spray twice or thrice at weekly intervals depending upon the weather conditions.

REFERENCES

Breen, T. H. (1985). *Tobacco Culture*. Princeton University Press. *Source on tobacco culture in eighteenth-century Virginia pp. 46–55*

Gately, Iain. *Tobacco: A Cultural History of How an Exotic Plant Seduced Civilization*. Grove Press, 2003.

Killebrew, J. B. and Myrick, Herbert (1909). *Tobacco Leaf: Its Culture and Cure, Marketing and Manufacture*. Orange Judd Company. *Source for flea beetle typology.*

Price, Jacob M. "Tobacco Use and Tobacco Taxation: A battle of Interests in Early Modern Europe". *Consuming Habits: Drugs in History and Anthropology*. Jordan Goodman, et al. New York: Routledge, 1995 166-169.

Bibliography

Achaya, K.T. 1990. *Oilseeds and oilmilling in India: a cultural and historical survey*. New Delhi, India, Oxford and IBH.

Alderks, O.K. 1948. Cooking of meats and recovery of the oil. *In* A.E. Bailey, ed. *Cottonseed and cottonseed products: their chemistry and chemical technology*, p. 615-650. New York, Interscience Publishers.

Alexander H. Tullo (September 8, 2008). "A Nutty Chemical". *Chemical and Engineering News* 86 (36): 26–27.

Arndt, C., Hertel, T., Dimaranam, B., Huff, K. & McDougall, R. 1997. China in 2005: implications for the rest of world. *Journal of Economic Integration*, 505-547.

Bailey, L. H. and Bailey, E. Z. 1976. *Hortus Third: A Concise Dictionary of Plants Cultivated in the United States and Canada*. MacMillan Publishing Company, New York.

Boddey, et.al., "Biological nitrogen fixation associated with sugar cane", *Plant and Soil*, Volume 137, Number 1 / November, 1991.

Bose, N.K. 1975. *The structure of Hindu society*, trans. André Beteille. Hyderabad, India, Orient Longman.

Breen, T. H. (1985). *Tobacco Culture*. Princeton University Press. *Source on tobacco culture in eighteenth-century Virginia pp. 46–55*

Campbell-Falck D, Thomas T, Falck TM, Tutuo N, Clem K (2000). "The intravenous use of coconut water". *Am J Emerg Med* **18** (1): 108–11.

Chaudhuri, J.L. & Selvaraj, K. 1985. Technological developments in ghani oil industry. *Khadigramodyog*, Oct./Nov.: 75-84.

Claud Bald: Indian Tea. A Textbook on the Culture and Manufacture of Tea. Fifth Edition. Thoroughly Revised and Partly Rewritten by C.J. Harrison. Thacker, Spink & Co., Calcutta 1940 (first edition, 1933).

Collinson, M. (editor): *A History of Farming Systems Research*. CABI Publishing, 2000.

Crosby, Alfred W.: *The Columbian Exchange: Biological and Cultural Consequences of 1492*. Praeger Publishers, 2003 (30th Anniversary Edition).

Dalby, Andrew (2002). *Dangerous Tastes*. Berkeley: University of California Press. http://books.google.com/books?id=7IIIcZ21dyjwC.

Davidson, Alan (2002). *Wilder Shores of Gastronomy: Twenty Years of the Best Food Writing from the Journal Petits Propos Culinaires*. Berkeley: Ten Speed Press.

Davis, Donald R., and Hugh D. Riordan (2004) Changes in USDA Food Composition Data for 43 Garden Crops, 1950 to 1999. Journal of the American College of Nutrition, Vol. 23, No. 6, 669-682.

Dieter Fischer. "History of Indonesian coffee". Specialty Coffee Association of Indonesia. http://www.sca-indo.org/history-of-indonesia/. Retrieved on 2008-08-13.

Dong et al., "A Nitrogen-Fixing Endophyte of Sugarcane Stems" (A New Role for the Apoplast), *Plant Physiology*, 1994, Vol 105, Issue 4 1139-1147.

Fife, Bruce (2005). *Coconut Cures*. Piccadilly Books, Ltd.. pp. 17. http://books.google.com/books?id=JW8RtCJqZ8MC. Retrieved on 2008-04-04.

Francois, J.F., MacDonald, B.J. & Nordström, H. 1996. *Trade liberalisation and capital accumulation in the GTAP model*. GTAP Technical Paper No. 7.

Friedland, William H. and Amy Barton (1975) Destalking the Wily Tomato: A Case Study of Social Consequences in California Agricultural Research. Univ. California at Sta. Cruz, Research Monograph 15.

Gately, Iain. *Tobacco: A Cultural History of How an Exotic Plant Seduced Civilization*. Grove Press, 2003.

Gujarathi, P.V. 1982. Design and efficiency of power ghani system. *In* V.H. Potty, S. Hariprasad & M.R. Mukunda, eds. *Village oil industry*, p. 91-100. Mysore, India. Central Food Technological Research Institute.

Harler, C.R.: The Culture and Marketing of Tea. Second edition. Oxford University Press, New York and Bombay, Reprinted 1958 (First edition 1933, second edition 1956).

Hertel, T.W. & Martin, W. 1999. *Would developing countries gain from inclusion of manufactures in the WTO negotiations?* GTAP working paper, Purdue University.

Hertel, T.W. (ed). 1997. *Global trade analysis: modelling and applications*. Cambridge University Press.

Huang, J., Rozelle, S. & Chang, M. 2004. Tracking distortions in agriculture: China and its accession to the World Trade Organization. *World Bank Economic Review*, 18(1): 59-84.

Jaffee, Steven (2004). "Delivering and Taking the Heat: Indian Spices and Evolving Process Standards". *An Agriculture and Rural Development Discussion Paper*. Washington: World Bank.

James A. Duke. "Coffea arabica L.". Purdue University. http://www.hort.purdue.edu/newcrop/duke_energy/Coffea_arabica.html#Ecology. Retrieved on 2007-07-20

John C. Evans (1992): *Tea in China: The History of China's National Drink*,Greenwood Press.

Killebrew, J. B. and Myrick, Herbert (1909). *Tobacco Leaf: Its Culture and Cure, Marketing and Manufacture*. Orange Judd Company. *Source for flea beetle typology*.

Lester Packer, Choon Nam Ong, Barry Halliwell (2004): *Herbal and Traditional Medicine: Molecular Aspects of Health*, CRC Press.

Mazoyer, Marcel; Roudart, Laurence (2006): *A history of world agriculture : from the Neolithic Age to the current crisis*, New York, NY : Monthly Review Press.

McGee, Harold (2004). "Black Pepper and Relatives". *On Food and Cooking (Revised Edition)*. Scribner. pp. 427–429.

Monier-Williams, M. 1899. *A Sanskrit-English dictionary*, Delhi, India, Motilal Banarsidass. Reprinted 1963.

Nag, T.K. 1982. *Village oil industry. Power-ghani installation: a handbook*. Bombay, India, Khadi and Village Industries Commission.

Patel, J.P. 1943. *Oil extraction*. Maganwadi, Wardha, India, All-India Village Industries Association. 3rd ed.

Pendergrast, Mark (1999). *Uncommon Grounds: The History of Coffee and How It Transformed Our World*. Basic Books. ISBN 0-465-05467-6.

Pillai, Rajmohan and Santha, P. *The World Çashew Industry* (Rajan Pillai Foundation, Kollam, 2008).

Price, Jacob M. "Tobacco Use and Tobacco Taxation: A battle of Interests in Early Modern Europe". *Consuming Habits: Drugs in History and Anthropology*. Jordan Goodman, et al. New York: Routledge, 1995 166-169.

Salvesen, David. "The Grind Over Sun Coffee". Smithsonian National Zoological Park. http://nationalzoo.si.edu/Publications/ZooGoer/1996/4/suncoffee.cfm. Retrieved on 2007-09-24.

Turner, Jack (2004). *Spice: The History of a Temptation*. London: Vintage Books. OCLC 61213802.

Van Tongeren, F. & Huang, J. 2004. *China's food economy in the early 21st century*. Report #6.04.04. The Hague, Agricultural Economics Research Institute (LEI).

Walmsley, T.L. 1998. *Long-run simulations with GTAP: Illustrative results from APEC trade liberalization.* GTAP technical paper No. 9.

Walmsley, T.L., Betina, V.D. & Robert, A.M. 2000. *A base case scenario for the dynamic GTAP model*. West Lafayette, Center for Global Trade Analysis, Purdue University.

Watson, Andrew. *Agricultural innovation in the early Islamic world. Cambridge University Press.* p.26-7.

Werth, E. 1933. Distribution, Origin and Cultivation of the Coconut Palm (in periodical: *Ber. Deutschen Bot. Ges.*, vol 51, pp. 301–304) (article translated into English by Dr. Child, R. (Director, Coconut Research Scheme, Lunuwila))

Winchester, Simon (2003). *Krakatoa: The Day the World Exploded, August 27, 1883*. HarperCollins. pp. 223–224.

Yamamoto, T; Kim, M; Juneja, L R (1997). *Chemistry and Applications of Green Tea*. CRC Press. .

Zhang, J., Huss, V.A.R., Sun, X., Chang, K. and Pan, D. 2008. Morphology and phylogenetic position of a trebouxiophycean green algae (Chlorophyta) growing on the rubber tree, *Hevea brasiliensis*, with the description of a new genus and species. *Eur. J. Phycol.* 43(2): 185 - 193.

Index